The Power of Networks

The Power of Networks describes a typology of network-based research practices in the historical disciplines, ranging from the use of quantitative network analysis in cultural, economic, social or political history or religious studies, to novel approaches in the Digital Humanities.

Network data visualisations and calculations have proven to be useful tools for the analysis of mostly textual sources containing relational information, offering new perspectives on complex historical phenomena. Including case studies from antiquity to contemporary history, the book provides a clear demonstration of the opportunities historical network research (HNR) provides for historical studies. The examples presented within the pages of this volume are arranged in a way to highlight three central typological pillars of HNR: (re-)construction and analysis of historical networks, computational extraction of network data and infrastructures for data collection and exploration.

The Power of Networks outlines the history and current state of research in HNR and points towards future research frontiers in the wake of new digital technologies. As such, the book should be essential reading for academics, students and practitioners with an interest in digital humanities, history, archaeology and religion.

Florian Kerschbaumer is Project Manager at the Danube University Krems and Lecturer at the University of Klagenfurt, Austria.

Linda von Keyserlingk-Rehbein is Curator and Head of the Document Department in the Military History Museum, Dresden, Germany.

Martin Stark is Senior Researcher at the ILS- Research Institute for Regional and Urban Development, Dortmund, Germany.

Marten Düring is Assistant Professor/Senior Research Scientist at the Luxembourg Centre for Contemporary and Digital History (C²DH) at the University of Luxembourg.

Digital Research in the Arts and Humanities
Series Editors: Marilyn Deegan, Lorna Hughes, Andrew Prescott, Harold Short and Ray Siemens

Digital technologies are increasingly important to arts and humanities research, expanding the horizons of research methods in all aspects of data capture, investigation, analysis, modelling, presentation and dissemination. This important series covers a wide range of disciplines with each volume focusing on a particular area, identifying the ways in which technology impacts on specific subjects. The aim is to provide an authoritative reflection of the 'state of the art' in technology-enhanced research methods. The series is critical reading for those already engaged in the digital humanities, and of wider interest to all arts and humanities scholars.

The Historical Web and Digital Humanities
The Case of National Web Domains
Edited by Niels Brügger and Ditte Laursen

Postdigital Storytelling
Poetics, Praxis, Research
Spencer Jordan

Humans at Work in the Digital Age
Forms of Digital Textual Labor
Edited by Shawna Ross and Andrew Pilsch

Feminist War Games?
Mechanisms of War, Feminist Values, and Interventional Games
Jon Saklofske, Alyssa Arbuckle, and Jon Bath

The Power of Networks
Prospects of Historical Network Research
Edited by Florian Kerschbaumer, Linda von Keyserlingk-Rehbein, Martin Stark and Marten Düring

For more information about this series, please visit: www.routledge.com/Digital-Research-in-the-Arts-and-Humanities/book-series/DRAH

The Power of Networks

Prospects of Historical Network Research

Edited by Florian Kerschbaumer, Linda von Keyserlingk-Rehbein, Martin Stark and Marten Düring

Routledge
Taylor & Francis Group
LONDON AND NEW YORK

First published 2020
by Routledge
2 Park Square, Milton Park, Abingdon, Oxon OX14 4RN

and by Routledge
605 Third Avenue, New York, NY 10017

First issued in paperback 2021

Routledge is an imprint of the Taylor & Francis Group, an informa business

Publisher's Note
The publisher has gone to great lengths to ensure the quality of this reprint but points out that some imperfections in the original copies may be apparent.

British Library Cataloguing-in-Publication Data
A catalogue record for this book is available from the British Library

Library of Congress Cataloging-in-Publication Data
A catalog record for this book has been requested

ISBN 13: 978-1-03-223667-4 (pbk)
ISBN 13: 978-1-138-73130-1 (hbk)

DOI: 10.4324/9781315189062

Typeset in Times New Roman
by Apex CoVantage, LLC

Contents

Figures

Tables

Contributors

Hans Blomme (Ghent) MA in History. Scientific Collaborator in TIC Collaborative and Policy Officer for Historical Cartography and Historical Data Processing at the Department of History at Ghent University, Belgium. Research interests: history of the 19th and 20th centuries, cultural history, social history, cartography, Digital Humanities.

Eberhard Crailsheim (Madrid) Dr. Phil., thesis on French and Flemish Merchant networks in Seville (1570–1650). Designated researcher at the Institute of History, Spanish National Research Council (CSIC) in Madrid. Research interests: Early Modern period, economic and social history, cultural history, network analysis, political communication (Spanish Empire, Atlantic and Pacific).

Thomas D'haeninck (Ghent) Doctoral Researcher on the intellectual mobility of Belgian and Dutch reformers in transnational networks of social reform (1850–1914). Scientific Collaborator in the Department of History, University Ghent, Belgium. Research interests: social history since 1750.

Marten Düring (Luxemburg) Dr. Phil., thesis on Covert Support Networks for Persecuted Jews during the Holocaust. Assistant Professor/Senior Research Scientist at the Luxembourg Centre for Contemporary and Digital History (C^2DH) at the University of Luxembourg, Luxembourg. Research interests: digital history and historiography, memory studies, network analysis, text analysis, data.

Kimmo Elo (Turku) Dr. Pol. Sci., Adjunct Professor. Senior Researcher in European studies at the Centre for Parliamentary Research at University of Turku, Finland. Research interests: European integration since 1945 as a political phenomenon, Germany and Europe, European policy networks, Cold War and Post-Cold War intelligence history, transatlantic relations from European perspective, network analysis and visualisations, Digital Social Sciences.

Frederik Elwert (Bochum) Dr. Phil., thesis on Religion as Resource and Restriction in the Process of Integration of Ethnic German Immigrants. Digital Humanities Coordinator at the Center for Religious Studies, Ruhr

University Bochum, Germany. Research interests: religion and migration, network analysis, computational text analysis, Digital Humanities.

Christine Fertig (Münster) Jun. Prof. Dr., Junior Professor for Social History at the University of Münster, Germany. Thesis about social networks and inequality in rural society. Research Interests: social and economic history (18.–20.c.), history of family and demography, rural history, history of material culture, Digital Humanities.

Robert Gramsch-Stehfest (Jena) PD Dr., Postdoctoral thesis on "The Empire as network of Princes. Political structures during the double kingship of Friedrich II. and Heinrich (VII.)". Academic Council at the Chair of Medieval History at University of Jena, Germany. Research interests: social network analysis and complexity theory, elites in the High and Late Middle Ages, Roman Curia, German political and constitutional history, medieval historiography, cultural history.

Florian Kerschbaumer (Krems/Klagenfurt) Mag. Phil., Florian Kerschbaumer is project manager at the Danube University Krems and lecturer at the University of Klagenfurt, Austria. Research interests: social history, civic and political education, history of social movements.

Geert Kessels (Den Haag) MA in history (Research Master) at the University of Amsterdam, Netherlands. Researcher and software developer (Digital Humanities). Co-Founder of LAB1100, a research and software development firm, which launched Nodegoat, a web-based research environment specifically focused on research in the humanities.

Malte Rehbein (Passau) Prof. Dr., Chair of Digital Humanities, University of Passau, Germany. Research interests: computational and digital history, quantitative methods in historiography/historical data studies, digitization of cultural assets, data modelling, text modelling and digital edition, ethics and critique of science and society, resistance and resistence.

Christian Rollinger (Trier) Dr. Phil., thesis on Roman elite society and friendship and the role of the role of *amicitia* networks as a social factor among the Roman republican elite. Lecturer in ancient history at the University of Trier, Germany. Research interests: economic and cultural history of the late Roman republic, late antique imperial ideology and ceremonial, seafaring in the ancient world, Classical Reception Studies.

Martin Stark (Aachen) Dr. Phil., thesis about the Social Embeddedness of a Rural Credit Market in the 19th Century. Senior Researcher at the ILS-Research Institute for Regional and Urban Development, Dortmund, Germany. Research interests: social and economic history of the 19th and 20th centuries, climate protection and climate change in urban development, network and governance research.

Pim van Bree (Den Haag) MA in New Media at the University of Amsterdam, Netherlands. Researcher and software developer (Digital Humanities).

Co-Founder of LAB1100, a research and software development Firm, which launched Nodegoat, a web-based research environment specifically focused on research in the humanities.

Charles van den Heuvel Prof. Dr., Professor of Digital Methods in Historical Disciplines at University of Amsterdam, Netherlands. Head of Department of History of Science and Scholarship, Huygens ING, Royal Netherlands Academy of Arts and Sciences. Research interests: history of science, art, history of architecture and urban planning, history of library and information studies, history of the World Wide Web, Digital Humanities.

Ingeborg van Vugt (Utrecht) PhD, thesis on "The structure and dynamics of scholarly networks between the Dutch Republic and the Grand Duchy of Tuscany in the 17th century". Postdoc in the SKILLNET project at Utrecht University, Netherlands. Research interests: intellectual and cultural history of Early Modern Europe, network analysis in historical research, republic of letters.

Christophe Verbruggen (Ghent) Prof. Dr., Director of the Ghent Centre for Digital Humanities and Associate Professor at the Research Unit Social History since 1750 at Ghent University, Belgium. Research interests: social history of intellectuals and cultural mobility in the 19th and 20th centuries, history of science and technology, environmental history and the use of prosopography, network analysis in historical research.

Linda von Keyserlingk-Rehbein Dr. Phil., thesis about the Network of the Attempted Coup D'état of July 20 1944 against the Nazi Regime. Curator and Head of the Document Department in the Military History Museum, Dresden, Germany. Research interests: social network analysis, German history of resistance, history of the Baltic Sea region, cultural history.

Bernd Wurpts (Lucerne) PhD, thesis on "Networks into Institutions or Institutions into Networks? Evidence from the Medieval Hansa". Postdoc in the Department of Sociology, University of Lucerne, Switzerland. Research interests: comparative historical sociology, economic sociology, organizations, social networks, theory.

Acknowledgments

We would finally like to express our gratitude towards the NeDiMAH network and most notably Lorna Hughes for their encouragement and generous financial support, which made the publication of this book possible. Annina Mathar, Lorna Schatzmayr and Maria Zwischenberger provided indispensable support in the preparation of the manuscript; we thank them for their endurance and precision. We would also like to thank our editors at Ashgate and Routledge Publishing for their continued support, here especially Elizabeth Risch, Heidi Lowther, Robert Langham and more. First and foremost, however, we wish to thank our authors for their contributions and patience.

1 Introduction

Almost four decades ago, the historian Wolfgang Reinhard was the first to recognise the potential added value of network analysis methods for social historiography in German-speaking countries. Using the term "entanglement analysis", he imported theoretical and formal concepts developed within the emerging field of social network analysis and tried to apply them to research in early modern history. The resulting "entanglement history" produced some valuable work on social entanglement ("Verflechtung") in Upper German cities and the early modern Papal State, but it failed to become a recognised method in historical studies. Instead, the impetus for historical network research (HNR) came from sciences that lend themselves more to quantitative approaches. Accordingly, some of the most widely received empirical network analyses on the basis of historical data were produced by American historical sociologists. Their research often focused on political and social upheavals and the resulting social tensions. In recent years, network approaches in history have been rediscovered and adapted by historians of all eras for their research objectives. The resulting research landscape is very heterogeneous, and its development is highly dynamic. Methods range from the mere adaptation of the theoretical concepts of SNA to the application of complex mathematical models. Accordingly, there is still no consensus on how social science network methods can be adequately applied to historical topics and sources and what added value network theory and network analysis methods can really offer to historical research.

The editors of this anthology were directly involved in this rediscovery of network approaches in history through their own doctoral theses. They met for the first time in 2008 at a social science-orientated methodological summer school for social network analysis at Trier University in Germany. The idea soon emerged that they should regularly exchange ideas on the specific challenges that arise in adapting network analysis theories and methods to history. Starting in 2009, this led to the establishment of a series of workshops on historical network research, initially in German, which took place every six months, with a first event in Cologne. These events in Germany, Austria and Switzerland, mostly organised by local third parties, offered newcomers the opportunity to discuss their project ideas and learn about other projects and solutions. Separate method training was subsequently added. In 2019, the thirteenth workshop in this series, “Networks Across Time and Space: Methodological Challenges and Theoretical Concerns of Network Research in the Humanities”, took place at the Academy of Sciences and Literature in Mainz, Germany. For the first time, the workshop was held entirely in English, and the programme also took into account the needs of advanced users, in an attempt to reflect the existing research landscape. These events were and continue to be supplemented by the Historical Network Research website, which provides a bibliography, a calendar of events, researcher profiles and other resources and the email newsletter of the same name.

In 2013, the European Network for Digital Methods in the Arts and Humanities (NeDiMAH) gave us the opportunity to organise a first international conference on historical network research in Hamburg. This was followed by conferences in Belgium (Ghent 2014), Portugal (Lisbon 2015), Finland (Turku 2017) and the Czech Republic (Brno 2018), run by local organisers. From 2013 onwards, we were also able to hold sessions on historical networks at the International Sunbelt Conferences of the International Network for Social Network Analysis (INSNA), and from 2014 on at the corresponding European Regional Conferences (EUSN), together with various colleagues, which drew a great response. We would like to thank all our colleagues who have accompanied us on this journey over the past few years.

On the basis of all these events and activities, we published the German “Handbuch Historische Netzwerkforschung” (Handbook on Historical Network Research) together with our Cologne-based colleague Ulrich Eumann in 2016. This handbook was a first attempt to bring together existing knowledge about the emerging field of historical network research. The aim was not to canonise this knowledge but rather to look back on what has been achieved so far and establish a foundation for further development. Questions that have arisen in recent years, from our own and external research projects, were addressed: What are the consequences of heterogeneous sources? What do centrality measures and network visualisations ultimately tell us? The universality of the network approach is accompanied by diverse possibilities for implementation and evaluation, which can be adapted to the specifics of questions. This means that no two projects are alike. Readers of the handbook are thus given an overview of the potentials, problems and challenges of historical network research, enabling them to design their own projects on this basis.

The year 2017 saw the publication of the first issue of the online, open access *Journal of Historical Network Research*. The idea behind the journal as an international publication solely devoted to the study of networks (social or otherwise) from a specifically historical perspective is to bring together research that seems relevant but is currently dispersed, thereby contributing to the consolidation of the field. The journal also promotes an exchange between different areas of historical research (in the broadest sense), the (digital) humanities, social and computer sciences and various research traditions and disciplines, as well as attempting to strengthen the dialogue between network research and traditional historical research.

This anthology follows the fundamental ideas behind the handbook and the journal, with a different layout. It is intended to give English-speaking readers an overview of the field of historical network research. Since a mere repetition would serve no purpose, the anthology is not structured like a handbook. Instead, selected thematically sorted research essays are used to introduce the reader to essential pillars of historical network research in a first attempt at consolidation. In addition, a glossary enables readers with little or no specific prior knowledge to orient themselves in this new and exciting field of research.

The enormous popularity of networks as a central object in historical research[1] has led to a rapidly growing interest in data-driven historical network research (HNR) across the historical disciplines.[2] The concept of analysing individuals not only in an isolated state but embedded in a larger social context with their resulting interdependent relationships represents an essential part of both social sciences and humanities.[3] As a result, systematic research on the influence of social structures on agency and the impact of agency on social structures is crucial for the field of history and may be of particular value for network-oriented perspectives. As Claire Lemercier has pointed out, social network analysis (SNA) is not a neutral process; it enables researchers to perceive their objects of research from a new perspective.[4] SNA is inherently biased towards a worldview that attributes explanatory power to social relationships. The gradual appropriation of SNA methods from the social sciences into historical research workflows has been accompanied by three distinct areas of tension.

The first synergistic field concerns the ongoing popularity of the term "network"[5] and its metaphorical use within the academic world as well as within other societal domains. The versatile implementation of the term may not always correspond to the accuracy and adequacy[6] called for by Max Weber. Wolfgang Reinhard, the earliest adopters of SNA for historical research in Germany, observed that the term network has turned, along with discourse, into another empty phrase.[7] Nevertheless, recent research (including the chapters in this book) contributes fundamentally to critically reflect on and to sharpen the definition of the network concept in historical research by taking on a non-metaphoric approach to build a sound methodological and theoretical basis for HNR.

A second synergistic dimension is the strained relationship between quantitative and hermeneutical approaches in historical research.[8] Quantitative methods have a long tradition within this field. Several innovative initiatives stemming from areas like cliometrics and new institutional economics (NIE) have produced a variety of results, including the Nobel-Prize-winning work of Robert Fogel and Douglas North.[9] By linking source-based data collection and quantitative analysis with traditional source criticism and reflective interpretation, many projects in HNR effectively bridge the gap between qualitative and quantitative approaches.

In recent decades, digitization has created a third synergistic dimension.[10] Increasing use of digitized sources and digital tools in the humanities has brought about profound changes, and the impact of digitization – from the search for materials to the analysis of sources to the communication of research results – in the field of history can be observed today.[11] This widely discussed approach initially emerged in the middle of the 20th century[12] and became known as "humanities computing", later as "digital humanities" from which later emerged the term "digital history". These processes not only raise awareness of the substantially growing digital toolbox available to historians; they also point towards new research methods and different ways of thinking about and perceiving research content. Digital history involves large-scale collections of digitized sources, ever-growing metadata catalogues and numerous efforts to interlink existing repositories. The data models used for these endeavours are not always relevant or even useful for network analysis research questions (nor should they be). Nevertheless, free access to a variety of datasets opens up exciting new opportunities, especially when combined with collaborative approaches such as crowdsourcing[13] or semi-automatic processes for the extraction of actors and relationships using methods applied in computer linguistics.[14] Moreover, different methods for visualizing networks are continuously providing new ways of interactively exploring and communicating findings.[15]

Despite all these technical achievements, the premise put forward by Konrad Jarausch and Kenneth Hardy still remains valid: "hardware and software are only tools, not ends in themselves. In historical research the intellectual question must always be dominant".[16]

The impact of the earlier-mentioned synergistic fields has turned HNR into a research area that is characterised by a high degree of self-reflection. With this book we hope to initiate a broader discussion about the applicability of social network analysis and digital humanities methodologies in historical research. To this end we have sought to represent the wide spectrum of network-based history research.

About this book

This edited volume presents best practices of HNR as a complement to existing methodological and theoretical manuals. This includes approaches inspired by and adopted from quantitative social network analysis but also more recent efforts to apply network analysis principles to the computational as well as crowd-based creation and curation of research data. We hope that this book will serve as a source of inspiration and a template for new network-based research projects. To guide readers through the diversity of approaches we suggest the following typology of research practices in historical network research:

1 (Re-)construction of historical networks and their analysis
2 Computational extraction of network data from large corpora
3 Infrastructures for data collection and exploration

We will use this typology to break down the case studies presented here:

Computational extraction of network data

Today most applications of network analysis principles in the historical disciplines are inspired by the methodological and theoretical toolkits offered by social network analysis as practiced in the quantitative social sciences. From a historical perspective, the perhaps most influential article in this field remains "Robust Action and the Rise of the Medici" by John Padgett and Christopher Ansell published in 1993.[17] The authors construct a model of multiple types of social ties including information on intermarriages between members of the Medici and other Florentine elite families based on data derived from historiographical works. Thereafter, they used blockmodeling, a technique to detect structural patterns in a network dataset. They come to the conclusion that the Medici family's preference for outgroup marriages led to a dominant position in their network. Their adoption of network analysis methods is characterised by a clear research question, a large but relatively simple dataset and advanced quantitative methods and yields a straightforward answer to explain a complex historical phenomenon.

This section illustrates how methods borrowed from the quantitative social sciences continue to be fruitfully applied to answer different kinds of research questions posed to diverse types of data. We observe a spectrum that ranges from the mere visual exploration of complex social relations to their statistical analysis and/or theory-driven interpretation.

Bernd Wurpts' presents an original empirical study on Hanseatic trade partnerships and kinship and reconstructs the development of contract enforcement institutions in mediaeval Lübeck before and after the Black Death in 1350. His article discusses the insights gathered from linking historical and social science approaches and demonstrates the added value of network perspectives for economic history coming from the viewpoint of historical sociology. *Eberhard Crailsheim* focuses on transatlantic trade in early modern Seville. He applies network analysis methods to study the significance of social networks for the success of the mercantile city. *Christian Rollinger* uses network-based methods to study the concept of friendship in Ancient Rome and to provide a new perspective on intrapersonal networks and their effect on societal cohesion and political history. *Christophe Verbruggen*, *Hans Blomme* and *Thomas D'haeninck* link network-based methods with work in the sociology of ideas and history of science to study the intellectual networks that led to the emergence of the field "history of science". In his work on mediaeval power relations, *Robert Gramsch-Stehfest* shows the interplay of network analytical modelling and interpretation of complex political processes in pre-modern societies. He describes the historical research process as a sequence of modelling, analysing, visualization and interpretation. *Christine Fertig* adopts a regional perspective in her study, which examines rural social networks in 19th-century Westphalia. Using a variety of sources (parish registers, land records, etc.) she identifies the crucial role of kinship networks for transactions of land and credit.

Computational extraction of network data

The rapidly growing availability of digitized collections of primary sources and other documents of interest to historians is raising new challenges: How can we effectively explore collections that are so large that it becomes impossible to inspect them manually? How can we identify higher-level patterns within them? This section contains two examples of how network analysis tools can be used to gather information about large datasets that are decidedly not based on (a model of) social exchanges or affiliations. Crucial for both is the usage of network analysis as a means of source or corpus criticism: Understanding of the conditions under which their data was created and their inherent biases is instrumental for the assessment of their value for historical research.

In his chapter, *Kimmo Elo* combines network analysis with conceptual considerations to study "intelligence cycles" based on a database that was created by the East German intelligence service. His approach contributes to our understanding of covert political or military activities in the Baltic region during the Cold War. *Frederik Elwert* investigates dynamic relations between religious traditions by analysing text structures in order to build semantic networks. In these semantic networks words are seen as nodes, and the grammatical structure indicates their interrelations.

Infrastructures for data collection and exploration

The digitization of historical documents has encouraged experiments that deviate both from traditional forms of historical scholarship and how historians used to engage with network analysis methods: Historians now work in mixed teams that combine expertise in history, software development, design and computer science. In and around the Digital Humanities a number of these collaborations have demonstrated the value of network principles beyond the idea of simplified models of social interactions: Recent efforts seek to incorporate the rich and not necessarily relational information we have about past societies into data models and thereby accentuate the historical interest in the specific as opposed to the social scientific interest in the generalizable. Other projects experiment with novel ways to automatically generate network data from unstructured text, using crowdsourcing techniques for network data curation and network visualisations for the general-purpose exploration of data.

We have selected three examples that describe such approaches.

Charles van den Heuvel, Ingeborg van Vugt, Pim van Bree and Geert Kessels use their research on the early modern history of science to explore the past and future of rich network data, presented in the form of interactive interfaces that capture the complexity of historical records and correspond more closely to the questions scholars are seeking to answer. *Marten Düring* reviews a number of applications that use network visualizations as gateways for the enrichment and exploration of digitized source collections.

Malte Rehbein's programmatic chapter concludes our book. With a view to the past, present and future, he discusses the interdependence of historical network research, digital history and digital humanities. Using the examples of digitization, automatic annotation, data linkage, visualization and publication, he shows that current developments in digital humanities could be an accelerator for historical network research, significantly increasing its research potential in the foreseeable future.

With this book we seek to capture the diversification of how historians appropriate the idea of networks in their research beyond the mere metaphorical usage. We hope that our attempt to capture the distinct trends can serve as a guide and a useful complement to the existing theoretical literature. It goes without saying that our selection is by no means exhaustive, complete or definitive, but we hope that it will provide impetus for the future development of the field.

Notes

1 Among numerous publications on this subject, one key work is Niall Ferguson, *The Square and the Tower: Networks and Power, from the Freemasons to Facebook* (New York: Allen Lane, 2018).

2 A comprehensive bibliography on historical network research is available at http://historicalnetworkresearch.org/bibliography/ (accessed 28 February 2020).

3 Hartmut Rosa, David Strecker and Andrea Kottmann, *Soziologische Theorien*, 2nd ed. (Konstanz: UTB, 2013).

4 Claire Lemercier, "Formale Methoden der Netzwerkanalyse in den Geschichtswissenschaften: Warum und Wie?," *Österreichische Zeitschrift für Geschichtswissenschaften* 23 (2012): pp. 16–41, 20.

5 Alexander Friedrich, *Metaphorologie der Vernetzung zur Theorie kultureller Leitmetaphern* (Paderborn: Wilhelm Fink, 2015); Sebastian Gießmann, *Die Verbundenheit der Dinge. Eine Kulturgeschichte der Netze und Netzwerke* (Berlin: Kadmos, 2014).

6 Max Weber, *Soziologische Grundbegriffe*, 6th ed. (Tübingen: Mohr Siebeck, 1984), p. 17.

7 Wolfgang Reinhard, "Kommentar: Mikrogeschichte und Makrogeschichte," in *Nähe in der Ferne. Personale Verflechtung in den Außenbeziehungen der Frühen Neuzeit*, ed. Hillard von Thiessen and Christian Windler, Zeitschrift für Historische Forschung 36, pp. 135–44, 135 (Berlin: Duncker & Humbolt, 2005).

8 Marten Düring and Florian Kerschbaumer, "Quantifizierung und Visualisierung. Anknüpfungspunkte in den Geschichtswissenschaften," in Handbuch Historische etzwerkforschung, ed. Marten Düring, Ulrich Eumann, Martin Stark and Linda von Keyserlingk pp. 31–43 (Münster: Lit, 2016).

9 See Robert W. Fogel, *Railroads and American Economic Growth: Essays in Econometric History* (Baltimore: Johns Hopkins Press, 1964); Douglass North, *Transaction Costs, Institutions, and Economic Performance* (San Francisco: ICS Press, 1992).

10 Peter Bearman, "Big Data and Historical Social Science," *Big Data & Society*, https://journals.sagepub.com/doi/full/10.1177/2053951715612497: pp. 1–5 (accessed 28 February 2020).

11 See Fotis Jannidis, Hubertus Kohle and Malte Rehbein, eds., *Digital Humanities: Eine Einführung* (Stuttgart: J.B. Metzler, 2017).

12 Manfred Thaller, "Geschichte der Digital Humanities," in Digital Humanities: Eine Einführung, ed. Fotis Jannidis, Hubertus Kohle and Malte Rehbein, pp. 3–12 (Stuttgart: J.B. Metzler, 2017).

13 Daniele Guido, Lars Wieneke and Marten Düring, *Histograph: Graph-Based Exploration: Crowdsourced Indexation, Version 0,7* (Luxembourg: CVCE, 2016), http://histograph.eu/ (accessed 28 February 2020).

14 Ryan Cordell, "Computational Methods for Uncovering Reprinted Texts in Antebellum Newspapers: Viral Texts," http://viraltexts.org/2015/05/22/computational-methods-for-uncovering-reprinted-texts-in-antebellum-newspapers/ (accessed 5 September 2018);

Christopher Warren et al., "Six Degrees of Francis Bacon: A Statistical Method for Reconstructing Large Historical Social Networks," *Digital Humanities Quarterly* 10, no. 3 (2016), http://digitalhumanities.org/dhq/vol/10/3/000244/000244.html (accessed 28 February 2020).

15 Stefan Jänicke et al., "Visual Text Analysis in Digital Humanities," *Computer Graphics Forum* 36, no. 6 (2016), https://onlinelibrary.wiley.com/doi/abs/10.1111/cgf.12873 (accessed 12 December 2018); Stefan Jänicke, Greta Franzini, Muhammad Faisal Cheema, and Gerik Scheuermann, "On Close and Distant Reading in Digital Humanities: A Survey and Future Challenges", in *Eurographics Conference on Visualization (EuroVis) - STARs*, ed. Rita Borgo, Fabio Ganovelli and Ivan Viola. (The Eurographics Association, 2015), https://diglib.eg.org/handle/10.2312/eurovisstar.20151113.083-103 (accessed 28 February 2020).

16 Konrad Jarausch and Hardy Kenneth, *Quantitative Methods for Historian: A Guide to Research, Data, and Statistics* (Chapel Hill, London: University of North Carolina Press, 1991), p. XIII.

17 John F. Padgett and Christopher K. Ansell, "Robust Action and the Rise of the Medici," *American Journal of Sociology* 98, no. 6 (1993): pp. 1259–319.

Bibliography

Bearman, Peter. "Big Data and Historical Social Science." *Big Data & Society*. https://journals.sagepub.com/doi/full/10.1177/2053951715612497 (accessed 28 February 2020).

Cordell, Ryan. "Computational Methods for Uncovering Reprinted Texts in Antebellum Newspapers: Viral Texts." http://viraltexts.org/2015/05/22/computational-methods-for-uncovering-reprinted-texts-in-antebellum-newspapers/ (accessed 5 September 2018).

Düring, Marten, Ulrich Eumann, Martin Stark, and Linda von Keyserlingk, eds. *Handbuch Historische Netzwerkforschung*. Schriften des Kulturwissenschaftlichen Instituts Essen (KWI) zur Methodenforschung 1. Münster: Lit, 2016.

Düring, Marten, and Florian Kerschbaumer. "Quantifizierung und Visualisierung. Anknüpfungspunkte in den Geschichtswissenschaften." In Handbuch Historische Netzwerkforschung. Edited by Marten Düring, Ulrich Eumann, Martin Stark and Linda von Keyserlingk, pp. 31–43. Münster: Lit, 2016.

Ferguson, Niall. *The Square and the Tower: Networks and Power, from the Freemasons to Facebook*. New York: Allen Lane, 2018.

Fogel, Robert W. *Railroads and American Economic Growth: Essays in Econometric History*. Baltimore: Johns Hopkins Press, 1964.

Friedrich, Alexander. *Metaphorologie der Vernetzung zur Theorie kultureller Leitmetaphern*. Paderborn: Wilhelm Fink, 2015.

Gießmann, Sebastian. *Die Verbundenheit der Dinge. Eine Kulturgeschichte der Netze und Netzwerke*. Berlin: Kadmos, 2014.

Guido, Daniele, Lars Wieneke, and Marten Düring. *Histograph: Graph-Based Exploration: Crowdsourced Indexation, Version 0,7*. Luxembourg: CVCE, 2016. http://histograph.eu/ (accessed 28 February 2020).

Jänicke, Stefan, Greta Franzini, Muhammad Faisal Cheema, and Gerik Scheuermann. "On Close and Distant Reading in Digital Humanities: A Survey and Future Challenges." In *Eurographics Conference on Visualization (EuroVis) – STARs*. Edited by Rita Borgo, Fabio Ganovelli and Ivan Viola. The Eurographics Association, 2015. https://diglib.eg.org/handle/10.2312/eurovisstar.20151113.083-103 (accessed 28 February 2020).

———. "Visual Text Analysis in Digital Humanities." *Computer Graphics Forum* 36, no. 6 (2016). https://onlinelibrary.wiley.com/doi/abs/10.1111/cgf.12873 (accessed 12 December 2018).

Jannidis, Fotis, Hubertus Kohle, and Malte Rehbein, eds. *Digital Humanities. Eine Einführung*. Stuttgart: J.B. Metzler, 2017.

Jarausch, Konrad, and Hardy Kenneth. *Quantitative Methods for Historian: A Guide to Research, Data, and Statistics*. Chapel Hill, London: University of North Carolina Press, 1991.

Lemercier, Claire. "Formale Methoden der Netzwerkanalyse in den Geschichtswissenschaften: Warum und Wie?" *Österreichische Zeitschrift für Geschichtswissenschaften* 23 (2012): pp. 16–41.

North, Douglass. *Transaction Costs, Institutions, and Economic Performance*. San Francisco: ICS Press, 1992.

Padgett, John F. and Christopher K. Ansell, "Robust Action and the Rise of the Medici," *American Journal of Sociology* 98, no. 6 (1993): pp. 1259–319.

Reinhard, Wolfgang. "Kommentar: Mikrogeschichte und Makrogeschichte." In *Nähe in der Ferne. Personale Verflechtung in den Außenbeziehungen der Frühen Neuzeit*. Edited by Hillard von Thiessen and Christian Windler. Zeitschrift für Historische Forschung 36. Berlin: Duncker & Humbolt, 2005: pp. 135–44.

Rosa, Hartmut, David Strecker, and Andrea Kottmann. *Soziologische Theorien*. 2nd ed. Konstanz: UTB, 2013.

Thaller, Manfred. "Geschichte der Digital Humanities." In Digital Humanities: Eine Einführung. Edited by Fotis Jannidis, Hubertus Kohle and Malte Rehbein, pp. 3–12. Stuttgart: J.B. Metzler, 2017.

Warren, Christopher et al. "Six Degrees of Francis Bacon: A Statistical Method for Reconstructing Large Historical Social Networks." *Digital Humanities Quarterly* 10, no. 3 (2016), http://digitalhumanities.org/dhq/vol/10/3/000244/000244.html (accessed 28 February 2020).

Weber, Max. *Soziologische Grundbegriffe*. 6th ed. Tübingen: Mohr Siebeck, 1984.

2

(Re-)construction of historical networks and their analysis

2.1 Networking the *res publica*

Social network analysis and Republican Rome

Christian Rollinger[1]

Roman society was permeated by and structured along well-understood personal relationships, ties of kinship, marriage, friendship, and patronage. "Patronage" in the modern sense is an inadequate concept for explaining all aspects of social cohesion and day-to-day functioning of the Roman aristocracy, both of which can better be understood to have been dominated by *amicitiae*. This perspective, along with the conceptualisation of Roman personal relationships as networks offers us new insights to better understand already well-known phenomena. Given the right source material – typically bodies of correspondence or documentary records – the ancient historian has enough data at his disposal to undertake network research and to visualise the results of his labour in the form of graphs and quantitative measures. The article shows that the SNA may reveal apparent contradictions within former interpretations in the field of ancient history; it helps to frame new questions and to take a fresh look at oftentimes well-known evidence, to systematise and conceptualise it in a new and different way. Since the proof of networks is neither revolutionary nor their existence, at this point, a surprise, Historical Network Research is about what we do *after*. The results of SNA, be they network graphs, centrality measures, or even the simple re-thinking of existing presuppositions, have to be combined with, preceded, and followed by a careful interpretation of historical context, sources and source biases.

For ancient historians (as for all others), the main challenge in undertaking Historical Network Research lies in combining two seemingly disconnected areas: SNA, as a quantitative research tool dependent on algorithms and complicated calculations, has to be related to traditional interpretative hermeneutics of historical research.[2] This is indispensable for the study of historical problems from a network perspective. Painstaking historical research and interpretation is essential in two respects: first, in establishing criteria for a research sample and assembling a database amenable to network research and, second, in interpreting the networks constructed. The latter is a particularly essential task: SNA is not and should not be *ars gratia artis*. It is not self-sufficient; in and of itself, it has little meaning. It is rather a starting point for further analysis and interpretation.[3]

In what follows, my research project on elite networks in Late Republican Rome will be used to show how essential heuristic criteria for network research can be developed specifically for research into ancient history from the available source material and, combined with traditional hermeneutics as well as prosopographic research, be utilised to conduct network research.[4]

Elite networks in Late Republican Rome between patronage and friendship

"Data! Data! Data! ... I can't make bricks without clay."[5]

Roman society was permeated by and structured along well-understood (although sometimes ill-defined) personal relationships, ties of kinship, marriage, friendship, and patronage. This is particularly true for the most important subset of Roman aristocracy; those families (*gentes*) whose members had reached the highest office of consul and thus belonged to the nobility (*nobilitas*).[6] For a long while, ancient historians have seen these relationships mostly as either part of larger marriage and alliance strategies or as expressions of patronage, which for the last decades has been understood in sociologically inspired terms, as proposed by Richard Saller in an important 1982 study, i.e., as involving a personal but asymmetrical relationship, "in the sense that the two parties are of unequal status and offer different kinds of goods and services" in an exchange that was nevertheless reciprocal by its very nature and definition.[7]

But "patronage" in the modern sense is an inadequate concept for explaining the social cohesion and day-to-day functioning of the Roman aristocracy, both of which can better be understood to have been dominated by *amicitiae*.[8] *Amicitia* in the Roman sense is not to be confused with modern notions of friendship, for "the range of *amicitia* is vast. From the constant intimacy and goodwill of virtuous or at least like-minded men, to the courtesy that etiquette normally enjoined on gentlemen, it covers every degree of genuinely or overtly amicable relation."[9] It seems logical, then, that in order to better understand Roman elite society by means of network analysis, the relationships between senators and knights, as dictated by the conventions of *amicitia* and the requirements of *gratia* and *fides*, are the ideal starting point. As a general remark, this is (hopefully) readily obvious. The difficulty in applying SNA to the field of Roman republican elites lies in gathering enough data to conduct meaningful analyses.

Therefore, to begin at the beginning: how can general remarks about the importance of *amicitia* and the exchange of favours, a truism, in any case, for most historical and contemporary societies, be transformed into a meaningful heuristic tool for establishing network criteria? The solution lies not in a priori mathematical abstraction or the reliance on models but rather in a close reading of the available sources. If *amicitia* was vital for elite society, then it stands to reason that the services, actions, and favours (*beneficia*) associated with it should be interpreted as a manifestation of *amicitiae*. Thus, in looking at the specific forms of favours granted to or expected among friends, network criteria can be established.

But before this is possible, the problem of the sources must be addressed. As one can well imagine, the availability and nature of historical sources are a major problem for any historian attempting network research. Those sources that most readily lend themselves to network analysis, i.e., closed bodies of process-produced serial sources, such as tax lists or name registers, both of which would allow a near-complete representation of the networks contained within them,[10] are exceedingly rare (if not unheard of) for the ancient world or indeed much of the (early) medieval period.[11] This challenge for ancient historians can only be overcome by applying traditional hermeneutic processes of historical research: close reading of available sources and interpretation *lege artis*.

In the case of this chapter, a relatively wide array of sources available for the late Roman republic can be used. They include works of historiography,[12] biographies of late republican figures,[13] as well as a number of varied literary works by polymaths such as Macrobius, Aulus Gellius, Valerius Maximus, and Pliny the Elder.[14] Of particular importance are the letters, speeches, and philosophical works of Cicero, which are invaluable sources (particularly the *Letters*) but have to be carefully vetted for information. Differences in literary and historiographical genres, in terms of reliability, as well as authors' cultural and temporal distance have to be accounted for. Thus, to state the obvious; while Cicero's *Letters* are by a large margin our best source for the late Republic, they have to be read carefully, and constant regard should be paid to the fact that the *res publica* depicted in his correspondence is always viewed through the massively non-impartial eyes of Cicero himself.

Taking into account the wide variety of source genres and authors, the criteria for the inclusion of actors in our network analysis have to be as specific and near-objective as possible. To recapitulate, friendly favours (*beneficia*) were seen as both an amical duty and a manifestation of friendship among the Roman elite. They were fundamentally reciprocal in nature in that social norms and conventions required favours to be followed by counter-favours at least of equal weight. *Beneficia* of different kinds can, for our intents and purposes, be regarded as equivalent and, given the effect of peer pressure (the shibboleth of *noblesse oblige* springs to mind), as equidirectional, that is to say that a *beneficium* given by one party to another binds both parties together.[15] In order to visualise the network of Roman aristocratic society, any of a number of *beneficia* may be included, provided they are exchanged between two named male individuals belonging to the same social class; that is either the *ordo senator-ius* or the *ordo equester*.[16] Thus, an *amicitia*-relationship will be coded between two actors A and B if

- an invitation to a dinner party (*convivium*) is extended to A by B or A's presence at a dinner party at B's house is attested;
- A's attendance at B's morning greeting (*salutatio*) is attested;
- A serves as a military tribune, as *praefectus fabrum* or legate of B, while the latter was acting as commanding general;[17]
- A acts as a lawyer (*patronus*), supporter (*advocatus*), or character witness (*laudator*) for B;

- A loans money to B or acts as guarantor of a loan;
- A acts as B's agent in financial or private matters;
- A is mentioned as a beneficiary in B's last will and testament or A is listed as a witness of the foresaid will;
- A is named as tutor of any descendant of B;
- A is attested as the author of a letter of recommendation or introduction addressed to C, in which B is the commended.[18]

While this selection of criteria should allow enough data for a meaningful analysis to be gathered, it should by no means be understood as complete, either in the sense of types of relationships/*beneficia* considered or in the sense that available sources provide a complete picture. This can be illustrated by the example of a relationship that has been consciously excluded from the network in this chapter: communication.

While the writing of letters among the educated and wealthy elite of republican Rome was no doubt considered as a *beneficium* in and of itself,[19] extreme caution is necessary. The only extant body of correspondence from this period is the collection of Cicero's *Letters*. From a total of 946 individual epistles, 490 letters written by Cicero himself are organized according to their addressee: 437 to his friend, Atticus (while not a single one of Atticus' letters to Cicero has survived); 26 to M. Iunius Brutus, one of the latter-day murderers of Caesar; and 27 to his brother, Quintus. The remaining 432 letters *ad familiares* comprise letters written by Cicero, letters written to him, and copies of letters written to or by third parties, which either Cicero or his correspondent included. Thus, through Cicero's *Letters* we may identify roughly 50 of his correspondents from the period of 68–43 BC. The vast majority of these correspondents belonged to the aristocracy (including the *ordo equester*). Considering that the members of the highest *ordines* numbered among the thousands, this is but an extremely small extract. It is also known that Cicero was a prolific letter writer and that other collections of letters have been lost, such as his correspondence with Caesar.

However, the real danger lies not in the incomplete nature of our information but rather in its extreme bias. If "communication by letter" was included as a criterion for inclusion in the network, this would invariably and dramatically skew the results, since Cicero's correspondence alone has survived. Thus, only his own correspondents would figure (with very few additions) in the resulting dataset.[20] This is a general problem, given the predominance of information gathered from the contents of Cicero's letters, whereby all the information is almost irredeemably skewed in favour of Cicero. This has to be taken into account from the start: within the network, Cicero will occupy a position of undoubtedly exaggerated importance, and measures have to be undertaken to reduce this exaggeration as much as possible.

The resultant network will follow a strictly binary coding: if a relationship conforming to the criteria laid out earlier is attested in the sources, a social relationship will be coded as an undirected and unweighted tie between two nodes.

Thus, the network will be reduced to the bare essentials: ties are undirected because it is assumed that social *mores* would, in the absence of contradicting factors, lead to reciprocal relationships. On the other hand, ties are unweighted because of the bias of our sources: while it would be possible to weight ties according either to the number of *beneficia* or to the different kinds of *beneficia* they represent and thus open up new perspectives, this would produce no reliable results. Since Cicero's letters are our main source, the distortion would be extreme, in that Cicero would be at the center of an extraordinarily well-connected cluster of multiplex ties. This would be true and false at the same time: true, in the sense that, as has been seen, relationships between senators or senators and knights were rarely strictly one-dimensional but instead encompassed a multitude of different aspects and *beneficia*. Still, such an image would be a decidedly imperfect representation of late Republican society as it is known from historical studies, since Cicero simply did not occupy as dominant a position as would result from our source limitations. His own views, expounded at length in his writings, notwithstanding, Cicero was, at best, a middle-sized fish in a very large pond. Such an approach would only be worthwhile if enough alternative sources were available to trace similarly dense clusters of multi-faceted relationships for other significant actors, above all for Caesar, Pompey, or Cato, who were among the decisive figures in late Republican history. However, such sources are not available.

Equally possible in theory but impractical in application would be the use of linguistic cues, such as those Adam Schor has put at the center of his own reconstructed network. The language of amity in Late Republican Rome is notoriously and consciously vague, precisely because *amicitia* was such a vast concept, ranging from extremely close personal friendships to mere acquaintances. But since *amicitia*, at its philosophical core, was a selfless connection to an alter ego, borne out of love (*amor*),[21] not utility (*utilitas*), any reference to personal connections was couched in deliberately emotional terms. Accordingly, it is not rare to find two people who were dedicated political enemies still addressing each other not only as *amici* but *amicissimi*, literally the "friend-iest of friends." While a differentiated vocabulary developed to navigate the vast oceans of individual relationships, with terms such as *necessarius* or *familiaris* usually denoting a closer personal connection, this vocabulary is not only exceedingly vague and difficult to pin down but also hotly contested among scholars.[22]

A final remark must be made on the temporal dimension. While it is, of course, theoretically possible to produce dynamic network graphs that recreate the temporal development of networks, this is not feasible in the case of this chapter. The primary reason for this, again, lies with the source materials and the lack of sufficient data, which make any such task difficult (though not impossible). Thus, the network presented later is an aggregated network spanning the period from 78–43 BC; a total of 35 years or, to put it another way, roughly a generation of Roman politics, from the death of Sulla to the beginning of the so-called second Triumvirate. While this was a decidedly restive period (and included the civil war between Pompey and Caesar), it was also the last period in republican history, in which the system of *amicitiae* and *beneficia* as understood earlier was able to function freely and unhindered by anomalies such as the all-dominating position of, e.g. the Triumvirs Caesar Octavian and Mark Antony.

Networking the *res publica*

In looking at the networks resultant from the coding used for the relationships defined earlier, the problems just described become immediately obvious. As an intermediate step in attempting to reconstruct a whole network, (directed) network graphs for all individually coded types of *beneficia* were drawn up. These individual relationships serve well to illustrate the difficulties inherent in the sources utilised. As an example, a closer look can be taken at the relationship defined as invitations to *convivia* and/or attendance at morning greeting ceremonies. Every instance of such an occurrence mentioned in the utilised sources was coded according to the criteria laid out previously. The coding process itself was divided into two steps: first, general information (names of actors, social status of actors, nature of ties, date of occurrence, source references) was collected in tabular form, then this data was coded into UCINET data language (DL) format, specifically as an edge list. This information was then fed into UCINET, and a network graph was drawn up using NetDraw.[23]

For this individual graph, ties were directed: the host of a *convivium* was coded as "giving," the guest as "receiving" a favour. Correspondingly, the visitor at a morning *salutatio* was coded as "giving" a favour (his obeisance), while the host was coded as "receiving" a *beneficium*. The resultant network is relatively small, with a total of 43 actors, almost all of whom belong to the Roman elite classes (Figure 2.1.1).[24]

As is obvious at first glance, Cicero is the dominant actor in this network. In fact, if the relatively few isolated dyads and triads not connected to the main component were eliminated, the result would be an ego-network of Cicero. The reason for this is simply the inherent bias of the applied sources. Except for a few isolated instances,[25] all information on attendance at dinner parties during the late republic originates from Cicero's correspondence, and whereas he was certainly not above gossiping with his friends about who had attended whose dinner party, most of the guest constellations mentioned in his correspondence are in the context of discussions of *convivia* he had himself attended and witnessed. While there is nothing inherently implausible in the relations represented in this graph (after all, Cicero did dine with Pompey, as did Pompey with Lucullus), it must always be understood to be a mere fragment. By no means should we assume that, e.g. Cicero and Brutus never dined together or, for that matter, Caesar, Pompey, and Crassus. The only thing that can be inferred from their lack of attestation in the sources is that either Cicero did not know of them, or he did not *choose to discuss* them in *the letters which have survived.* Even if complete access to every single letter Cicero had ever written was readily available there would still be information lacking about important or

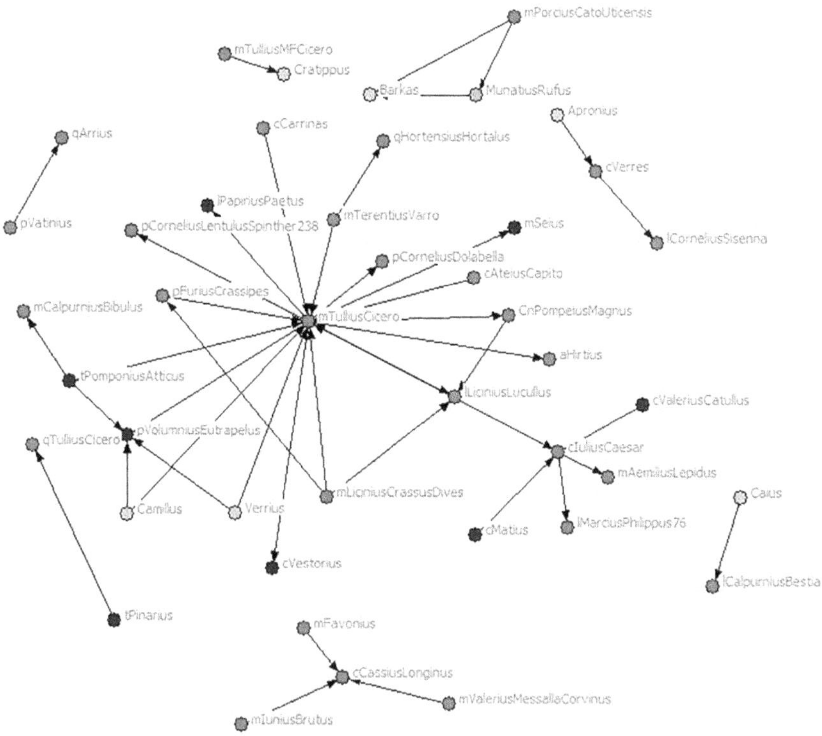

Figure 2.1.1 "Convivial" network[26]

socially relevant dinner invitations, for the simple reason that these things were often discussed orally (or not at all).

Thus, this particular graph is no more than a visualisation of what little information sources have yielded on the topic, and its interpretative potential is therefore limited. The same can unfortunately be said for most (if not all) representations of individual relationships (as laid out earlier), as the fundamental problem, that is reliance on Cicero as a source, remains the same.[27] There is one exception to this, namely the case of military relationships; service as an officer in the provinces whereby the availability of more diversified sources is a definite advantage including but (importantly) not exclusively limited to Cicero's. Many officers are mentioned in biographical and historiographical works, including for instance Caesar's *Gallic War* and *Civil War*. While this is a valuable addition to Cicero's letters, wherein, too, such matters are discussed, the fundamental problem of source distortion remains the same, as shown by Table 2.1.1.

Ancient historians naturally focused on the spectacular cases of Roman expansions, such as Caesar's conquest of Gaul and Pompey's suppression of Mediterranean piracy or his reorganization of much of the Hellenistic east as Roman provinces or client kingdoms. In addition, Caesar's own works provide two

Table 2.1.1 List of (selected) known generals and officers allotted to them

General	*Number of officers attested*	*Campaigns fought*	*Years active (dates B.C.)*
C. Iulius Caesar	79	Governorship of Spain	61–60
		Civil War	49–44
Cn. Pompeius Magnus	41	Civil War	82–80
		Civil War	49–48
M. Iunius Brutus	16	Governorship of Cisalpine Gaul	46–45
		Civil War	44–42
M. Tullius Cicero	15	Governorship of Cilicia	52–50
L. Licinius Lucullus	11	First Mithridatic War	88–81
		Third Mithridatic War	74–66
M. Licinius Crassus	10	War against Spartacus	72–71
		Parthian campaign	54–53

very detailed accounts of his campaigns over more than a decade. Consequently, historians are comparatively well informed about the composition of Caesar's officer corps and, to a lesser degree, that of Pompey as well. In the cases of Brutus, Cicero, Lucullus, and Crassus, the situation is more difficult. The similar number of officers attested for them is by no means an accurate reflection of their success as generals but rather of the source bias. Thus, Brutus' governorship of Cisalpine Gaul (northern Italy) was a fairly quiet affair, as was Cicero's tenure of Cilicia.[28] But since both his own stay in Cilicia and his dealings with Brutus in general take up a significant part of Cicero's correspondence, there is more information about them than about the dozens of other low-key provincial governorships. On the other hand, Lucullus led a very substantial army against some of Rome's most determinate and successful foes for almost a decade; these were Mithridates of Pontus and his ally Tigranes of Armenia, over whom Pompey was later to achieve his greatest triumph. His campaigns saw him reach the outliers of the Caucasus and bring both rulers to their knees. Likewise, Crassus' campaign against Spartacus included the walling-off by fortifications of a substantial part of southern Italy, and in 55 BC he had command of a punitive expedition against the Parthians (which famously ended in disaster) conferred upon him.[29] But in both cases, it is necessary to rely on historiographical sources for information. No letters to or from Crassus or Lucullus have survived that could shed light on the composition of their officer corps, and thus we are restricted to a number of off-hand remarks concerning their campaigns, made by biographers and historians, who, understandably, tend to focus on other aspects. Still, taking this into account, the relative importance of Caesar and Pompey in the resultant network graph are in all likelihood accurate reflections of their exceptional status as Rome's most successful and important generals.

In this particular graph (Figure 2.1.2), the different shading of node clusters denotes subgroups identified by using the Girvan-Newman algorithms available in UCINET.[30] The two largest subgroups (Nos. 1 and 2) are centered around Caesar and Pompey respectively. In general terms, the impression given by

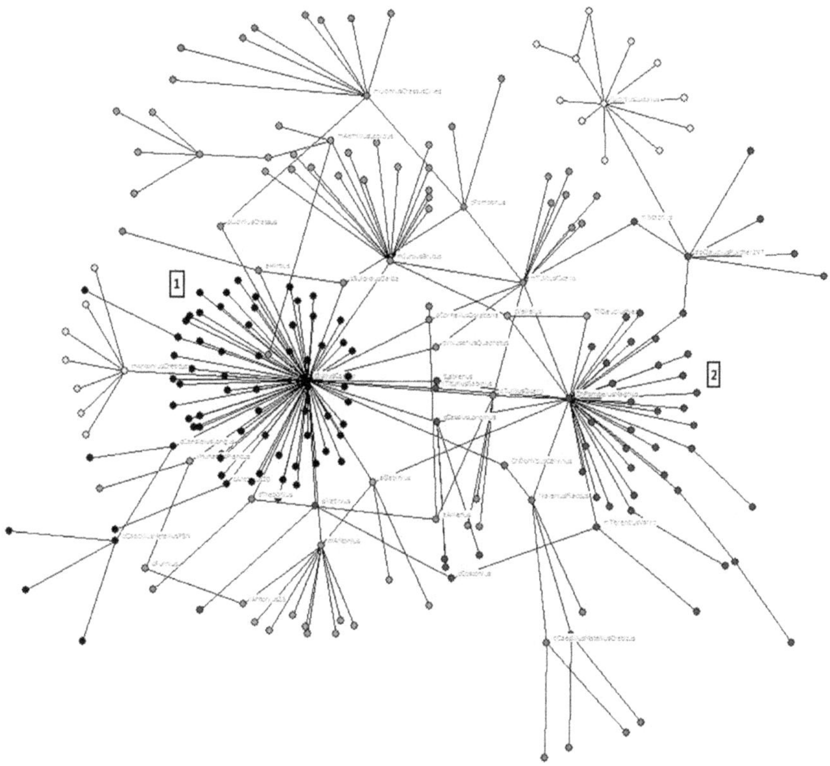

Figure 2.1.2 "Military" network[31]

the graph may be accurate, in that Caesar and Pompey were by all accounts able to accumulate a large military clientele of loyal soldiers and officers bound to them by numerous *beneficiae* obtained from their commanders. While the size of these clienteles was unquestionably unusual, the phenomenon in itself was not, as other provincial commanders were also in a position to gather an entourage around them and to ferment loyalty. Cicero himself is a good example of this. Though he was a decidedly unmilitary man, he was nevertheless forced to act as governor of Cilicia from the end of the 50s BC In this capacity, he commanded military forces and assembled an entourage of experienced commanders such as C. Pomptinus and his brother, Quintus.[32] Since his governorship, in contrast to the many other regular governorships, is well attested in his *Letters*, Cicero's "military" network also figures prominently in this graph (compare with the officers attested in Table 2.1.1). With a total size of seven nodes, his subgroup here rivals that of Crassus (8) and Mark Antony (9). As the former led the campaign against Spartacus and the latter enjoyed a long and distinguished military career and, at the end, command authority over dozens of legions, this is clearly a preposterously prominent position for Cicero to occupy. Thus,

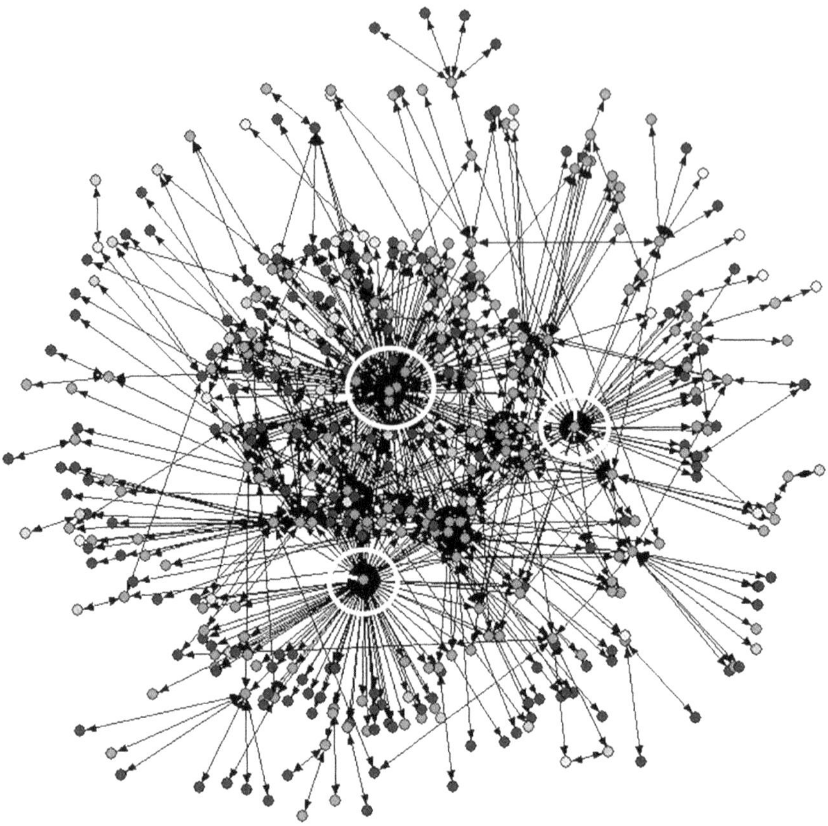

Figure 2.1.3 Whole-network (main component, n = 457)

again, the analytical value of this graph is, on its own, limited (though it usefully shows at first glance the relative strengths and importance of military clienteles).

But while in and of themselves graphs based on individual relations as attested by available sources are not conclusive, they are indispensable stepping stones, as their very lack of expressiveness demonstrates the necessity of a different approach which will, in turn, lead to better results (Figure 2.1.3).

Indeed, the only viable way to gain a glimpse of the basic structural characteristic of the Roman elite from a network perspective is to aggregate individual and difficult-to-interpret discrete networks into a general, whole-network with undirected and unweighted ties. Thereby, it is possible to gain visual representation of the aristocratic network of 78–43 BC, insofar as is presently achievable. This network encompasses 490 nodes connected by a total of 842 ties and represents connections of *amicitia* between senators (220) and knights (191) during the last 35 years of the *res publica libera*. The remaining 79 nodes represent members of different classes of Roman society, whose social status is hard to specify.[33] It consists of one main component of 457 nodes (Figure 2.1.3), as well as 13 smaller (dyadic and triadic) components and a single component with four nodes (the latter two are not shown in this graph). As is readily apparent, Cicero appears as the proverbial spider in the web, both in the graph where Cicero appears at the very center and in the quantitative measures taken.[34] The overall network structure is obvious: aside from Cicero, two other nodes are at the center of a large number of connections (all marked by circles in the graph). They are, naturally, Julius Caesar (right) and Cnaeus Pompey (bottom left), the two most influential politicians in the final stages of the late republic. While each of these three nodes appears at the nexus of its own well-connected network, most other nodes are only connected by single ties. In the case of Caesar and Pompey, this is unproblematic; both are justly famous for their roles in Roman history, and their prominent position in this network is evidence of this. The same, however, cannot be said for Cicero. As an ex-consul and famed orator, he was no doubt a well-known and influential politician, but as far as influence and political power went, he was significantly less important than the other two. His position at the center of the network is due to his overbearing significance in utilized source materials which, it must be remembered, consist mostly of his writings.

To minimize Cicero's distorting influence and so as to get a glimpse of the underlying structure of the network, Cicero is eliminated as a node in the next graph, which also contents itself with undirected ties, as the depiction of directed (but equidirectional) ties offers no analytic advantage but rather complicates visual interpretation (Figure 2.1.4). This figure, with the influence of Cicero eliminated, depicts an aristocratic network still heavily dependent on a number of nodes with a large number of connections, i.e., hubs. While Caesar and Pompey again feature as the most prominent of these hubs (particularly in terms of the sheer number of connections, i.e., in network jargon, their degree centrality), it is now also possible to determine several second-tier connectors as part of the central cluster nodes with a high (if not exceptionally high) degree centrality.

Although it is impossible to delve into a complete analysis of this network,[35] some general observations can be indicated. First, the network graph tends to confirm the notion of Roman society as dependent on personal brokerage and patronage in the distribution of social resources. As the relationships coded for analysis purposes are manifestations of *amicitia*, nodes with the highest degree centrality are those actors at the centre of a large nexus of interpersonal relationships. Given the social conventions governing *amicitia*, it is a reasonable assumption that any two actors connected by a *beneficium* were in some way bound together by the actuality or the potential of a future exchange of *beneficia*. Thus, in theory, since *amicitiae* could be used to further personal economic and political interests, the number of *amicitiae* a single individual could accrue, as represented by the number of connections of any node in the previous graph, may also serve as an indicator of that individual's status and influence within the republican elite.

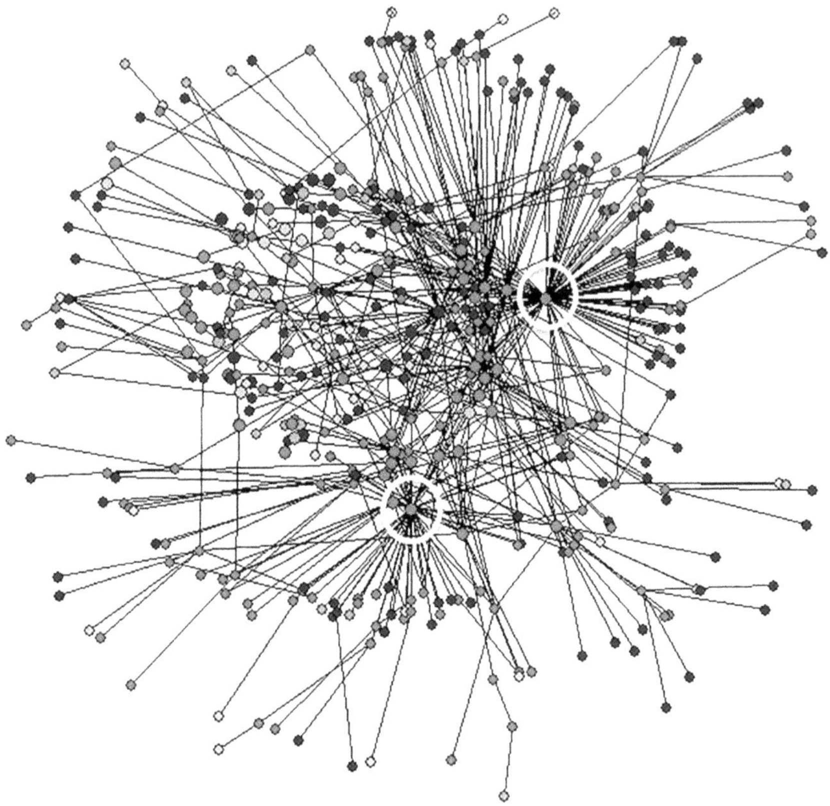

Figure 2.1.4 Whole-network (excluding Cicero)[36]

That this is generally true is confirmed by the fact that the results from SNA correlate largely with the image of Roman elite society traditionally presented based on classical hermeneutic interpretation. Thus, a comparison of centrality measures with received wisdom about the Republican elite seems perfectly reasonable and is suited to reinforce confidence in the general accuracy of the network model constructed as previously explained. Actors with the highest *degree* centrality (given in rounded figures, higher being better) are in descending order: Caesar (104), Pompey (59), Atticus (35), Brutus (32), Crassus (20), Hortensius (19), Antony (18) Aemilius Scaurus (16), Licinius Lucullus (15), Ap. Claudius Pulcher (14), Cassius Longinus (12), Cato (7). By contrast, Cicero's degree centrality is 209, almost twice that of Caesar's. The average degree centrality across the network is three.[37]

Figures for *betweenness* centrality, which attempts to measure the influence of individual nodes by calculating the number of connections passing through them and thus focusing on their role as brokers, largely reflect the same hierarchy: Caesar (30,793), Pompey (18,619), Atticus (5,048), Crassus (5,010), Brutus

(4,891), Lucullus (4,466), Aemilius Scaurus (4,386), Hortensius (4,049), Antony (3,793), Q. Caecilius Metellus Pius Scipio Nasica (3,120), Ap. Claudius Pulcher (2,828), Cassius Longinus (2,346), Cato (1,149).

Still, that being said, it is important to not fall prey to a crude functionalism. In fact, quantitative measures should always be interpreted carefully and with due regard to the theories and suppositions underlying the algorithms used by software solutions such as UCINET. Centrality measures are a good example of this.[38] Whilst *degree* centrality simply measures the number of nodes attached to any node (x), *betweenness* centrality is a reflection of how many ties between other nodes (y, z) use (x) as an intermediary. Another approach to centrality measures taken by Philipp Bonacich called *Bonacich power measures* gives different results. It measures embeddedness, i.e., not only the number of connections of any given node but their quality as well. In Bonacich's scale, negative values are given to those actors with a majority of ties to other equally well-connected actors. Positive values are assigned to those actors whose ties are mostly to isolated, less well connected nodes. The idea behind this is simple: an actor with ties to many isolated nodes is likely to be more influential in his own network, as those nodes tend to rely on him; he is less dependent. In contrast, those nodes connected to other well-connected nodes in a local network are more deeply embedded, that is, their own position in relation to other nodes is weaker, because those nodes do not have to rely on him, and thus he himself is in no position to exert uncontested dominance. Thus, he is more dependent.[39]

Taking *Bonacich power measures* of our network renders interesting results: as might be expected, Caesar (9.737) and Pompey (5.959) have high positive values. While by no means meaning that those figures were not "embedded," that is connected to other important members of the elite, it is a reflection of their special status, based largely on their successful military commands and the concomitant opportunities for patronage. Both had a host of relatively junior supporters, whose history of service with the great generals bound them together. Aemilius Scaurus, on the other hand, while appearing surprisingly important (for a senator who never achieved the consulate) in other *degree* measures, here has a negative score (-1.708), meaning he appears more dependent, whereas Caesar and Pompey clearly dominate within their own networks. This correlates much better with reality than the high *degree* and *betweenness* centrality scores of Scaurus using other measures, for one simple reason: the large number of ties he possesses in our network is a result not of power but of weakness. They are, in actuality, instances of support given to him by *amici*, specifically of *amici* acting as lawyers for Scaurus, who, as far as is reflected in the available admittedly meager sources, has to have been one of the most sued (as well as one of the richest) senators of his day: he was prosecuted for extortion and bribery and was forced to go into exile, even though his legal team included the greatest advocates of his day, Cicero and Hortensius.

There is also a general caveat that must be kept in mind: each tie depicted in the graph stands for a relationship of *amicitia*, extrapolated from any number of *beneficia* that was exchanged between two actors. Thus, while individual ties are proven, the network graph as a whole offers only a model of social embeddedness according to the Roman notions of friendship expounded upon earlier. It is less a map of *actual*, observed resource flow (though that is the basis for the graph) than it is a *model of potential exchange*.[40]

To magnify this point, one example will be briefly considered. Pompey's seemingly unassailable position at the centre of a large personal network, as reinforced both by the graphical visualisation and quantitative measures of *degree* and *betweenness* centrality, has to be reconciled with historical facts, which seemingly argue against any personal dominance. This is best illustrated by the fact that, even during his alliance with Caesar and Crassus in the so-called (first) triumvirate, Pompey came nowhere near the central position of power he aspired to. In fact, it has become a truism that, even with Caesar in Gaul and with Crassus' influence diminished (not to speak of his early death in 53 BC), Pompey remained somewhat surprisingly unable to bring his own influence to bear on matters close to him. This was famously the case during a period in the mid-50s, when opposing aristocratic circles launched an all-out judicial attack on important allies and supporters of the "triumvirate."[41] Despite Pompey's efforts, some of the triumvirate's most important collaborators were successfully prosecuted and went into exile. The reasons for this political defeat are varied: it was certainly a reminder of how powerful senatorial opposition to triumviral politics still was and how much combined clout these opponents still had at their disposal. From a network perspective, however, Pompey's defeat is even more surprising. If we are to take the network model graph as a map of potential support, it is not easy to explain. This in turn must lead to the consideration of other avenues. One possible explanation would be that Pompey, while enjoying a maximum of *possible* ties of support, failed to *realise* these ties in actuality. This ties in with another truism: Pompey has long been seen as a somewhat inept politician, particularly when compared to Caesar. Looking at his actions from the perspective of a network generated by and based on individual favours and *beneficia*, on mutual and exhaustive efforts for friends,[42] and not least on a well-understood but difficult to pin down code of conduct (*humanitas – urbanitas*) expected of a Roman "gentleman" (*vir bonus*) and friend (*bonus amicus*), it may simply be that Pompey was deficient in this particular regard and that he was also, in that sense, a "bad" friend rather than simply a bad politician. While Pompey's politics, his supporters, and his supposed inability to rely on them has recently been the subject of an exhaustive study,[43] the social aspect of his relationship with friends, allies, and supporters has not (or hardly) been considered. Judging from the importance of such "soft" skills evident in the construction and maintenance of personal networks, it may well be time to remedy this, in order to come to a fuller understanding of Pompey's failures.[44] Here (and elsewhere), the combination of social and cultural history, of Social Network Analysis and traditional source criticism, should prove to be helpful.

In other words, in drawing up network graphs, interest should lie less in "mapping" individual favours or services and thus using the graphs mostly for purposes of illustration, and more on generally observing the structure of the network and thus using the graph as a heuristic tool.[45] That being said, two pertinent general remarks can be made about the aristocratic network presented in this chapter.

First, we are clearly dealing with a rather highly centralised network. The vast majority of connections are not direct ties between individual actors exchanging all kinds of services and favours. Instead, most connections are to important hubs, influential aristocrats with a vast number of connections. These individuals are brokers, instrumental in allocating social and material resources.

Second, while the network is not very dense,[46] the average distance between any given nodes is only ~3, which means that, on average, any two actors can be connected to each other by three intermediaries (thus there are, here, a mere three degrees of separation).[47] In this respect, even the least-well connected senator or knight in republican Rome had access to the network, as he was invariably connected by friendship or patronage to members of the network. According to the logic of preferential attachment (or what is widely known as the *Matthew Effect*), all actors, especially those of relatively lower status, seek attachment to influential brokers and to hubs. Thus, senators such as Caesar, Pompey or Crassus invariably attracted a large number of adherents, as they offered the greatest potential advantages.

One problem with this is that preferential attachment as a concept is largely based simply on the number of connections. If this were true, older and established actors would invariably be beyond the reach of younger, aspiring ones, since they had a distinct advantage simply by the fact that they had had 40-plus years to gather adherents and to amass favours. By this logic, to illustrate with an example from the late Republic, the young Octavian, named Caesar's adopted heir by testament, would never have stood a chance against established politicians and Caesarians such as Mark Antony. No one would have followed him, as they had little to gain and much to lose from it. Likewise, while it is obvious to anyone who has read Cicero's letters or any of the other sources available from the years 59–44 BC that Caesar and Pompey were the dominating individuals of their time, this point is further reinforced in the face of the network graphs presented in this chapter (Figures 2.1.3 & 2.1.4), in which they gather the largest number of connections around them. If a large number of connections was to be naively equated with concrete power, Caesar's and Pompey's massively dominant position within the network would have prevented any significant opposition, of which, however, there was a fair amount. How can this be reconciled? And how does the use of SNA/HNR fit into the greater scheme of things?

Conclusion

The apparent contradiction helps with the framing of new questions, and it is in this that the real value of SNA/HNR perspectives for Ancient History lie, namely in forcing historians to take a new look at seemingly long-answered questions.

As has been shown, there are periods and subjects within the field of ancient history, where Historical Network Research is a viable research interest. Given the right source material – typically bodies of correspondence or documentary records – the ancient historian has enough data at his or her disposal to undertake network research and visualise the results of his or her labour in the form of graphs and quantitative measures. This is the easy part.

The difficult part is knowing whether one should attempt any such undertaking and what, precisely, to do afterwards. By its very methodology, SNA forces the historian to take another look at frequently well-known evidence and to systematise and conceptualise it in a new and different way. In drawing up network graphs as visual representation of the evidence (the significance of that step and its potential for, literally, eye-opening re-imaginations should not be underestimated), the historian may start anew and cast a different light on long-known facts. But what historians have to remember in dealing with network graphs aspiring to represent fundamental truths about ancient societies, is that, in fact, they do not. Graphs are mere models, abbreviated, abstract representations of social relations. This is the case for the network described in this chapter, which can be interpreted in two different ways. A priori, it represents an infrastructure for the allocation of scarce resources, whether material (e.g. money) or social (e.g. influence). A network perspective on the late republic has shown it to be a decidedly small world: even from the fragmentary and sometimes contradictory sources at our disposal, we can catch a glimpse of a very well-connected aristocratic society, where personal favours and interactions were *de rigueur* for getting ahead. In and of itself, this is hardly new. Historians of ancient Rome have known for a long time about the importance of personal connections in what may best be termed a face-to-face society. Likewise, the notion of patronage and personal, *fides*-based relationships has been around for a while. Still, the fresh perspective associated with conceptualising personal relationships as networks (with all the theoretical and methodological baggage that this implies) offers new perspectives, which may help historians to better or more fully understand already well-known phenomena. Therein lies the promise (and the future) of Historical Network Research. In and of itself – and this point cannot be stressed enough – SNA seldom produces results. Or, to put it another way, the results of SNA, whether they are network graphs, centrality measures, or even the simple re-thinking of existing presuppositions, have to be combined with, preceded, and followed by a careful interpretation of historical context, sources, and source biases. This final heuristic step is of utmost importance. Since the proof of networks is neither revolutionary, nor their existence at this point (although deeply meaningful) a surprise, Historical Network Research has to be in fact all about what is done *afterwards*, if it is to become more than just a fad.

Notes

1 The following abbreviated case study draws on previously published work, in particular Christian Rollinger, *Amicitia sanctissime colenda. Freundschaft und soziale Netzwerke in der Späten Republik*, Studien zur Alten Geschichte 19 (Heidelberg, Neckar: Verlag Antike, 2014), where a fuller and more detailed argument is available. I wish to

thank the editors of this volume for the opportunity to present my research in this publication.

2 Readers should bear in mind that "SNA" refers to the variety of theories and methodologies commonly subsumed under the moniker of "Social Network Analysis" and thus describes a paradigm not limited to historical research, while "Historical Network Research" is a description that is steadily gaining in acceptance and refers specifically to the application of this paradigm to questions and problems relating to past societies. In what follows I shall use "SNA" to refer to the methodology as such and "HNR" for the developing field in (digital) history.

3 Marten Düring and Linda V. Keyserlingk, "Netzwerkanalyse in den Geschichtswissenschaften. Historische Netzwerkanalyse als Methode für die Erforschung von historischen Prozessen," in *Prozesse: Formen, Dynamiken, Erklärungen*, ed. Rainer Schützeichel and Stefan Jordan, pp. 337–50 (Wiesbaden: Springer VS, 2015), p. 347f.

4 For a review of previous research into social networks in ancient history, see Christian Rollinger and Christian Nitschke, "Network Analysis Is Performed: Die Analyse sozialer Netzwerke in den Altertumswissenschaften: Rückschau und aktuelle Forschungen," in *Knoten und Kanten*, ed. Markus Gamper et al., Sozialtheorie, pp. 213–60 (Bielefeld: Transcript Verlag, 2010–2015) and Christian Rollinger, "Historical Network Research and Ancient History: Some Methodological Prolegomena," in *The Ties That Bind: Network Analysis and Ancient Politics*, ed. Wim Broekaert, Elena Köstner and Christian Rollinger (forthcoming).

5 Sherlock Holmes to Dr John Watson in *The Adventure of the Copper Beeches*, first published in the June 1892 edition of the *Strand Magazine*.

6 Matthias Gelzer, *Die Nobilität der römischen Republik*, 2nd ed. (Stuttgart: B. G. Teubner Verlag, 1983) and Friedrich Münzer, *Römische Adelsparteien und Adelsfamilien* (Stuttgart: J.B. Metzler, 1920).

7 Richard P. Saller, *Personal Patronage Under the Early Empire* (Cambridge: Cambridge University Press, 1982), p. 1.

8 For a more developed version of this argument, see Rollinger, *Amicitia sanctissime colenda. Freundschaft und soziale Netzwerke in der Späten Republik*, especially pp. 52–121 for Roman *amicitia* and pp. 133–352 for the actions and duties associated with it. On *amicitia* in general, see Peter A. Brunt, "*Amicitia* in the Late Roman Republic," in *The Fall of the Roman Republic and Related Essays*, ed. Peter A. Brunt, pp. 351–81 (Oxford: Clarendon Press, 1988); David Konstan, *Friendship in the Classical World*, Key Themes in Ancient History (Cambridge: Cambridge University Press, 2005); David Konstan, "Friendship and Patronage," in *A Companion to Latin Literature*, ed. Stephen J. Harrison, Blackwell Companions to the Ancient World. Literature and Culture, pp. 345–60 (Malden, MA: Blackwell Publishers, 2005); Michael Peachin and Maria L. Caldelli, eds., *Aspects of Friendship in the Graeco-Roman World: Proceedings of a Conference Held at the Seminar für Alte Geschichte Heidelberg, on 10–11 June, 2000*, Journal of Roman Archaeology. Supplementary Series 0043 (Portsmouth: Journal of Roman Archaeology, 2001). Koenraad Verboven, "Friendship Among the Romans," in *The Oxford Handbook of Social Relations in the Roman World*, ed. Michael Peachin, pp. 404–21 (Oxford [etc.]: Oxford University Press, 2014); Koenraad Verboven, *The Economy of Friends: Economic Aspects of Amicitia and Patronage in the Late Republic*, Collection Latomus 269 (Bruxelles: Latomus. Revue d'Études Latines, 2002); Craig A. Williams, *Reading Roman Friendship* (Cambridge: Cambridge University Press, 2012); Christian Rollinger, "Beyond Laelius: The Orthopraxy of Friendship in Late Republican Rome," *Ciceroniana On Line*, no. 1 (2) (2017b): pp. 344–67.

9 Brunt, "*Amicitia* in the Late Roman Republic," pp. 381, 351–81.

10 Matthias Bixler, "Wurzeln der Historischen Netzwerkforschung," in Düring; Eumann; Stark; Keyserlingk, *Handbuch Historische Netzwerkforschung*, pp. 53, 45–62.

11 One notable exception is the presence of individual archives preserved in papyrus *corpora*. Cf. Giovanni Ruffini, *Social Networks in Byzantine Egypt* (Cambridge: Cambridge University Press, 2008).

12 The *Histories* of Sallust, Appian, Diodorus Siculus, Cassius Dio, and Velleius Paterculus, as well as the commentaries on the Gallic and Civil War by Caesar.

13 The *Parallel Lives* of Plutarch, as well as the biographies of Caesar and Octavian/Augustus by Suetonius.

14 As the oldest surviving encyclopedia of Western history, the *Natural History* of Pliny the Elder hardly needs an introduction. The *Memorable Deeds and Sayings* of Valerius Maximus are a collection of anecdotes relating to and structured along the lines of common virtues and vices, while the *Saturnalia* of Macrobius and the *Attic Nights* of Aulus Gellius may best be described as literary cabinets of curiosities.

15 This notion is amply attested to within the ancient sources; see e.g. Ter. Eun. 148. Dion. Hal. ant. Rom. 8.34.1–3. Sen. ben. 1.2.5. Plin. paneg. 85.8.

16 For the full rationale behind the selection of which *beneficia* to include, cf. Rollinger, *Amicitia sanctissime colenda. Freundschaft und soziale Netzwerke in der Späten Republik*, pp. 133–352. For *beneficia* in general, see now also Antje Junghanß, *Zur Bedeutung von Wohltaten für das Gedeihen von Gemeinschaft: Cicero, Seneca und Laktanz über "beneficia"*, Palingenesia 109 (Stuttgart: Franz Steiner Verlag, 2017). In determining to which *ordo* individuals belonged, specialized prosopographies are used, such as Thomas Robert Shannon Broughton and Marcia L. Patterson, *The Magistrates of the Roman Republic, Vol. 1: 509 B.C.-100 B.C.; Vol. 2: 99 B.C.-31 B.C.* (New York: American Philological Association, 1951–1952); Timothy Peter Wiseman, *New Men in the Roman Senate: 139 B.C. – A.D. 14* (London [etc.]: Oxford University Press, 1971); Claude Nicolet, *L'ordre équestre à l'époque républicaine (312–43 av. J.-C.). 2, 2* (Paris: De Boccard, 1974); Israël Shatzman, *Senatorial Wealth and Roman Politics*, Collection Latomus 142 (Bruxelles: Latomus. Revue d'Études Latines, 1975); Klaus Zmeskal, *Adfinitas: Die Verwandtschaften der senatorischen Führungsschicht der römischen Republik von 218–31 v. Chr*, ed. Armin Eich (Passau: Stutz, 2009).

17 The reigning assumption (cf. Jaakko Suolahti, *The Junior Officers of the Roman Army in the Republican Period: A Study on Social Structure*, Annales Academiae Scientiarium Fennicae, Ser. B. 97 (Helsinki: Suomalainen Tiedeakatemia, 1955)) is that those officer ranks served at the pleasure of their general and were not career officers assigned to individual legions. Thus, a Roman general could exert patronage as a matter of course, as the example of Pompey shows: for his campaign against the pirates, he was given authority to personally name 24 legates (Plut. *Pomp*. 25–26. App. *Mithr*. 25). The *tribuni militum* were a partial exception: 24 of these officers were elected by the *populus* and distributed among the regular consular levies (6 tribunes per *legio*). As the need for more troops grew constantly during the middle and late republic, these tribunes were gradually supplemented, and generals were authorized to appoint six military tribunes per additional legion. But due to the enormous benefits a junior officer could reap from his connection to a powerful general, relationships of *amicitia* and (in modern terms) patronage invariably occurred even between commanders and elected (not appointed) tribunes. Those individuals for whom only service in general is attested, without an explicit mention of their commanding officer, are excluded.

18 In this case, all three individuals are taken to be connected by *amicitia*, since, if the addressee (C) acquiesces to the wishes expressed by the author (A), the latter is henceforth morally obligated to C, as is the beneficiary of the letter (B), who is indebted both to A and C. Letters of recommendation thus create a triadic relationship.

19 Amanda Wilcox, *The Gift of Correspondence in Classical Rome: Friendship in Cicero's Ad familiares and Seneca's Moral epistles*, Wisconsin Studies in Classics (Madison: The University of Wisconsin Press, 2012).

20 Thus, in terms of network analysis, any study of Cicero's communication networks almost invariably has to be more or less qualitative in nature, i.e., a study of his ego-network. That is indeed what Michael C. Alexander and James A. Danowski, in "Analysis of an Ancient Network: Personal Communication and the Study of Social Structure in a Past Society," *Social Networks* 12, no. 4 (1990): pp. 313–35 have undertaken.

21 For the philosophical grounding of the Roman "theory" of friendship as it applies to our theme, see Rollinger, *Amicitia sanctissime colenda. Freundschaft und soziale Netzwerke in der Späten Republik*, pp. 52–78.

22 Jean Hellegouarc'h, *Le vocabulaire Latin des relations et des partis politiques sous la République* (Paris: Société d'Édition Les Belles Lettres, 1963).

23 Stephen P. Borgatti, Martin Everett and Lin Freeman, *Ucinet for Windows: Software for Social Network Analysis* (Harvard, MA: Analytic Technologies, 2002). All network graphs used in this chapter were drawn up using this software package.

24 The exceptions are Cratippus, Barkas, Apronius, Caius, Verrius, and Camillus, which cannot be identified with certainty.

25 Graph drawn up with UCINET/NetDraw and reprinted from Rollinger, *Amicitia sanctissime colenda. Freundschaft und soziale Netzwerke in der Späten Republik*, p. 530 with kind permission of *Verlag Antike* (2017). The different shadings of individual nodes indicate their social status (senatorial, equestrian, other) but are of no importance for the general thoughts presented in this chapter. For a discussion of the social composition of these networks and its implications, readers are referred to ibid.

26 Such as, for instance, the triad *mPorciusCatoUticensis-MunatiusRufus-Barkas*, which represents the reconciliation of Cato and Munatius Rufus at a dinner party hosted by Barkas and is attested independently of Cicero, i.e., in Plutarch's biography of Cato (Plut. *Cat. min.* 37.7–10).

27 This is not the place for their discussion, but see Rollinger, *Amicitia sanctissime colenda. Freundschaft und soziale Netzwerke in der Späten Republik*, pp. 319–402.

28 Though he did fight some smaller actions (cf. David Engels, "Cicéron comme proconsul en Cilicie et la guerre contre les Parthes," *Revue Belge de Philologie et d'Histoire*, no. 86 (2008): pp. 23–45).

29 On the scope of Crassus' commands, cf. generally Bruce A. Marshall, *Crassus: A Political Biography* (Amsterdam: Hakkert, 1976), pp. 139–70 and Allen Mason Ward, *Marcus Crassus and the Late Roman Republic* (Columbia, MO: University of Missouri Press, 1977), pp. 83–98. See now J.F. Vervaet, "Crassus Command in the War against Spartacus (73–71 BCE): His Official Position, Forces and Political Spoils," *KLIO*, no. 96 (2014): pp. 607–44 for the war against Spartacus and Katharina Weggen, *Der lange Schatten von Carrhae: Studien zu M. Licinius Crassus*, Studien zur Geschichtsforschung des Altertums 22 (Hamburg: Kovač, 2011) for a special emphasis on the Parthian campaign. On Lucullus, see Arthur Keaveney, *Lucullus: A life*, Classical Lives (London [etc.]: Routledge, 1992), pp. 75–128.

30 Reprinted from Rollinger, *Amicitia sanctissime colenda. Freundschaft und soziale Netzwerke in der Späten Republik*, p. 536 with kind permission of *Verlag Antike* (2017).

31 Michelle Girvan and Mark E.J. Newman, "Community Structure in Social and Biological Networks," *Proceedings of the National Academy of Sciences of the United States of America*, no. 99 (2002): pp. 7821–6.

32 See Engels, "Cicéron comme proconsul en Cilicie et la guerre contre les Parthes," pp. 23–45 and Christian Rollinger, "Ciceros *supplicatio* und aristokratische Konkurrenz im Senat der Späten Republik," *KLIO*, no. 99 (2017a): pp. 192–225 for accounts of his proconsulship.

33 Judging from their names and associations, 30 individuals probably also belong to the upper ranks of society, but their status is impossible to prove conclusively. Forty-nine

actors belong to the remaining ranks of Roman society (slaves, freedmen, citizens and non-citizens).

34 In Figure 2.1.3, the central circled node represents Cicero. The other two circles indicate the position of Caesar (right) and Pompey (bottom) within the network. Cicero's undue influence is also reflected in a number of quantitative measurements (taken for the main component), such as *betweenness* centrality (Cicero = 67321, Caesar = 30793, Pompey = 18619, Crassus = 5010, network mean = 473) or *degree* centrality (Cicero = 209, Caesar = 104, Pompey = 59, Crassus = 20). Interestingly, Bonacich power measures (see later) tend to negate Cicero's dominant position (see Rollinger, *Amicitia sanctissime colenda. Freundschaft und soziale Netzwerke in der Späten Republik*, pp. 393–6).

35 Reprinted from Rollinger, *Amicitia sanctissime colenda. Freundschaft und soziale Netzwerke in der Späten Republik*, p. 544 with kind permission of *Verlag Antike* (2017).

36 For this, the reader is again referred to ibid.

37 All measures were taken with UCINET.

38 Cf. Stephen P. Borgatti, Kathleen M. Carley and David Krackhardt, "Robustness of Centrality Measures under Conditions of Imperfect Data," *Social Networks*, no. 28 (2006): pp. 124–36; Martin Stark, "Netzwerkberechnungen. Anmerkungen zur Verwendung formaler Methoden," in Düring; Eumann; Stark; Keyserlingk, *Handbuch Historische Netzwerkforschung*, pp. 155–72.

39 Philipp Bonacich, "Power and Centrality: A Family of Measures," *American Journal of Sociology*, no. 92 (1987): pp. 1170–82.

40 Cf. Claire Lemercier, "Formal Network Methods in History: Why and How?," in *Social Networks, Political Institutions, and Rural Societies*, ed. Georg Fertig, Rural History in Europe 11, pp. 281–310 (Turnhout: Brepols Publishers, 2015), p. 286. On networks as "maps," see ibid., p. 290: "It is in fact possible to "draw maps" of networks, but only if we remember that the map is not the territory: it concentrates on some precisely defined phenomena, momentarily forgetting everything else." In a similar vein, see Marten Düring, *Verdeckte soziale Netzwerke im Nationalsozialismus: Die Entstehung und Arbeitsweise von Berliner Hilfsnetzwerken für verfolgte Juden* (Berlin, Boston: De Gruyter, 2015), p. 70: "Ihr Zweck ist es nicht, die Komplexität der Wirklichkeit so genau wie möglich zu kopieren, sondern durch die zielgerichtete Reduktion von Informationen Orientierung zu ermöglichen."

41 See Elaine Fantham, "The Trials of Gabinius in 54 B.C.," *Historia*, no. 24 (1975): pp. 425–43 and especially Erich S. Gruen, "Pompey, the Roman Aristocracy, and the Conference of Luca," *Historia*, no. 18 (1969): pp. 71–108 and Erich S. Gruen, *The Last Generation of the Roman Republic* (Berkeley: University of California Press, 1974), pp. 260–357.

42 What has in untranslatable German been called a mutual "Sichaufreiben für die Belange anderer" (Jan Wolkenhauer, *Senecas Schrift De beneficiis und der Wandel im römischen Benefizienwesen*, Freunde – Gönner – Getreue 10 (Göttingen: Vandenhoeck & Ruprecht, 2014), p. 277).

43 Matthias Dingmann, *Pompeius Magnus: Machtgrundlagen eines spätrepublikanischen Politikers*, Osnabrücker Forschungen zu Altertum und Antike-Rezeption 12 (Rahden: Verlag Marie Leidorf, 2007).

44 There are in fact some striking indications of Pompey's unfamiliarity with and inability (or unwillingness?) to follow senatorial-aristocratic conventions, but this is not the place to discuss them. See instead my paper on this subject, Christian Rollinger, "Die kleinen Freunde des großen Pompeius: *amicitiae* und Gefolge in der Späten Republik," in *Gnaeus Pompeius Magnus – Ausnahmekarrierist, Netzwerker und Machtstratege. Beiträge der Heidelberger Pompeius-Tagung am 24. September 2014*, ed. Georg-Philipp Schietinger, Pharos 43, pp. 93–138 (Rahden: Verlag Marie Leidorf, 2019).

45 Cf. now Ulrich Eumann, "Heuristik. Hypothesenentwicklung und Hypthosentest," in Düring; Eumann; Stark; Keyserlingk, *Handbuch Historische Netzwerkforschung*, pp. 123–38; Marten Düring and Florian Kerschbaumer, "Quantifizierung und Visualisierung. Anknüpfungspunkte in den Geschichtswissenschaften," in Düring; Eumann; Stark; Keyserlingk, *Handbuch Historische Netzwerkforschung*, pp. 31–44.
46 Only 0.7% of possible ties are realized.
47 The largest single distance is seven.

Bibliography

Alexander, Michael C., and James A. Danowski. "Analysis of an Ancient Network: Personal Communication and the Study of Social Structure in a Past Society." *Social Networks* 12, no. 4 (1990): pp. 313–35.

Bixler, Matthias. "Wurzeln der Historischen Netzwerkforschung." In Düring; Eumann; Stark; Keyserlingk, *Handbuch Historische Netzwerkforschung*, pp. 45–62.

Bonacich, Philipp. "Power and Centrality: A Family of Measures." *American Journal of Sociology*, no. 92 (1987): pp. 1170–82.

Borgatti, Stephen P., Kathleen M. Carley, and David Krackhardt. "Robustness of Centrality Measures under Conditions of Imperfect Data." *Social Networks Social Networks*, no. 28 (2006): pp. 124–36.

Broekaert, Wim, Elena Köstner, and Christian Rollinger, eds. *The Ties That Bind: Network Analysis and Ancient Politics*. forthcoming.

Broughton, Thomas Robert Shannon and Marcia L. Patterson. *The Magistrates of the Roman Republic. Vol. 1: 509 B.C.-100 B.C.; Vol. 2: 99 B.C.-31 B.C.* New York: American Philological Association, 1951–1952.

Brunt, Peter Astbury. "Amicitia in the Late Roman Republic." In *The Fall of the Roman Republic and Related Essays*. Edited by Peter A. Brunt, pp. 351–81. Oxford: Clarendon Press, 1988.

———, ed. *The Fall of the Roman Republic and Related Essays*. Oxford: Clarendon Press, 1988.

Dingmann, Matthias. *Pompeius magnus: Machtgrundlagen eines spätrepublikanischen Politikers*. Osnabrücker Forschungen zu Altertum und Antike-Rezeption 12. Rahden: Verlag Marie Leidorf, 2007.

Düring, Marten. *Verdeckte soziale Netzwerke im Nationalsozialismus: Die Entstehung und Arbeitsweise von Berliner Hilfsnetzwerken für verfolgte Juden*. Berlin, Boston: De Gruyter, 2015.

Düring, Marten, Ulrich Eumann, Martin Stark, and Linda von Keyserlingk, eds. *Handbuch Historische Netzwerkforschung: Grundlagen und Anwendungen*. Schriften des Kulturwissenschaftlichen Instituts Essen (KWI) zur Methodenforschung 1. Berlin: LIT, 2016.

Düring, Marten, and Florian Kerschbaumer. "Quantifizierung und Visualisierung. Anknüpfungspunkte in den Geschichtswissenschaften." In Düring; Eumann; Stark; Keyserlingk, *Handbuch Historische Netzwerkforschung*, pp. 31–44.

Düring, Marten, and Linda von Keyserlingk. "Netzwerkanalyse in den Geschichtswissenschaften. Historische Netzwerkanalyse als Methode für die Erforschung von historischen Prozessen." In Schützeichel; Jordan, *Prozesse: Formen, Dynamiken, Erklärungen*, pp. 337–50.

Engels, David. "Cicéron comme proconsul en Cilicie et la guerre contre les Parthes." *Revue Belge de Philologie et d'Histoire*, no. 86 (2008): pp. 23–45.

Eumann, Ulrich. "Heuristik. Hypothesenentwicklung und Hypthosentest." In Düring; Eumann; Stark; Keyserlingk, *Handbuch Historische Netzwerkforschung*, pp. 123–38.

Fantham, Elaine. "The Trials of Gabinius in 54 B.C." *Historia*, no. 24 (1975): pp. 425–43.

Fertig, Georg, ed. *Social Networks, Political Institutions, and Rural Societies*. Rural History in Europe 11. Turnhout: Brepols Publishers, 2015.

Gamper, Markus, Linda Reschke, Michael Schönhuth, and Marten Düring, eds. *Knoten und Kanten III*. Sozialtheorie. Bielefeld: Transcript Verlag, 2010–2015.

Gelzer, Matthias. *Die Nobilität der römischen Republik*. 2nd ed. Stuttgart: B. G. Teubner Verlag, 1983.

Girvan, Michelle, and Mark E. J. Newman. "Community Structure in Social and Biological Networks." *Proceedings of the National Academy of Sciences of the United States of America*, no. 99 (2002): pp. 7821–26.

Gruen, Erich S. "Pompey, the Roman Aristocracy, and the Conference of Luca." *Historia*, no. 18 (1969): pp. 71–108.

———. *The Last Generation of the Roman Republic*. Berkeley: University of California Press, 1974.

Harrison, Stephen J., ed. *A Companion to Latin Literature*. Blackwell Companions to the Ancient World. Literature and Culture. Malden, MA: Blackwell Publishers, 2005.

Hellegouarc'h, Jean. *Le vocabulaire Latin des relations et des partis politiques sous la République*. Paris: Société d'Édition Les Belles Lettres, 1963.

Junghanß, Antje. *Zur Bedeutung von Wohltaten für das Gedeihen von Gemeinschaft: Cicero, Seneca und Laktanz über "beneficia"*. Palingenesia 109. Stuttgart: Franz Steiner Verlag, 2017.

Keaveney, Arthur. *Lucullus: A Life*. Classical Lives. London [etc.]: Routledge, 1992.

Konstan, David. "Friendship and Patronage." In *A Companion to Latin Literature*. Edited by Stephen J. Harrison, pp. 345–60. Blackwell Companions to the Ancient World. Literature and Culture. Malden, MA: Blackwell Publishers, 2005.

———. *Friendship in the Classical World*. Digital Printing. Key Themes in Ancient History. Cambridge: Cambridge University Press, 2005.

Lemercier, Claire. "Formal Network Methods in History: Why and How?" In *Social Networks, Political Institutions, and Rural Societies*. Edited by Georg Fertig, pp. 281–310. Rural History in Europe 11. Turnhout: Brepols Publishers, 2015.

Marshall, Bruce A. *Crassus: A Political Biography*. Amsterdam: Hakkert, 1976.

Münzer, Friedrich. *Römische Adelsparteien und Adelsfamilien*. Stuttgart: J.B. Metzler, 1920.

Nicolet, Claude. *L'ordre équestre a l'époque républicaine (312–43 av. J.-C.). 2, 2*. Paris: Boccard, 1974.

Peachin, Michael, ed. *The Oxford Handbook of Social Relations in the Roman World*. Oxford [etc.]: Oxford University Press, 2014.

Peachin, Michael, and Maria L. Caldelli, eds. *Aspects of Friendship in the Graeco-Roman World: Proceedings of a Conference Held at the Seminar für Alte Geschichte Heidelberg, on 10–11 June, 2000*. Journal of Roman Archaeology. Supplementary Series 0043. Portsmouth: Journal of Roman archaeology, 2001.

Rollinger, Christian. *Amicitia sanctissime colenda. Freundschaft und soziale Netzwerke in der Späten Republik*. Studien zur Alten Geschichte 19. Heidelberg, Neckar: Verlag Antike, 2014.

———. "Ciceros *supplicatio* und aristokratische Konkurrenz im Senat der Späten Republik." *KLIO*, no. 99 (2017a): pp. 192–225.

———. "Beyond Laelius: The Orthopraxy of Friendship in Late Republican Rome." *Ciceroniana On Line*, no. 1 (2) (2017b): pp. 344–67.

———. "Die kleinen Freunde des großen Pompeius: amicitiae und Gefolge in der Späten Republik." In *Gnaeus Pompeius Magnus – Ausnahmekarrierist, Netzwerker und Machtstratege. Beiträge der Heidelberger Pompeius-Tagung am 24. September 2014*. Edited by Georg-Philipp Schietinger, Pharos 43, pp. 93–138. Rahden: Verlag Marie Leidorf, 2019.

———. "Historical Network Research and Ancient History: Some Methodological Prolegomena." In *The Ties That Bind: Network Analysis and Ancient Politics*. Edited by Wim Broekaert, Elena Köstner and Christian Rollinger, forthcoming.

Rollinger, Christian, and C. Nitschke. "'Network Analysis Is Performed': Die Analyse sozialer Netzwerke in den Altertumswissenschaften: Rückschau und aktuelle Forschungen." In *Knoten und Kanten III*. Edited by Markus Gamper et al., pp. 213–60. Sozialtheorie. Bielefeld: Transcript Verlag, 2010–2015.

Ruffini, Giovanni. *Social Networks in Byzantine Egypt*. Cambridge: Cambridge University Press, 2008.

Saller, Richard P. *Personal Patronage under the Early Empire*. Cambridge: Cambridge University Press, 1982.

Schietinger, Georg-Philipp, ed. *Gnaeus Pompeius Magnus – Ausnahmekarrierist, Netzwerker und Machtstratege. Beiträge der Heidelberger Pompeius-Tagung am 24. September 2014*. Pharos 43. Rahden: Verlag Marie Leidorf, 2019.

Schützeichel, Rainer, and Stefan Jordan, eds. *Prozesse: Formen, Dynamiken, Erklärungen*. SpringerLink Bücher. Wiesbaden: Springer VS, 2015.

Shatzman, Israël. *Senatorial Wealth and Roman Politics*. Collection Latomus 142. Bruxelles: Latomus. Revue d'Études Latines, 1975.

Stark, Martin. "Netzwerkberechnungen. Anmerkungen zur Verwendung formaler Methoden." In Düring; Eumann; Stark; Keyserlingk, *Handbuch Historische Netzwerkforschung*, pp. 155–72.

Suolahti, Jaakko. *The Junior Officers of the Roman Army in the Republican Period: A Study on Social Structure*. Annales Academiae Scientiarium Fennicae, Ser. B. 97. Helsinki: Suomalainen Tiedeakatemia, 1955.

Verboven, Koenraad. *The Economy of Friends: Economic Aspects of Amicitia and Patronage in the Late Republic*. Collection Latomus 269. Bruxelles: Latomus. Revue d'Études Latines, 2002.

———. "Friendship among the Romans." In *The Oxford Handbook of Social Relations in the Roman World*. Edited by Michael Peachin, pp. 404–21. Oxford [etc.]: Oxford University Press, 2014.

Vervaet, J.F. "Crassus' Command in the War against Spartacus (73–71 BCE): His Official Position, Forces and Political Spoils." *KLIO*, no. 96 (2014): pp. 607–44.

Ward, Allen Mason. *Marcus Crassus and the Late Roman Republic*. Columbia, MO: University of Missouri Press, 1977.

Weggen, Katharina. *Der lange Schatten von Carrhae: Studien zu M. Licinius Crassus*. Studien zur Geschichtsforschung des Altertums 22. Hamburg: Kovač, 2011.

Wilcox, Amanda. *The Gift of Correspondence in Classical Rome: Friendship in Cicero's Ad familiares and Seneca's Moral Epistles*. Wisconsin studies in classics. Madison: The University of Wisconsin Press, 2012.

Williams, Craig A. *Reading Roman Friendship*. Cambridge: Cambridge University Press, 2012.

Wiseman, Timothy Peter. *New Men in the Roman Senate: 139 B.C. – A.D. 14*. London [etc.]: Oxford University Press, 1971.

Wolkenhauer, Jan. *Senecas Schrift De beneficiis und der Wandel im romischen Benefizienwesen*. Freunde – Gönner – Getreue 10. Göttingen: Vandenhoeck & Ruprecht, 2014.

Zmeskal, Klaus. *Adfinitas: Die Verwandtschaften der senatorischen Führungsschicht der römischen Republik von 218–31 v. Chr.* Edited by Armin Eich. Passau: Stutz, 2009.

2.2 Community detection and structural balance

Network analytical modelling of political structures and actions in the Middle Ages

Robert Gramsch-Stehfest

During recent years, historical network research has experienced a considerable upswing.[1] This also applies to Medieval Studies, although the source material comes with a lot of restrictions.[2] Nevertheless, recent publications and ongoing projects have shown that social network analysis (SNA) can be successfully, not only metaphorically, applied in several research fields of medieval history.[3] At the same time, however, there are many difficulties and open questions. Some authors, for instance, explicitly asked for "The limits of the network", beyond which SNA can only produce an illusion of certainty.[4] The discussion about this, as well as other questions, is still in progress, which is urgently needed for the formation of a new branch of historical research.

Historical network research basically consists of four aspects: 1. the extraction of network data from historical sources (data gathering, modelling), 2. the analysis of this data by means of appropriate network analytic methods, 3. the adequate historical interpretation of the results of SNA and 4. if necessary, the formulation of further questions for a more profound investigation. This "workflow" can be understood as a circular process; although it begins with the aspect of historical interpretation and the formulation of adequate research goals and questions, it also needs to simultaneously consider the problems of the availability of sources and the selection of appropriate network analytic concepts and techniques. Hence, all three aspects cannot be considered in isolation. The search for rich data material can lead, for instance, to the research field of political history. In this section of Medieval Studies, an almost inexhaustible number of sources is available, including thousands of charters and letters, chronicles, genealogical information and so on. The aim of the following chapter is to demonstrate how the research process, as a sequence of modelling, analysing, visualisation and interpretation of political networks, can be successfully designed.

After several "turns" in modern historical research, a certain amount of "political return" (Oliver Auge), that is a new interest in political history,[5] can be observed. One reason for this rediscovery of a relatively "old-fashioned" field of research is the availability of new explanatory models, for instance the concept of medieval "consensual leadership" ("konsensuale Herrschaft").[6]

Furthermore, new research methods provide the tools for a deeper understanding of political actions and structures. Social network analysis is one of these new instruments. In the following pages, I would like to introduce into my professorial dissertation „Das Reich als Netzwerk der Fürsten" ("The Empire as a Network of Princes"), which gives an example of the use of SNA in medievist research.[7] This study pays attention to one of the most unfortunate kings of the German Middle Ages, Henry (VII), who lived from 1211 to 1242. The following remarks will focus on methodological questions, which are discussed in the book in greater detail. After a short discussion of some general aspects of SNA in Medieval Studies, I will ask how the medieval German Empire can be modelled as a network and will explain a special analytical technique to clarify its structure and dynamic evolution. At the end, the relation between abstract analytical models and historical social and political reality will be discussed.

First it is necessary to raise the question: which network analytical techniques are particularly promising in the investigation of medieval politics? Unfortunately, in historical network research, this problem is frequently not treated with the necessary diligence. Many researchers are satisfied with the application of standard algorithms "out of the textbook" from common software. In fact, it is always necessary to ask whether these algorithms are suitable for one's own research goals and to allow an adequate and fruitful historical interpretation of the findings. A careful and competent choice is the first precondition for the final success of the work![8] Eventually, historians even should develop their own algorithms for the analysis of specific network data.

It is crucial to keep in mind that the fundamental aim of SNA is the uncovering of hidden structures within a given network scenario. However, a lot of SNA studies content themselves with the reconstruction and visual presentation of historical networks. This is obviously a necessary first step, but in many cases it is insufficient for a clear understanding. This problem shall be illustrated with the help of a simple fictitious example (Figure 2.2.1).

The displayed network with 18 actors and a relatively slight density of 0,21 (this means 21% of the possible relations are really set) seems to be quite intricate.[9] It is impossible to see any structural pattern within it. Hence the SNA offers diverse analytical methods to examine this structure. But which techniques are appropriate? One well-established method of SNA is to measure the importance of the several actors by the calculation of certain network analytic parameters like degree and centrality. But in a historical context these parameters are all too often not very helpful. In many cases they only reflect an already well-known social hierarchy: a king, for instance, should normally be the actor with the greatest degree and closeness, because he is the person with the strongest "gravitational force" in the medieval political system and appears extraordinarily often in written records. In other cases, these parameters are too undetermined to characterize the actors in a clear and distinct way. This is also the problem in the given example. Some actors have a degree centrality of four or five (A, B and C); the actor D is the least linked with a degree of only two. If we compute the closeness centrality, which also takes indirect contacts into consideration, the pattern

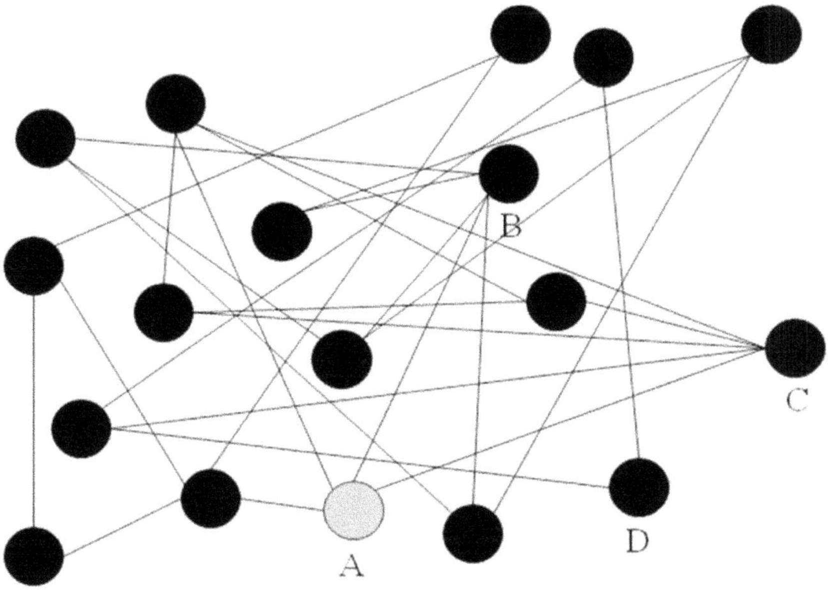

Figure 2.2.1 A fictitious network with 18 actors (D = 0,21)[10]

changes a little. Now actor A takes the first place with C_C = 0,51which means he – or she – can reach any other actor at an average path length of two. Actors B and C follow tightly behind (C_C = 0,44 and 0,45), and actor D has the smallest closeness centrality again (C_C = 0,26, that means he – or she – needs on average four steps to reach any other actor).

As can be seen, the network is centralised relatively high and some actors seem to be much better linked and therefore more important than others. Such an individual "ranking" of actors may be useful for the historical interpretation. But the SNA has more to offer. The displayed network contains a very distinct structure, which is much more significant regarding the historical interpretation. That is why it is necessary to focus on another SNA technique, which deals with the structure of overall networks; cluster analysis or community detection. There are diverse mathematical procedures to identify clusters in a network. I relied on a simple algorithm, which is solely orientated on the parameter of the interior density of a cluster.[11] In this hypothetical case, the result of the cluster analysis is very clear and easy to interpret (Figure 2.2.2).

In this sociogram, which differs from Figure 2.2.1 only in the spatial arrangement of the 18 actors, four clusters with an interior density = 1 or marginal below can immediately be identified. They are separated by structural holes and linked together by the relations of the broker A. His – or her – prominent position has already been proved by the calculation of the closeness centrality. But now his – or her – really outstanding position within the network can be directly

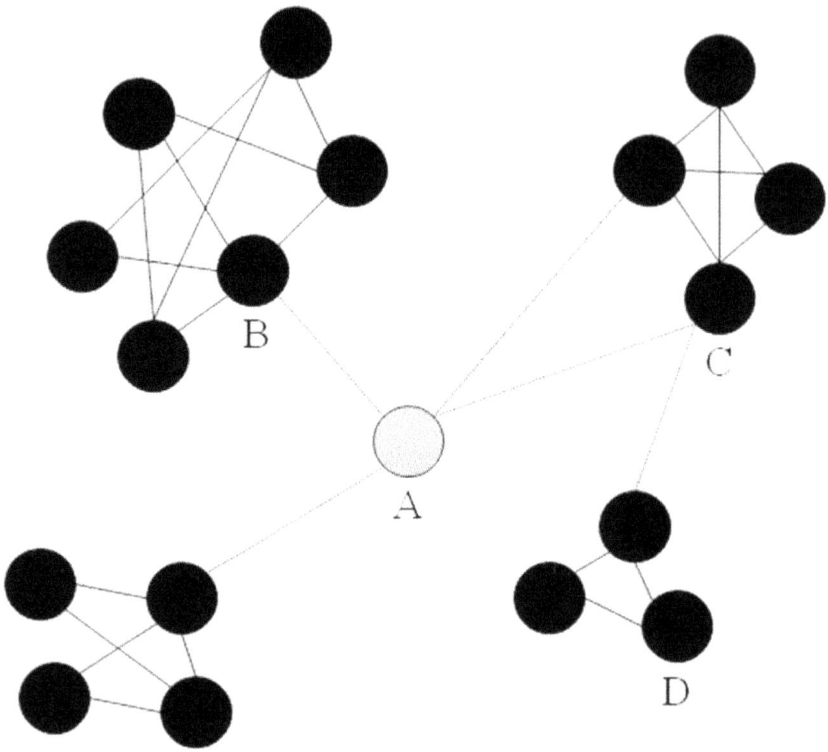

Figure 2.2.2 The same fictitious network as in Figure 2.2.1, arranged in clusters

seen. Between the actors B and C, who are totally equal with regard to centrality parameters, it can be clearly distinguished now. The consequence is that it would be very easy, at this point, to interpret the whole network scenario, depending on the individual historical context.

This simple example shows what SNA should supply. The task is not only the compilation of a list of abstract individual network parameters but rather the unveiling of hidden network structures, which can be displayed in optimised network graphs with the aim of improving the understanding of a given scenario. The exploration of clusters, structural holes and brokers within a network can particularly stimulate the process of historical interpretation in a very productive way. Of course, it must be underlined that this example is only a fictitious one. Even though it could occur in historical reality, it would probably be recognised without the help of analytical procedures: such distinct clusters would emerge clearly enough in historical sources, and it is quite probable that they also have an institutionalised form. But there are a lot of historical networks of much greater complexity that can be successfully analysed with the help of community detection algorithms.[12] Hereinafter, a new technique of community detection shall be introduced, which is a very efficient tool for the analysis of political networks but also offers the key to another important question of historical network research, namely the modelling and understanding of historical dynamics and political change.

In most cases, analytical models of social networks are exclusively based upon positive relations between actors. At first glance this seems reasonable: negative relations, for instance a declared hostility, can merely cause a *lack* of contacts and therefore the *absence* of networks. But there is a great and fundamental difference between missing and hostile links. In many cases, social networks are highly affected by the impact of negative relations, and that is why conflicts should not be ignored in network models! This can be demonstrated by using the example of the ideal network in Figure 2.2.2: the massive structural holes between the four clusters can best be explained by the existence of hidden conflicts. It is a clear "*divide et impera*" situation: In this scenario, the central actor A can keep his – or her – privileged broker position best by playing the various groups off against each other. Failing this, the constellation would not be stable; without conflicts, the holes would rapidly disappear with the mediation of the broker him – or her – self! Consequently, negative relations (conflicts) have the desired property of sharpening the cluster structure of a social network. In addition, there is another important point: if conflicts are systematically taken into consideration, the evolution of networks can be better understood. This problem, the modelling of network dynamics, is one of the great challenges of SNA.[13] Hence I would like to provide a short introduction into an appropriate data model and analytical technique to identify clusters and their dynamics by means of negative relations.[14]

The analytical model of the "Network of Princes" is based upon the theory of structural balance by the Austro-American social psychologist, Fritz Heider (†1988).[15] He pointed out that conflicts can induce dynamic processes into networks by polarising them. The central argument of his theory can best be explained using the example of a triad, which is made up by three actors. They are connected by positive or negative relations in four possible ways (Figure 2.2.3).

The crucial fact is that only balanced triads can be stable while non-balanced triads have the tendency to decay or to change into a balanced status. Heider

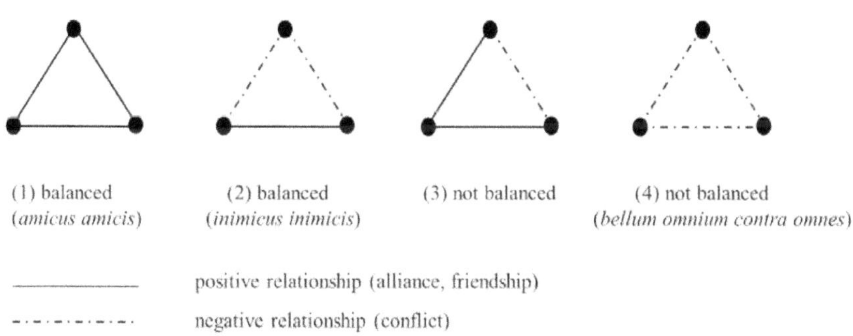

Figure 2.2.3 Cognitive (structural) balance in triads (Heider 1958, Gramsch 2013)

postulated that in non-balanced triads forces occur that determine the change of the triad. For instance, in a triad of type (3), there is a tendency that the actor who is a friend of both other actors will try to reconciliate them (type (1)) or he/she has to take one side against the other (type (2)). Of course, this is not inevitable; the actor could actively resist these forces. But the mechanism works with clear statistical evidence.[16]

This fundamental law of structural balance can also be applied to larger network scenarios because every social network can be understood as a composition of triads. Therefore, the outbreak of a conflict can induce the changing of the involved triads and consequently modify the whole system step by step. A very clear and well-known example of such a chain reaction is provided by the beginning of First World War in 1914.[17] Reciprocally, the ending of a conflict can lead to a chain reaction of general pacification too.

In this perspective, political systems can be understood as networks of many actors that are connected by positive, negative and ambivalent (neutral) relations (dyads). Only in a few cases are such systems simple and bipolar like Europe in 1914. A complex political system with dozens or hundreds of actors contains a lot of balanced and non-balanced triads and is in a permanent state of change. There are, of course, a lot of other parameters that determine these dynamics, but the influence of the network structure itself can best be explained by using the principles of structural balance. However, how can we operationalise these principles? I have developed an algorithm that detects communities of positively linked actors within a given network with friendly and hostile links.[18] To explain its functionality, I will use a simple and well-known fictitious example from a medieval epic poem, "*The Lay of the Nibelungs*" („*Das Nibelungenlied*"). A key scene in this drama is the quarrel between the two queens, Kriemhild and Brunhild, at the front gate of the cathedral of Worms: Kriemhild's words gravely insult Brunhild, and Brunhild swears revenge against Kriemhild's husband, King Siegfried. The following verses describe the reaction of the vassals of King Gunther, Brunhild's husband:

> *By this dispute were many fair women kept apart.*
> *Brunhilda still the matter so sorely to heart*
> *That needs must Gunther's warriors feel pity for the dame.*
> *Then Hagen, knight of Tronjé, unto his lady came.*
> *He bade her say what ail'd her, finding her weeping sore.*
> *Then told she him the story, and unto her he swore*
> *That either Kriemhild's husband must for the lie repent*
> *Or he himself thereafter would never live content.*
> *Ortwein and also Gernot, in council join'd the twain;*
> *And there the heroes plotted how Siegfried should be slain.*[19]

The conflict between the queens consequently divides the Burgundian court into two factions; on the one side, Siegfried and his wife Kriemhild, on the other side, Gunther, Brunhild and their vassals who intend to murder Siegfried.

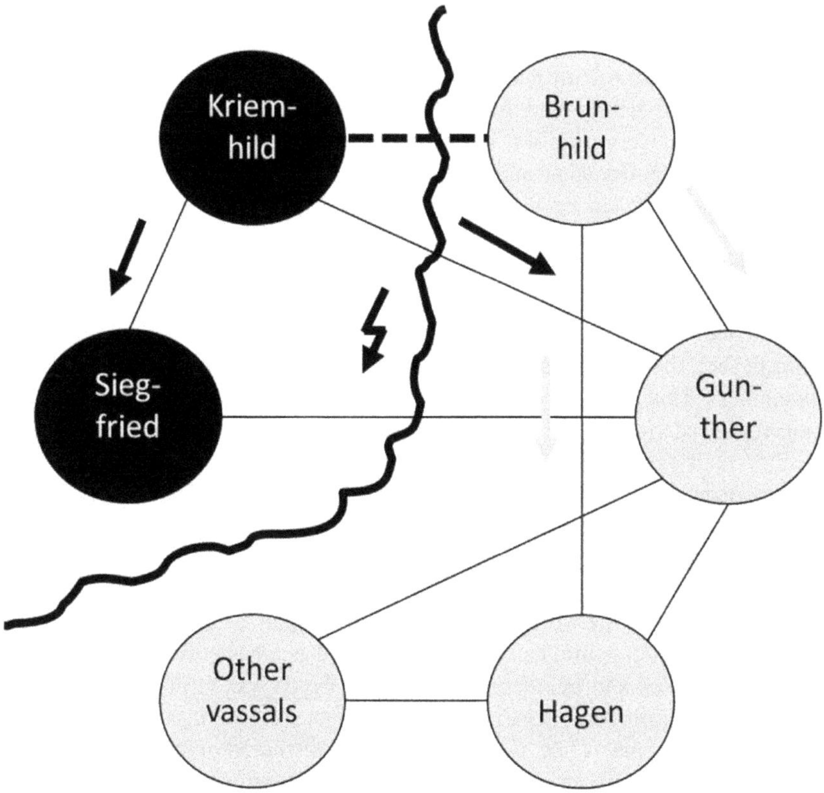

Figure 2.2.4 The Burgundian court and the conflict between Kriemhild and Brunhild (Nibelungenlied, v. 863–865)

It is very easy to translate this plot into a network schema (Figure 2.2.4). The different grey shades visualise the splitting of the network in two opposite groups, which are separated by a line of conflict. The cluster distribution has been calculated by the computer program ("*opposing groups detection*").

How does the algorithm work to determine the opposing groups within the network? As a first step, it identifies the two women as the centres of antagonistic clusters, because these actors are directly connected by a hostile link (dotted line in Figure 2.2.4). They both try to activate other actors as party supporters (arrows along the positive links). Kriemhild wins her husband, King Siegfried (black cluster). Brunhild is in a more complicated situation: she counts on her husband, King Gunther, but Gunther is concurrently Kriemhild's brother and tries to remain neutral and mediate between the queens.[20] That is why the "strategic pact" between Brunhild and the King´s vassal, Hagen, is decisive: Hagen becomes a part of Brunhilds cluster. He influences the decision of the other vassals too, and, all together, they convince (or constrain) Gunther to join Brunhild's party (grey cluster). In the same steps, remarkable concordant to the narration of the poem, the "*opposing groups detection*" algorithm calculates how the clusters are composed.

The sociogram illustrates not only the confrontation of the two clusters but also the ambivalent role of King Gunther who is situated in the broker position: he is Kriemhild's brother and Siegfried's best friend and political partner but his tragedy is that he has to follow the "bad advice" of his vassals. Gunther is by no means free in his decision; an accurate allegory for the political reality in the Middle Ages which has been characterised by Bernd Schneidmüller as „konsensuale Herrschaft" ("consensual leadership")![21]

Today, it is widely acknowledged in Medieval Studies that in the middle ages monarchic power was restricted in many ways. Neither in the German Empire nor, for instance, in medieval England did the command belong exclusively to the king. In fact, the consensus of the princes was essential for a successful royal government. This principle can also be applied to the era of the Roman-German Emperor, Frederick II (1212–1250), and his son, King Henry (VII) (1220–1235):[22] their case is of special historical interest because Frederick II deposed and imprisoned his son in 1235, an outrageous event in German history, which has remained misunderstood to the present day. In my survey, I traced the events that led to this end and developed a new explanation for the fall of Henry. For this purpose, the political system of the medieval German Empire was modelled as a network of princely and comital actors subjected to the rules of Heider's structural balance. This "Empire as a Network of Princes" constituted a complex and interdependent system, which was formed by numerous interactions of positive or negative character traits, such as kinship, political alliances, personal contacts (e.g. document witness), territorial or status competition, legal and military conflicts and so on. This political environment also determined the royal activities: frictions and conflicts between the princes had a negative impact on the relation of both Staufian sovereigns. In this way, the circumstances of the conflict between father and son could be presented in a completely new light.

The first task of the investigation was to reconstruct these most changeable social relationships as exactly as possible and to establish a database for the computer-aided analysis based on sources and literature. One short example may illustrate the technique of data gathering and network analytical modelling: the following two sociomatrices show a small part of the network, consisting of the Bishop of Utrecht, the Archbishop of Bremen, the Counts of Geldern and Oldenburg and the Lords of Lippe (Figure 2.2.5).

The upper sociomatrix represents the network status in the summer of 1227 before the death of Bishop Otto of Utrecht in a conflict with Frisian rebels. Otto was a member of the Lippe family, and his brother was the Archbishop of Bremen. That is why a "1" can be found in the corresponding positions of the first line (2nd and 5th columns), which encodes a positive link. Furthermore, the Count of Geldern was an ally of Otto ("1" in the 3rd column). After Otto´s death, Bishop Wilbrand of Paderborn, a member of the family of the Counts of Oldenburg, was transferred to Utrecht.[23] Consequently, the system of alliances

```
12345
-----,
\11-1| 1  #BUtr.
 \N11| 2  #EBBrem.
  \-N| 3  #GrGeld.
   \-| 4  #GrOldenbg.
    \| 5  #HeLippe.
```

```
12345
-----,
\-11-| 1  #BUtr.
 \N11| 2  #EBBrem.
  \-N| 3  #GrGeld.
   \-| 4  #GrOldenbg.
    \| 5  #HeLippe.
```

Figure 2.2.5 The network of bishops and counts in northern Germany before and after the transfer of Wilbrand of Oldenburg from the bishopric in Paderborn to Utrecht (August 1227)

around the bishop´s seat of Utrecht changed (new first line in the lower sociomatrix): The links between Utrecht with respect to Bremen and Lippe are deleted (encoded by "-"), and a new link between Utrecht and the Counts of Oldenburg appears (4th column).[24] In this way, every historical event, as well as other relevant information about the configuration of this medieval political system, can be systematically translated into the "language" of a network model.

The data gathering started with about 150 actors, but only a set of 68 main actors was finally included in the analysis. The political relations and interactions between them over a period of ten years (from 1225 to 1235) are represented by about 3.000 dyads. Many of them are only repeated contacts between the same actors. But nonetheless: at every point in time, the whole network contains on average 350 or 400 dyads, which corresponds to a network density of about 16%. For such a large network, this is a very considerable value. The following sociogram shows the "naked" political relations at the starting point of the investigation, that is March 1225 (Figure 2.2.6):

In this picture, the actors are topographically located. The most prominent actors (kings, dukes, archbishops among others) are symbolised by quadrats, labelled with self-explanatory acronyms. Less important actors (counts, bishops) are marked by circles without identifiers (for reasons of readability). They are connected by thick, fine and dashed lines that symbolise hostile, friendly and neutral (ambivalent) relations.[25] The whole image, however, looks very confusing at first sight. The visualisation of such a complex historical network cannot give any ideas for an adequate comprehension. That is why it is necessary to use additional

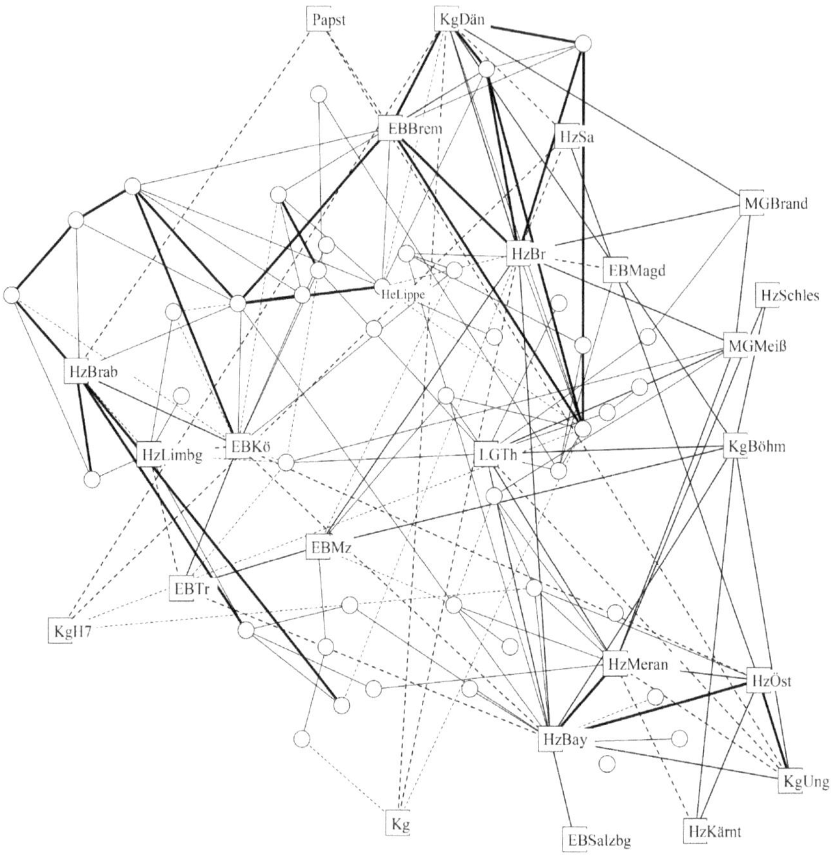

Figure 2.2.6 The German political network in March 1225 (Gramsch 2013, simplified presentation)

network analytic methods to support the historical interpretation. For this purpose, I used the "*opposing groups detection*" algorithm, which was described before in the example of "*The Lay of the Nibelungs*". The result can be seen in the following picture (Figure 2.2.7):

In this sociogram, the most relevant information consists of the distribution of actors in different clusters. In this way, the chaotic multiplicity of actors and relations is more easily comprehensible.[26] The conflicts give the network a relatively clear structure with three clusters (black, grey and striped) which cum grano salis can be regarded as political factions. They do not have to be densely cross-linked; it is mainly the absence of interior conflicts that connects them. The striped cluster, for instance, is a group of North German actors. It is particularly connected by the kinship ties of the Lords of Lippe[27] while the external border of the group is marked by military and political conflicts in northern Germany and

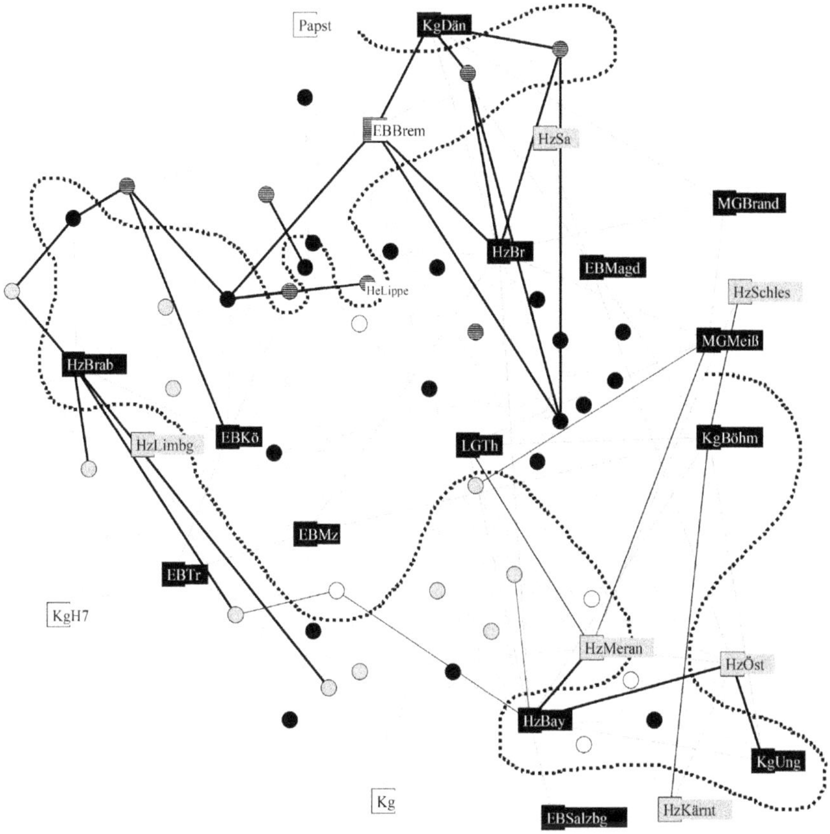

Figure 2.2.7 The cluster structure of the German political network in March 1225 (Gramsch 2013, simplified presentation)

in the Lower Rhine region against "black" actors (especially Denmark, the Lord of Brunswik and the Archbishop of Cologne).[28]

The most important result of the cluster analysis is the detection of a dualistic structure of the network of princes in 1225: we find a big black cluster in the geographic centre of the empire, which is surrounded by a striped cluster in the north and a grey cluster in the west and respectively south of Germany. Both parties are separated by a thick dotted line, which connects the various conflicts.[29] While one previously considered these hostilities in isolation, it is now possible to find out hidden relations between them and to identify the overall framework of imperial politics. It can be shown that Emperor Frederick II in 1225 predominantly collaborated with actors of the grey cluster while King Henry (VII) tended to form alliances with "black" actors. The later conflict between the Staufian father and son thus already casts its shadow!

Compared with the picture painted by earlier research, these results are absolutely new and original. SNA does not provide, of course, "hard facts", but it offers an insight into the general structural characteristics of a given historical scenario, which can be used as a hypothesis for further examination. Interesting findings are also provided with regard to the position of individual actors. In "*The Lay of the Nibelungs*", for example, King Gunther plays a specific role because he holds the broker position between the opposite groups, thanks to his contacts to Kriemhild and Siegfried. Such "bridge ties" between members of opposite clusters are highlighted in Figure 2.2.7 by thin black lines, while the ties within the clusters are grey. For instance, the Landgrave Ludwig of Thuringia ("LGTh", in the centre) had positive relations with his uncle, the Duke of Bavaria ("HzBay", black) as well as to the Duke of Merania ("HzMeran", grey) who was a relative of Ludwig's wife, the former Saint Elisabeth. Bavaria and Merania, however, were enemies at this time. This is a very important indication for historical interpretation, as the Landgrave who was not an active part in these conflicts had a strategic broker position, which he used very effectively. His defection from the black to the grey cluster, which took part in the summer of 1225, decided the affair of the marriage of King Henry (VII) – a first "stumbling block" in the relation of the two Staufian rulers.[30]

This example shows in detail how a network model can be used for historical explanation. The cluster analysis makes a complex scenario comprehensible and gives a guideline for the interpretation of sources as well as for the description of the history of events and the underlying political structures. In the given case, the history of 13th-century Germany, this approach is particularly fruitful because the multiplicity of relevant actors and relations prohibited, until recently, a clear understanding. Already, contemporary historiography has surrendered to the complexity of the matter. Not coincidentally do great historical "master-narratives" of the late Staufian era along the lines of Otto of Freisings "*Gesta Friderici*" not exist.[31] It was impossible for medieval chroniclers as well as for modern historians to report the history of the empire only in the form of a conventional biography of a king. Network analysis, however, focuses on the interrelations of the entire system and thus can suggest possible explanations of concrete events as well as uncover long-term structural patterns.

The analysis of such a network scenario, which is no more than a snapshot of a historical moment, is admittedly only a first step. To analyse a longer historical course it is necessary to repeat the described cluster analysis for other points in time. As shown before (Figure 2.2.5), every historical event causes the addition or the deletion of positive or negative relations in the network model with the consequence that the network structure changes. This change, the transformation of the clusters, is most notably apparent in the changing colouring of the actors.[32] Such a "movie clip" of an evolving network typically consists of about ten different sociograms per year. By the "crossfading" of these images, it is possible to identify the more stable parts of the network; groups of actors who stay together in one cluster for a longer period of time. These techniques open further possibilities of historical interpretation, which cannot be described here more in detail.

Finally, I would like to briefly discuss the question of the relation between abstract analytical models and the complex historical social and political reality. The clarity of the sociograms should not deceive us: in fact, the exact composition of the clusters often remains uncertain and erratic. Moreover, black, grey and striped nodes are seldom real political entities; in some cases, various actors of the same large cluster, which appears in the network model, may probably never have heard of each other. Needless to say, the whole network itself is only a very simple and reduced model with a lot of information gaps caused by a lack of sources and so on. And of course, there was no political player in the middle ages who possessed a coloured map of the political landscape that facilitated decision-making.

Nevertheless, the modelling of the German Empire as a "Network of Princes" proves to be very useful. It suits medieval social and political mentality very well: in a weakly institutionalised face-to-face society everybody was aware of the great importance of social relations and concrete interactions. Numerous examples demonstrate that the rules of political networks, the importance of friends and loyalty, active "networking strategies" (e.g. nepotism), structural balance and so on, were ubiquitous, and they really regulated political action.[33] Therefore, it is possible to model and analyse such a political system by means of SNA. The results of this network analysis obviously exceed the horizons of the contemporary actors as they are artificial and in some respects hypothetical. But they allow modern historians to realise the emergent effects of political networks, ideally much better than the medieval actors could. Emperor Frederick II, for instance, probably did not really understand why he was so successful in his fight against his son in 1235. But it is evidently a legitimate aim of historical research to find adequate explanations for his success.[34] The introduction of perspectives and methods of SNA into political history broadens the horizons of historians. But it is also fruitful for SNA itself; particularly in the field of political history, it is possible to observe dynamic network processes and to find ways for modelling and explaining them.

Notes

1 As an introduction into German historical network research see Marten Düring, Markus Gamper and Linda Reschke, *Knoten und Kanten III: Soziale Netzwerkanalyse in der Geschichts- und Politikforschung* (Bielefeld, 2015); Marten Düring, Ulrich Eumann, Martin Stark and Linda von Keyserlingk, eds., *Handbuch Historische Netzwerkforschung: Grundlagen und Anwendungen*, Schriften des Kulturwissenschaftlichen Instituts Essen (KWI) zur Methodenforschung 1 (Münster, 2016) (especially the paper of Christian Marx, "Forschungsüberblick zur Historischen Netzwerkforschung. Zwischen Analysekategorie und Metapher," in Düring; Eumann; Stark; Keyserlingk, *Handbuch Historische Netzwerkforschung: Grundlagen und Anwendungen*, pp. 63–84, which gives also an overview over the international research in this field).

2 Eva Jullien, "Netzwerkanalyse in der Mediävistik. Probleme und Perspektiven im Umgang mit mittelalterlichen Quellen," *Vierteljahrschrift für Sozial- und Wirtschaftsgeschichte*, no. 100 (2013): pp. 135–53; Robert Gramsch, "Zerstörte oder verblasste Muster? Anwendungsfelder mediävistischer Netzwerkforschung und

das Quellenproblem," in Düring; Eumann; Stark; Keyserlingk, *Handbuch Historische Netzwerkforschung: Grundlagen und Anwendungen*, pp. 85–99.

3 See for instance Mike Burkhardt, *Der Hansische Bergenhandel im Spätmittelalter. Handel – Kaufleute – Netzwerke*, Quellen und Darstellungen zur Hansischen Geschichte 60 (Köln, Wien, Weimar, 2009); Richard Engl, "Das Ende muslimischen Lebens im mittelalterlichen Süditalien. Netzwerkanalytische Überlegungen zu einer hundertjährigen Forschungsfrage," in *Gesellschaftliche Umbrüche und religiöse Netzwerke. Analysen von der Antike bis zur Gegenwart*, ed. Daniel Bauerfeld and Lukas Clemens, pp. 119–54 (Bielefeld, 2014); Johannes Preiser-Kapeller, "Mapping MEDieval CONflicts: A Digital Approach towards Political Dynamics in the Pre-Modern Period (Ongoing Project, Online [22 February 2017], https://oeaw.academia.edu/MappingMedievalConflict). Also, in international research there are a lot of interesting publications and projects in this field. To give a few examples: the doctoral thesis of Hervin Fernández-Acheves (University of Leeds) deals with the network of aristocratic elites in medieval Norman Sicily; Matthew Hammond (University of Glasgow) and Cornell Jackson (Kings College London) analysed Scotch charters of the „People of Medieval Scotland 1094–1314 (PoMS) Database" by means of network analytical methods (e-book, see the homepage of the project: www.poms.ac.uk); Isabelle Rosé (University of Rennes 2) retraced and analysed the personal network of Abbot Odo of Cluny, see Isabelle Rosé, "Reconstitution, représentation graphique et analyse des réseaux de pouvoir au haut Moyen Âge. Approche des pratiques sociales de l'aristocratie à partir del'exemple d'Odon de Cluny († 942)," *REDES – Revista hispana para el análisis de redes sociales*, no. 21 (2011): pp. 199–272.

4 Kerstin Hitzbleck and Klara Hübner, *Die Grenzen des Netzwerks 1200–1600* (Ostfildern, 2014).

5 Oliver Auge, *Handlungsspielräume fürstlicher Politik im Mittelalter. Der südliche Ostseeraum von der Mitte des 12. Jahrhunderts bis in die frühe Reformationszeit*, Mittelalter-Forschungen 28 (Stuttgart, 2009), p. 6.

6 Bernd Schneidmüller, "Konsensuale Herrschaft. Ein Essay über Formen und Konzepte politischer Ordnung im Mittelalter," in *Reich, Regionen und Europa in Mittelalter und Neuzeit*, ed. Paul-Joachim Heinig et al., Historische Forschungen 67, pp. 53–87 (Berlin, 2000).

7 Robert Gramsch, *Das Reich als Netzwerk der Fürsten. Politische Strukturen unter dem Doppelkönigtum Friedrichs II. und Heinrichs (VII.) 1225–1235*, Mittelalter-Forschungen 40 (Ostfildern, 2013) (see there for further information and evidences).

8 On the subject of SNA there exist a lot of useful introductory literature. A very influential German introduction is the textbook of Dorothea Jansen, *Einführung in die Netzwerkanalyse: Grundlagen, Methoden, Forschungsbeispiele*, 3rd ed. (Wiesbaden, 2006). See also Stanley Wasserman and Katharina Faust, *Social Network Analysis: Methods and Applications* (Cambridge, 1999); Wouter de Nooy, Andrej Mrvar and Vladimir Batagelj, *Exploratory Social Network Analysis with Pajek*, Structural Analysis in the Social Sciences 27 (Cambridge [u.a], 2009); Marc E.J. Newman, *Networks: An Introduction* (Oxford, 2012).

9 All the following sociograms have been created with the help of WORD (Version 6.0 and 97–2003) by the author.

10 The use of network analytical concepts and parameters like density, degree and closeness centrality is oriented on the textbook of Jansen, Einführung (wie Anm. 8).

11 The interior density is a parameter that characterises the connectivity of the involved actors: an interior density = 1 means that all actors of the cluster are connected to each other. See also Jansen, *Einführung in die Netzwerkanalyse: Grundlagen, Methoden, Forschungsbeispiele*, p. 200.

12 Community detection algorithms like the previously described or the Girvan-Newman algorithm can help to identify clusters in all kinds of social networks, for instance in

networks of scholars or kinship networks. For further examples see Robert Gramsch, "Zerstörte oder verblasste Muster? Anwendungsfelder mediävistischer Netzwerkforschung und das Quellenproblem," in Düring; Eumann; Stark; Keyserlingk, *Handbuch Historische Netzwerkforschung: Grundlagen und Anwendungen*, pp. 85–99 and Robert Gramsch, "Networks of Noblemen and Scholars: Social Groups, Group Consciousness and Their Institutionalization in the Late Middle Ages," in *Collective Identity in the Context of Medieval Studies*, ed. Michaela Antonín-Malaníkova and Robert Antonín, pp. 13–31 (Ostrava, 2016).

13 Jansen, *Einführung in die Netzwerkanalyse: Grundlagen, Methoden, Forschungsbeispiele*, p. 275. An interesting recent example of the discussion of dynamics and emergent effects of networks is the study of John F. Padgett and Walter W. Powell, *The Emergence of Organizations and Markets* (Princeton, 2012).

14 For more details see Gramsch, *Das Reich als Netzwerk der Fürsten. Politische Strukturen unter dem Doppelkönigtum Friedrichs II. und Heinrichs (VII.) 1225–1235*, especially chapter 1.2. and 1.3.

15 Fritz Heider, *The Psychology of Interpersonal Relations* (New York, 1958); Dorwin Cartwright and Frank Harary, "Structural Balance: A Generalization of Heider's Theory," *Psycological Review*, no. 63 (1956): pp. 277–93; Paul W. Holland and Samuel Leinhardt, "Social Structure as a Network Process," *Zeitschrift für Soziologie*, no. 6 (1977): pp. 368–402; Jansen, *Einführung in die Netzwerkanalyse: Grundlagen, Methoden, Forschungsbeispiele*, p. 40.

16 Gramsch, *Das Reich als Netzwerk der Fürsten. Politische Strukturen unter dem Doppelkönigtum Friedrichs II. und Heinrichs (VII.) 1225–1235*, p. 40.

17 See for instance Christopher Clark, *The Sleepwalkers: How Europe Went to War in 1914* (London, 2012) (in German: Die Schlafwandler: Wie Europa in den Ersten Weltkrieg zog, München 2013). An interesting thought experiment was made by Niall Ferguson, who discussed the possibility that the government of the British Empire had resisted the "attraction force" of the system of alliances and had abstained from the war. See Niall Ferguson, *The Pity of War: Explaining World War I* (London, 1998)(in German: Der Erste Weltkrieg und das 20. Jahrhundert, 3 Aufl. Stuttgart 1999).

18 The algorithm ("*opposing groups detection*") was programmed in WORD Basic (macro programming in WORD 6.0). At the present time (2017) I cooperate with Silvio Dahmen and Ana Bazzan (Porto Alegre, Brazil) to compare the software with a similar standard algorithm ("*spinglass algorithm*") and to create a new program version on the base of the programming language Python. See Silvio R. Dahmen, Ana L.C. Bazzan and Robert Gramsch, "Community Detection in the Network of German Princes in 1225: A Case Study," in *Complex Networks VIII: Proceedings of the 8th Conference on Complex Networks CompleNet 2017*, ed. Bruno Gonçalves et al., pp. 193–200 (Springer, 2017).

19 Alice Horton, *The Lay of the Nibelungs: Metrically Translated from the Old German Text*, Vols. 863–865 (London, 1901), p. 147s, www.archive.org/stream/nibelungslay00hortrich#page/146/mode/2up (accessed 21 February 2017).

20 The black and grey arrows in Figure 2.2.4 symbolise the attraction forces exerted on the actors who are directly linked with the hostile queens. These forces are not weighted – that's why the attraction forces between Kriemhild and her brother Gunther and between Brunhild and her husband Gunther are equal, and they neutralise each other.

21 Schneidmüller, "Konsensuale Herrschaft. Ein Essay über Formen und Konzepte politischer Ordnung im Mittelalter," pp. 53–87. The poet of the "*Nibelungenlied*" explicitly criticises this subjection of King Gunter under the will of his princes (v. 876): "*Unto his vassal Gunther in evil hour gave ear. With treason foul to tamper ere any grew aware*".

22 The following biographies and monographs may serve as an overview of the topic: Thomas C. van Cleve, *The Emperor Frederick II of Hohenstaufen. Immutator mundi* (London, 1972); Wolfgang Stürner, *Friedrich II (Gestalten des Mittelalters und der Renaissance)*, 2nd ed. (Darmstadt, 1992/2000); Hubert Houben, *Kaiser Friedrich II. (1194–1250). Herrscher, Mensch und Mythos*, Urban-Taschenbuch 618 (Stuttgart, 2008); Emil Franzel, *König Heinrich VII. von Hohenstaufen. Studien zur Geschichte des "Staates" in Deutschland*, Quellen und Forschungen aus dem Gebiet der Geschichte 7 (Prag, 1929); Christian Hillen, *Curia regis: Untersuchungen zur Hofstruktur Heinrichs (VII.) 1220–1235 nach den Zeugen seiner Urkunden*, Europäische Hochschulschriften, Reihe 3: Geschichte und ihre Hilfswissenschaften 837 (Frankfurt, 1999).

23 The historical background is shortly explained in Michael Matscha, *Heinrich I. von Müllenark, Erzbischof von Köln (1225–1238)*, Studien zur Kölner Kirchengeschichte 25 (Siegburg, 1992), p. 256 and Erwin (Hg.) Gatz, *Die Bischöfe des Heiligen Römischen Reiches 1198–1448: ein biographisches Lexikon* (Berlin, 2001), p. 824.

24 The sociomatrix contains also information about the contacts between the other actors. Some of these contacts are neutral/ambivalent, encoded by an „N" (negative links are encoded by a „0"). This part of the network (the lower part of the sociomatrix) remains unchanged, because it is not affected by the event (the transfer of the bishop).

25 The numerous positive contacts between the Staufian rulers, respectively the Pope and the princes, are not included in the sociogram.

26 For purposes of a better visualisation, positive links between members of the same cluster and all neutral/ambivalent links are plotted in a grey tone because these links are not so important for the understanding of the network scenario. Conflicts are plotted as black bold lines; positive links between different clusters ("bridge ties") are marked by black thin lines.

27 The whole cluster is composed of eight actors with 12 ties all in all (D = 0,43). The degrees of the Lords of Lippe amount to five; their closeness centralities within the green cluster amount to 0,78. See also their genealogical table in Detlev Schwennicke, *Europäische Stammtafeln. Stammtafeln zur Geschichte der europäischen Staaten, N.S. I.3* 335 (Marburg, 2000).

28 For more details see Gramsch, *Das Reich als Netzwerk der Fürsten. Politische Strukturen unter dem Doppelkönigtum Friedrichs II. und Heinrichs (VII.) 1225–1235*, chapter 2.5.

29 We can distinguish four trouble spots: 1.) The conflict in northern Germany between Denmark and Brunswik at the one and some northern German actors (the striped cluster) at the other side. 2.) The conflict in the Lower Rhine region with the Bishop of Utrecht in the centre. 3.) The inheritance battle after the death of Countess Gertrud of Dagsburg (†1225.III.19) in western Germany (Duke of Brabant and others). 4.) Competition and open conflicts in the southeast of Germany between the Dukes of Austria and Merania at the one and Bavaria-Hungaria at the other side. For more details see ibid., pp. 90–4 and following chapters.

30 For further details see ibid., chapter 2.2.

31 Otto von Freising and Rahewin, *Die Taten Friedrichs oder richtiger Cronica, hg. von Franz-Josef Schmale, übers. von Adolf Schmidt*, 3rd ed., Freiherr v. Stein-Gedächtnisausgabe 17 (Darmstadt, 1986) – the well-known chronicle of the first years of the reign of Frederick I Barbarossa.

32 It should be noted that the positions of the actors in the sociograms are fixed, because only in this way is a direct comparison of the different network scenarios and the visualisation of their change possible.

33 Only two studies on this topic shall be mentioned here: Gerd Althoff, *Verwandte, Freunde und Getreue: zum politischen Stellenwert der Gruppenbindungen im früheren Mittelalter* (Darmstadt, 1990) and Claudia Garnier, *Amicus amicis – inimicus*

inimicis. Politische Freundschaft und fürstliche Netzwerke im 13. Jahrhundert, Monographien zur Geschichte des Mittelalters 46 (Stuttgart, 2000).

34 A very well-known example of the potential of SNA to explain the success of political actors is the survey of John F. Padgett and Christopher K. Ansell, "Robust Action and the Rise of the Medici," *American Journal of Sociology*, no. 98 (1993): pp. 1259–319.

Bibliography

Althoff, Gerd. *Verwandte, Freunde und Getreue: zum politischen Stellenwert der Gruppenbindungen im früheren Mittelalter*. Darmstadt: Wissenschaftliche, Darmstadt, 1990.

Antonín-Malaníkova, Michaela, and Robert Antonín, eds. *Collective Identity in the Context of Medieval Studies*. Ostrava: University of Ostrava, 2016.

Auge, Oliver. *Handlungsspielräume fürstlicher Politik im Mittelalter. Der südliche Ostseeraum von der Mitte des 12. Jahrhunderts bis in die frühe Reformationszeit*. Mittelalter-Forschungen 28. Stuttgart: Thorbecke, 2009.

Bauerfeld, Daniel, and Lukas Clemens, eds. *Gesellschaftliche Umbrüche und religiöse Netzwerke. Analysen von der Antike bis zur Gegenwart*. Transcript: Bielefeld, 2014.

Burkhardt, Mike. *Der Hansische Bergenhandel im Spätmittelalter. Handel – Kaufleute – Netzwerke*. Quellen und Darstellungen zur Hansischen Geschichte 60. Köln, Wien, Weimar: Böhlau, 2009.

Cartwright, Dorwin, and Frank Harary. "Structural Balance: A Generalization of Heider's Theory." *Psychological Review*, no. 63 (1956): pp. 277–93.

Clark, Christopher. *The Sleepwalkers: How Europe Went to War in 1914*. London: Allen Lane, 2012.

Dahmen, Silvio R., Ana L.C Bazzan, and Robert Gramsch. "Community Detection in the Network of German Princes in 1225: A Case Study." In *Complex Networks VIII: Proceedings of the 8th Conference on Complex Networks CompleNet 2017*. Edited by Bruno Gonçalves et al., pp. 193–200. New York: Springer, 2017.

Düring, Marten, Ulrich Eumann, Martin Stark, and Linda von Keyserlingk, eds. *Handbuch Historische Netzwerkforschung: Grundlagen und Anwendungen*. Schriften des Kulturwissenschaftlichen Instituts Essen (KWI) zur Methodenforschung 1. Münster: Lit Verlag, 2016.

Düring, Marten, Markus Gamper, and Linda Reschke. *Knoten und Kanten III: Soziale Netzwerkanalyse in der Geschichts- und Politikforschung*. transcript: Bielefeld, 2015.

Engl, Richard. "Das Ende muslimischen Lebens im mittelalterlichen Süditalien. Netzwerkanalytische Überlegungen zu einer hundertjährigen Forschungsfrage." In *Gesellschaftliche Umbrüche und religiöse Netzwerke. Analysen von der Antike bis zur Gegenwart*. Edited by Daniel Bauerfeld and Lukas Clemens, pp. 119–54. transcript: Bielefeld, 2014.

Ferguson, Niall. *The Pity of War: Explaining World War I*. London: Basic Books, 1998.

Franzel, Emil. *König Heinrich VII. von Hohenstaufen. Studien zur Geschichte des "Staates" in Deutschland*. Quellen und Forschungen aus dem Gebiet der Geschichte 7. Prague: Deutsche Gesellschaft der Wissenschaften und Künste für die Tschechoslowakische Republik, 1929.

Freising, Otto von, and Rahewin. *Die Taten Friedrichs oder richtiger Cronica, hg. von Franz-Josef Schmale, übers. von Adolf Schmidt*. 3rd ed. Freiherr v. Stein-Gedächtnisausgabe 17. Darmstadt: Wissenschaftliche Buchgesellschaft, 1986.

Garnier, Claudia. *Amicus amicis – inimicus inimicis. Politische Freundschaft und fürstliche Netzwerke im 13. Jahrhundert*. Monographien zur Geschichte des Mittelalters 46. Stuttgart: Verlag Anton Hiersemann, 2000.

Gatz, Erwin (Hg.). *Die Bischöfe des Heiligen Römischen Reiches 1198–1448: ein biographisches Lexikon*. Berlin: Duncker und Humblot Verlag, 2001.

Gonçalves, Bruno, Ronaldo Menezes, Roberta Sinatra, and Vinko Zlatic, eds. *Complex Networks VIII: Proceedings of the 8th Conference on Complex Networks CompleNet 2017*. New York: Springer, 2017.

Gramsch, Robert. *Das Reich als Netzwerk der Fürsten. Politische Strukturen unter dem Doppelkönigtum Friedrichs II. und Heinrichs (VII.) 1225–1235*. Mittelalter-Forschungen 40. Ostfildern: Thorbecke, 2013.

———. "Networks of Noblemen and Scholars: Social Groups, Group Consciousness and Their Institutionalization in the Late Middle Ages." In *Collective Identity in the Context of Medieval Studies*. Edited by Michaela Antonín-Malaníkova and Robert Antonín, pp. 13–31. Ostrava: University of Ostrava, 2016.

———. "Zerstörte oder verblasste Muster? Anwendungsfelder mediävistischer Netzwerkforschung und das Quellenproblem." In Düring; Eumann; Stark; Keyserlingk, *Handbuch Historische Netzwerkforschung: Grundlagen und Anwendungen*, pp. 85–99.

Heider, Fritz. *The Psychology of Interpersonal Relations*. New York: Wiley, 1958.

Heinig, Paul-Joachim, Sigrid Jahns, Hans-Joachim Schmidt, Rainer Christoph Schwingers, and Sabine Wefers, eds. *Reich, Regionen und Europa in Mittelalter und Neuzeit*. Historische Forschungen 67. Berlin: Duncker & Humblot, 2000.

Hillen, Christian. *Curia regis: Untersuchungen zur Hofstruktur Heinrichs (VII.) 1220–1235 nach den Zeugen seiner Urkunden*. Europäische Hochschulschriften, Reihe 3: Geschichte und ihre Hilfswissenschaften 837. Frankfurt: Lang, 1999.

Hitzbleck, Kerstin, and Klara Hübner. *Die Grenzen des Netzwerks 1200–1600*. Ostfildern: Thorbecke, 2014.

Holland, Paul W., and Samuel Leinhardt. "Social Structure as a Network Process." *Zeitschrift für Soziologie*, no. 6 (1977): pp. 368–402.

Horton, Alice. *The Lay of the Nibelungs: Metrically Translated from the Old German Text*. London: George Bell and Sons, 1901.

Houben, Hubert. *Kaiser Friedrich II. (1194–1250). Herrscher, Mensch und Mythos*. Urban-Taschenbuch 618. Stuttgart: W. Kohlhammer GmbH, 2008.

Jansen, Dorothea. *Einführung in die Netzwerkanalyse: Grundlagen, Methoden, Forschungsbeispiele*. 3rd ed. Wiesbaden: VS Verlag für Sozialwissenschaften, 2006.

Jullien, Eva. "Netzwerkanalyse in der Mediävistik. Probleme und Perspektiven im Umgang mit mittelalterlichen Quellen." *Vierteljahrschrift für Sozial- und Wirtschaftsgeschichte*, no. 100 (2013): pp. 135–53.

Marx, Christian. "Forschungsüberblick zur Historischen Netzwerkforschung. Zwischen Analysekategorie und Metapher." In Düring; Eumann; Stark; Keyserlingk, *Handbuch Historische Netzwerkforschung: Grundlagen und Anwendungen*, pp. 63–84.

Matscha, Michael. *Heinrich I. von Müllenark, Erzbischof von Köln (1225–1238)*. Studien zur Kölner Kirchengeschichte 25. Siegburg, 1992.

Newman, Marc E.J. *Networks: An Introduction*. Oxford: Oxford University Press, 2012.

Nooy, Wouter de, Andrej Mrvar, and Vladimir Batagelj. *Exploratory Social Network Analysis with Pajek*. Structural Analysis in the Social Sciences 27. Cambridge [u.a]: Cambridge University Press, 2009.

Padgett, John F., and Christopher K. Ansell. "Robust Action and the Rise of the Medici." *American Journal of Sociology*, no. 98 (1993): pp. 1259–319.

Padgett, John F., and Walter W. Powell. *The Emergence of Organizations and Markets*. Princeton: Princeton University Press, 2012.

Rosé, Isabelle. "Reconstitution, représentation graphique et analysedes réseaux de pouvoir au haut Moyen Âge. Approche des pratiques sociales de l'aristocratie à partir del'exemple d'Odon de Cluny († 942)." *REDES – Revista hispana para el análisis de redes sociales*, no. 21 (2011): pp. 199–272.

Schneidmüller, Bernd. "Konsensuale Herrschaft. Ein Essay über Formen und Konzepte politischer Ordnung im Mittelalter." In *Reich, Regionen und Europa in Mittelalter und Neuzeit*. Edited by Paul-Joachim Heinig et al., pp. 53–87. Historische Forschungen 67. Berlin: Duncker & Humblot, 2000.

Schwennicke, Detlev. *Europäische Stammtafeln. Stammtafeln zur Geschichte der europäischen Staaten, N.S. I.3 335*. Marburg: Klostermann, 2000.

Stürner, Wolfgang. *Friedrich II (Gestalten des Mittelalters und der Renaissance)*. 2nd ed. Darmstadt: Wissenschaftliche Buchgesellschaft/Primus, 1992/2000.

van Cleve, Thomas C. *The Emperor Frederick II of Hohenstaufen: Immutator Mundi*. London: Clarendon Press, 1972.

Wasserman, Stanley, and Katharina Faust. *Social Network Analysis: Methods and Applications*. Cambridge: Cambridge University Press, 1999.

2.3 The value of network analysis in historical sociology: economic and social relations in medieval Lübeck[1]

Bernd Wurpts

In this chapter it is claimed that social network analysis offers a research perspective and toolset that should be taken up by historians and historically oriented social scientists. Important features of organization could be missed (or gotten wrong) if scholars are not taking a network perspective. The chapter starts with a discussion of features and benefits of social network research in an historical economic context.[2] It continues with short summaries of three selected best practice studies from historical analytical sociology. These studies investigate economic networks in various historical periods and exemplify how useful, flexible and widely applicable the network approach is. In addition, these applications are complemented by the author's own work related to the medieval Hansa, Hanse or Hanseatic League. The application includes an illustration of network concepts, methodology and an empirical case study related to the Hansa. Social network analysis is applied to one of the oldest systematic trade records from Northern Europe (Saß 1953, Cordes, Friedland, and Sprandel 2003) and inquiries reveal striking similarities to micro-level structures observed in medieval Genoa earlier in time when feudal warrior-cultures prevailed (van Doosselaere 2009). Thus, studies that focus on wealth and economic output may overlook important similarities between societies and overstate the distinctness of economic arrangements that are potentially associated with economic development. Ultimately, the chapter concludes with notes on challenges in network research applied to Hansa trade and suggestions for ways to get started with new research projects in historical networks more generally.

Features and benefits of the social network approach

Economic history is a particularly obvious field for social network analysis[3] because the economy produces many types of suitable data. Relational data are produced in the work process (as a by-product) but usually not thought of as network data. States collect tax data on income and estates, companies record their transactions and individual merchants often document who they trade with. These recording procedures were established many centuries ago and offer a rich arsenal of data for network researchers. Historians can use their case related knowledge and access to primary sources to extract various types of social relations from original documents. Many historical books also include tables and appendices with lists suitable for network analysis. Trade records, membership lists, event attendance data, etc. can be transformed into network data and analyzed with social network techniques.

New computer technology allows the extraction, storage and analysis of diverse types of relational information as well as unprecedented amounts of data. Entire text corpora can be turned into computer readable data, and all types of networks may be extracted from these automatically, depending on the qualities of written documentation. Handwritten original documents or complex printed materials are still very challenging to handle for computers and often require humans as coders. But, with increasing precision, printed documents can be digitized using scanners or modern OCR software (optical character recognition), which makes these much more readily accessible for modern data analysis and has already led to the creation of large historical databases. Leading network scholars claim that large digital collections of historical information "can revolutionize historical social sciences."[4] The merits may be particularly huge when it comes to identifying causal mechanisms and processes in the long run. Bearman argues that the usual work of historians, writing narrative sentences about the past, may be less affected by the data revolution related to digitization of large amounts of archival data.[5]

Nevertheless, economic historians who are looking for a fresh view of their historical cases may find their best alternative to established research procedures in social network analysis.[6] The network approach offers an additional perspective to common statistical aggregations in economic history. It shifts the focus from the characteristics of individuals or actors in the widest sense to the relations and structures among them. Social scientists who study economic history with methods of social network analysis underline the importance of social structure and relations for economic phenomena.

In the last decade, sociologists produced a number of high-quality studies of social networks in the economy and often focused on patterns of trade.[7] Social network analysis is a particularly suitable methodology for the analysis of exchange or transaction data in general and for studying relational processes like those related to flows of trade goods. In economic history, scholars who study trade volumes may shift their view from aggregating capital to what trade really is, namely flows of valued goods and resources. Focusing on amounts of capital gives a reduced picture of the economy and particularly of trade as an inherently relational process often connected to other domains like kinship and politics. The network approach is uniquely suited to analyze trade and questions related to flows, content and rates of flows.

Other main benefits of using social network analysis in economic history are related to the capabilities of compelling and elucidating visualization. Perhaps the most prominent and simple form of network visualization is a (network) graph depicting a set of nodes (persons, organizations, etc.) and a set of lines indicating existence of relations (kinship, trade, etc.) between them. The creation of network graphs and the use of network tools in general mean that researchers clearly define the set of actors (nodes) and types of relations (lines) that constitute the social network under investigation. This requirement of defining nodes and lines in social network analysis leads to potential gains in objectivity. Many traditional historical studies do not clearly define their network as the object of study and therefore remain metaphorical and vague.[8] Furthermore, social network analysis offers a set of standardized network measures that can be used for a formal, numeric summary of a network compared to mere visual interpretations of graphs and are particularly helpful for comparisons of networks. While many networks cannot be compared across studies, the comparison of network measures seems particularly fruitful for historical research when the same location or very similar structures are observed over time.

Two broad types of network analyses are connectivity and positional methods.[9] Both methods can be used on various levels of analysis to describe the network, which may be considered as a first step in the identification of patterns in social structure and/or social mechanisms.[10] Simple descriptive procedures on the individual node level include counting the number of connections of individuals (degree), how far they are in terms of relational distance (closeness) or the extent to which they are located on shortest paths between actors (betweenness). These centrality measures can be used to understand activity in networks, prestige, social influence and/or opportunities to control information flows.[11] Other connectivity methods look at larger units of analysis on the group-level and investigate phenomena like the cohesion within subgroups including cliques or k-cores.[12] At the total network level, connectivity methods include measures like centralization or network density to capture hierarchy and overall cohesion in a network.[13] On the other hand, positional methods look at more abstract role structures by studying nodes with similar connections. This approach focuses on the absence of ties rather than connections themselves. Methods like blockmodeling identify structurally equivalent actors.[14] These and many other descriptive measures can be used to inform researchers about structures of relations impossible to reveal with other methods. An increasing amount of network studies also combines descriptive network measures with statistical procedures such as regression models.[15]

While descriptions and visualizations may be one step in analysis, particularly at the exploratory stage, many network studies go further and point to specific effects particular network structures or relational processes have on various economic outcomes of interest.[16] In this way, the network perspective helps to make theoretical contributions, particularly those highlighting social structures and relational dynamics as causal agents. Most ambitious studies from historical sociology follow the paradigm of "analytical sociology," which is particularly interested in the mechanisms and processes that link the micro level to macro outcomes.[17] "An analytical historical sociology would focus on explaining why phenomena happen and how they happen, rather than relying only on description and interpretation."[18] This approach may share similarities with ontological theories in history, which also focus on underlying mechanisms.[19] Braudel recognized a "brotherly correspondence" between history and sociology but also noticed a lack in actual dialogue between sociologists and historians.[20] This trend, it seems, has changed partially due to a group of historical network scholars in sociology.

To illustrate this point, a groundbreaking study by Padgett and McLean can be referred to in which they argue that economic historians who studied the invention of the partnership system confuse explanations of origins and genesis with explanations of consequences.[21] The article demonstrates that the social network approach, including network theory and methods, may be used to tackle established arguments and get around functional explanations. Padgett and McLean show how the application of network thinking in combination with quantitative analyses of historical data may add to the understanding of key phenomena in economic history. Their unrivaled database of relational information from medieval Florence captures multiple relational domains that exist beside each other and overlap via occupations of multiple roles. The relative impact of network domains is analyzed systematically using quantitative techniques, e.g. multiple regression. Network scholars like Padgett and McLean emphasize the role of social relations and analyze how these unfold in the period leading up to the phenomenon of interest at a certain moment in history. Thereby social network methods may not only help solve theoretical puzzles but could also be used to answer more concrete questions in history. The network approach offers means and perspectives to develop new research questions but also answer old ones.[22]

Finally, network concepts can be very useful for studying social and economic change and/or stability. Concepts from sociology or economic history may be expressed in network terms and analyzed in network structures or meanings of ties. This allows researchers to objectively analyze patterns of social change using concrete network terms and measures. Historical change can be understood in terms of changes in "ideal types" of social organization, for instance. The great Max Weber stated that

> (t)he construction of abstract ideal-types recommends itself not as an end but as a *means*. Every conscientious examination of the conceptual elements of historical exposition shows however that the historian as soon as he attempts to go beyond the bare establishment of concrete relationships and to determine the *cultural* significance of even the simplest individual event in order to "characterize" it, *must* use concepts which are precisely and unambiguously definable only in the form of ideal types.[23]

A recent study of trade networks has, for instance, observed an institutional change in trade organization among traders to Bergen (Norway) during the Late Middle Ages. Observing concrete relationships among traders in selected time periods ranging from 1360 to 1510, Burkhardt finds that in the late 14th century trade partnerships existed predominantly among family members, which changed within the next century in favor of short-term transactions with many different, unrelated traders. Increases in the principle of "Treu und Glauben" (equity and good faith) show the cultural significance of changes in social networks.[24]

In the next section, a set of selected studies from sociology that use a network approach to study social change and economic development using network ideal-types as analytical tools will be described.

Selected best practice studies

In the early 2000s, Roger Gould published an important article on the state of the art in historical network research in sociology.[25] In order to avoid redundancy, best practice studies published in the decade thereafter have been selected. This selection includes three case studies from European and world history spanning a period of about 800 years. The studies differ by period of investigation, geographical location and in terms of the types of social relations at the center of analysis. It is striking that these studies differ in the types of network concepts or ideal types they emphasize in relation to the characteristic historical period. The first study uses the sociological concepts brokerage and closure in the context of industrialization; the second study focuses on legitimacy of ties in the context of globalization and the third study analyzes concepts from economic history such as patronage and feudalism through a social network lens. All these studies explain changes at the level of the larger economy (macro) through the application of lower level network structures and relational mechanisms (micro or meso). To commence, the historically most recent period will be examined.

Hillmann and Aven are interested in the question of how economic development is possible in the absence of strong public institutions.[26] The authors present a historical case study that focuses on entrepreneurship in late imperial Russia (1869–1913); a case of late industrialization due to a lack of reliable market-supporting institutions. Hillmann and Aven are interested in business networks and underline the fact that economies based on informal institutions such as reputation, associated with network closure, may stabilize but not develop. Network brokerage, on the other hand, may attract the resources and information necessary to bridge local economies. Empirical analyses are based on a complete list of share partnerships and joint-stock companies including information about characteristics of thousands of companies and individual founders. Network data are derived from shared affiliations of founders and are studied over time.

Using multivariate OLS regression analysis,[27] the authors find that previous business success had a positive effect on the amount of capital raised by a company. In other words, reputation matters for capital mobilization. In the second part of their analysis, the authors find that network fragmentation, a characteristic expected for emergent societies, did exist in the business network in Russia during this period. Finally, Hillmann and Aven go on to investigate whether the network location, core vs. periphery, is of importance for capital mobilization. In other words, they ask whether reputation works similarly well in closed and bridging network positions. The authors find that reputation effects are about twice as high in the network periphery compared to the network core. They also find that individual network reach and brokerage measured as constraint was higher in the core than in the periphery and that this was beneficial for capital mobilization on average. This study is best practice because concepts like reputation, brokerage and fragmentation are operationalized in a clear manner. Aspects of time in networks are considered carefully, and the authors also allow for various aspects that might affect causal interpretation, particularly aspects of individual heterogeneity. Overall, the quality and completeness of data are impressive and enable a complex investigation of the problem at hand.

Emily Erikson's book on the Early Modern English East India Company is another exemplary network-based study of an important period in economic history.[28] Erikson (and Bearman) analyzes how malfeasance of actors affected larger patterns of trade.[29] She argues that the opportunistic behavior of employees connected otherwise loosely integrated commercial regions and increased information flows between Asian ports and England. For data analysis, the author uses information about communication between ships and between captains. Information about voyages is directly drawn from the ship logs and includes reliable information about the paths of the ships. Overall, the data include 264 ports in the East Indies including 1,480 ships or 4,725 voyages for the period 1601–1835. Erikson analyzes voyages between ports and how malfeasant voyages affect the overall trade structure. She takes out the non-legitimate trade routes from the overall trade networks for the 58 periods of a length of four years. Taking out malfeasant voyages leads to a reduced network in each period and lowers network density. As network density reduces automatically when connections are taken out, she compares taking out the malfeasant voyages with taking out randomly selected legitimate voyages.

Erikson finds that taking out the private trade connections disconnects the entire network into separate components in periods around 1712, 1720 and 1728 while taking out the legitimate "matched" voyages does not disconnect the network at all. Moreover, to verify her findings using formal methods instead of network visualizations, Erikson uses a modified network integration measure to further assess different effects of the removal of private trade vs. legitimate trade from the network. She finds that private trade had a greater impact on network integration than legitimate trade and that the removal of private trade leads to network fragmentation. The English East India Company profited from private traders as they promoted communication in Asia and between Asia and England. This study can be considered as of highest quality because of the completeness of the data, which perhaps allows a full understanding of the macro phenomenon of a first global trade network using micro level relational logics. Erikson elegantly uses the different meanings of ties in her historical context building on legitimacy ideal types. Besides linkages between theory and data, the study includes methodological innovations.

The research of van Doosselaere on the transition from feudal to mercantile organization in medieval Genoa is a wonderful example of historical network research on the Middle Ages.[30] Van Doosselaere focuses on commercial dynamics and observes a change in trade partner selection starting at the end of the 13th century, which contributed to the rise of a mercantile oligarchy. For empirical analyses, the author uses data that are rougher than those used in the previous studies due to decentralized recording practices and problems of survival of records. However, the available data from medieval Genoa are among (if not) the oldest surviving trade records of this magnitude and degree of completeness. The author analyzes thousands of equity associations in the time period 1154 to 1315. These associations, called commenda(e), were limited to single voyages or arrangements of short duration in which resources were pooled among two or more partners. Van Doosselaere finds an increasing rarity of this type of contract at the end of his observation period in addition to increasing capital values and variation in monetary values of partnership ties.[31] In analyses of trader attributes, he confirms previous findings in history regarding a wide but declining integration of women in the medieval economy and disconfirms previous research by emphasizing the importance of changing "aristocratic elites" for the development of Mediterranean long-distance trade instead of "new men."[32] Van Doosselaere then shifts the focus from attributes to relations and studies network change of commendae over time.

Network graphs for selected periods of varying lengths (10–26 years) show large numbers of components because the connections between operators in the samples show low commercial connection. The author calculates degree centralization for each network and compares this with an ideal type of feudal social organization resembling a star-like shaped network. Network analyses show that network centralization declines over time in the period of observation. Large traders were mainly unconnected in the early period, which changed in later periods. Thus, during the 13th century, elite collaboration stretched beyond politics and military and increasingly included commerce as well. Another finding is that families who would dominate Genoa in the Renaissance were not very prominent in volumes of trade but in network betweenness centrality, thus network control, during the early periods of investigation. To analyze the integration or cohesion of the network over time, the author uses connectedness indices and interprets his empirical findings, "as feudal-like control over the network declines, the trade network's integration decreases sharply, before commercial ties knit the whole back together to form a more integrated – yet more decentralized – trade architecture."[33] Finally, van Doosselaere looks at status and occupational homophily in partner selection by coding for each pair possible combinations of binary attributes and calculating point correlation statistics for the selected periods. Findings are that status homophily in partner selection was relatively low until the end of the 12th century and then increased steadily by the early 14th century. In other words, aristocrats were more likely to be partners with other aristocrats at the end of the period of observation compared to the beginning. This study impresses due to its high ambition to study large scale social change in economic history.

To sum up briefly, the selected sociological studies add to our understanding of how social relations may have contributed to historical economic development and social change more generally. They carefully link the conceptual part with the empirical measures and data they have. All studies have in common that they tell history by forward looking, not backward looking, regarding concrete social structures and relational processes as main agents of change. All studies take a sociological perspective on historical developments by emphasizing concrete social ties and their potentially different cultural meanings. Instead of adding complexity to history and more detail to specific cases, the authors strive for abstraction, generalization and explanation. The best practice studies presented here benefit from detailed case-related knowledge and the use of primary and secondary data in combination with the creative applications of tools from social network analysis. By carefully balancing sociology and economic history these studies contribute to a better understanding of historical economies and the major social forces underlying them.

Application of social network analysis: case study of medieval Hansa trade

Illustration of network concepts and methodology

The analytical power of the social network approach should further be demonstrated by an example from the author's own research through an elaboration of a classic analysis of trade. One of the classical sociologists, Georg Simmel, argued that in the case of the medieval Hansa, traders traveled to sell their goods in other towns and thereby built connections between localities and overlaps in social circles.[34] More recently these ideas have been developed under the label of the brokerage concept, which underlines the advantages or disadvantages "brokerage" entails for individuals, groups and outcomes like economic development.[35] To illustrate the brokerage concept, network graphs are used in Figure 2.3.1 including connections between traders and markets.

The graph on the left is a "two-mode network" named after the different types of nodes it includes. Lines in the graph indicate relations between traders (primary nodes) and towns (secondary nodes).[36] The hypothetical example shows that four out of five Hansa merchants travel to one town only, e.g. traders A and B trade in town/market 1. Merchant C travels to both markets and therefore connects otherwise unconnected markets. This structural location of merchant C designates her as a "broker." We can imagine how brokers such as Hansards contributed to the integration of economies during the Middle Ages. While the graph in Figure 2.3.1 (left) illustrates an integrated economy, we can also imagine how taking out the broker, trader C, might swiftly create a local economy and disintegrate markets.

While the two-mode network shows how merchant C connects markets, the corresponding one-mode projection shows how shared affiliations may also create connections between individuals. Simple matrix multiplications between two-mode networks lead to one-mode networks characterized by one type of nodes.[37] The two-mode network of traders and towns may be projected into a

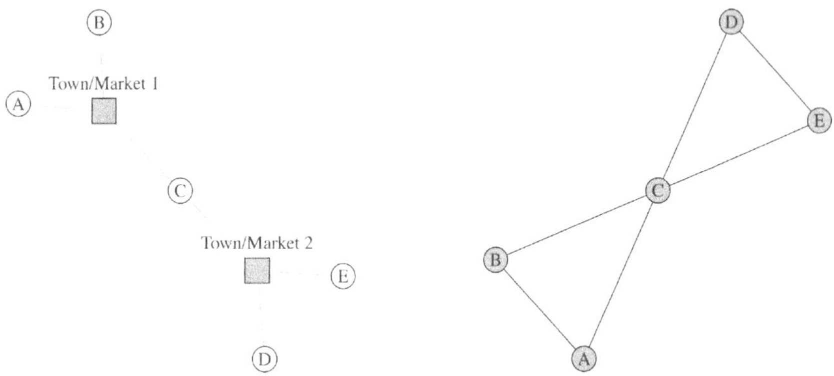

Figure 2.3.1 Left: Two-mode network of persons/traders (circles) linked to towns/markets (squares). Right: One-mode projection showing relations between traders through shared affiliations

Source: Figure produced in R using igraph[38] package and Fruchterman Reingold layout

trader network or town network. Projecting the two-mode network (left) into a one-mode network of traders (right) results in a single component in which all traders are connected allowing communication at least in principle.[39] One might infer how network brokers may promote transmission of information and enable groups to mobilize, coordinate and act together, such as the medieval Hansa did over centuries.[40] While the network graphs are based on an artificially constructed network dataset, scholars could use empirical data to evaluate the extent of overlaps in medieval social circles. Collaborations between medieval towns in the Hansa regions are observable in the records of the Hansa Diets (1356–1669) for instance and indicate political and economic ties between northern regions. In some of my own research, together with Katie Corcoran and Steve Pfaff, I show that historical embeddedness among Hansa elites likely contributed to the early adoption of Protestantism in Northern Europe.[41]

Empirical analysis of historical trade data

Instead of analyzing market integration based on social connections across towns, the following short study focuses on registered trade from a single town in Northern Europe. In certain (very modest) ways, the present analysis follows in the footsteps of Max Weber, one of the founders of sociology, who wrote his dissertation on Italian partnerships and could not extend it with a study on German commercial towns.[42] From contemporary research, the work of van Doosselaere on medieval trade in Genoa, as described earlier, seems to be an appropriate reference point and case of comparison.[43] Important institutional differences such as the mostly absent notary system in Northern Germany along with differences in recording practices of trade led to lower amounts of data in the North, but otherwise the two regions had very similar legal structures of trade partnerships during the 14th century.[44] In this application of network methodology I aim to answer whether the micro-level patterns of social relations and partner selection observed in medieval Genoa also formed the basis of economic organization in the northern German town Lübeck.

Lübeck was among the largest towns in the Holy Roman Empire during the Late Middle Ages and had status as an imperial town. In the 13th century, Lübeck was the most important loading place and supply port for the crusades to the European Northeast.[45] Here we can see another similarity to the Italian town Genoa, which also served as a major port in the Crusades that had started much earlier in Southern Europe. In Genoa, military expeditions caused growth in the local economy as well as involvement of the feudal nobility in commerce.[46]

Located close to Baltic and North Sea, the Northern German town was also the designated "head" of the medieval "Hanse" (German) or "Hanseatic League" (English).[47] The Hansa (Latin) was one of the most famous and long-living governance institutions in economic history (e.g. 1358–1669) and at the same time a large (perhaps decentralized) trade network across medieval borders. Hansa trade was predominantly in bulk goods but also in luxury goods, which were often transported via ships in the Baltic and North Sea regions. In 1350, Hansa regions including Lübeck were struck by the plague, Black Death, which caused tremendous death tolls and population decline. One of the economic effects was that the plague seems to have led to a concentration of wealth in the hands of a few. There was also a likely decline in demand for consumer goods whereas the demand for luxuries increased.[48] The plague seems to have had an impact on the development of early Hansa trade because it might have caused a breakdown in the trade with bulk goods.[49]

For data analysis, information from the "Societates" register as recorded in the town hall of Lübeck (Germany), 1311–1361, was extracted using the published version by Cordes et al. (2003) as well as an unpublished handwritten transcription by Saβ (1953) from the city archive in Lübeck.[50] The societates register recorded trade partnerships and included voluntary recognitions of obligations and cancellations of these debts. The register is among the earliest sources of systematically recorded trade in northern Europe. According to Cordes, it includes only one type of organization called "Widerlegung."[51] This German word refers to the merging of capital and also marked the foundation of a new association. The register was likely to have been initially created for a small group of capital providers in medieval Lübeck. This group instructed in almost every transaction a new capital leader who was then sent on trade voyages with vast amounts of capital, e.g. in Mark silver or bullion.[52]

Using traditional methods from economic history, Sprandel identifies 17 big investors with equity participation in four and more partnerships or extensions of partnerships.[53] The top four list of partnership participation is: H. Mornewech (22 partnerships), C. Attendorn (nine partnerships), J. Morkerke and E. Pape (both eight partnerships). At the level of partnerships, Sprandel finds that numbers of partnerships declined in the observed period from 1311–1361. In his analysis of capital size of partnerships by groups of small, big and very

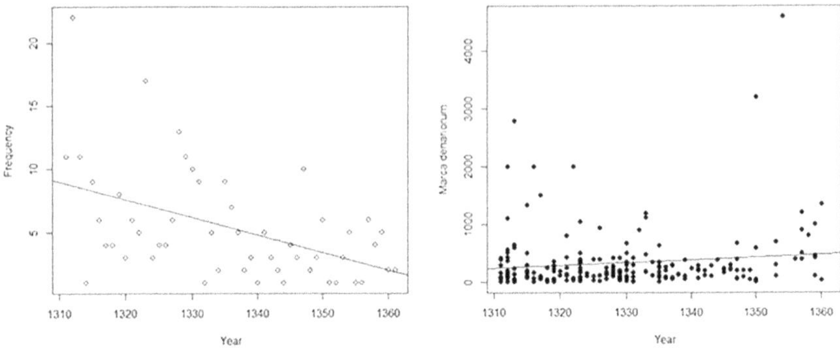

Figure 2.3.2 Frequencies of partnerships per year (left) and size of partnerships per year (right)

Source: Figures produced in the statistical programming software R

big investments, Sprandel finds that small partnerships declined first.[54] Over time, the relative size of entries shifts to larger associations. Ultimately, Sprandel argues that this recording practice declined due to the decline of use by important entrepreneurs and due to increasing written private book keeping. He supposes that the diffusion of merchant books in the Hansa towns may have been the reason why the societates register expired. The intense use of the register declined in the 1330s due to the resignation of the assumed initiators and a new generation of wholesalers with preferences for private book-keeping.[55] To illustrate the decline of societates and the increase of relative capital value, scatterplots have been shown, including fitted lines in Figure 2.3.2.

In a next step, traditional economic history is extended by using social network analysis. The data in the societates register has been treated as two-mode data similar to the example in Figure 2.3.1, where actors are linked to one another through joint partnerships. Analyses include all 279 entries in the societates register including 477 individuals connected by 597 partnership ties in the Cordes et al. dataset and 467 individuals linked by 605 ties in the Saβ (1953) dataset. While the start of the partnerships are recorded, the end of relations between traders is often unclear. The duration of ties varied between one-shot collaborations and multi-year associations. As the observation period spans 50 years, issues like mortality of merchants have to be considered for interpretation.

Furthermore, interpretations need to acknowledge potential changes in meanings of ties over time, e.g. different sets of obligations and opportunities. Van Doosselaere states that in Genoa agency relationships emerged in the 14th and 15th century and had relatively strict guidelines compared to previous cooperative partnerships in which travelers had substantial autonomy.[56] While his analysis of commendae ends in 1315 and cannot capture this process, the present dataset of societates records seems to reflect these changes as increases in the complexity of ties can be identified in the 1330s and 1340s.[57] Cordes also observes increases in numbers of cancellation receipts since about 1350 and relates this to potentially changing business practices as responses to the plague. The Black Death likely increased the need for written preservation of evidence.[58]

As node attributes, 32 traders who had "dominus"[59] status in the Cordes et al. data and 29 "domini" in the unpublished Saβ data were identified. Many traders in the societates register were born into old councilor families; some of them would become councilors later, but were not part of the city council themselves during the period of observation.[60] To capture these traders, a distinct category "council family background" was coded, including 83 nodes in Cordes et al. data and 79 nodes in Saβ data. Overall, 294 different last names in the datasets from Cordes et al. and Saβ were identified, thus, traders from almost three hundred different families used the register to secure trade partnerships. The cumulative merchant network of the societates register, 1311–1361, is depicted in Figure 2.3.3. Node

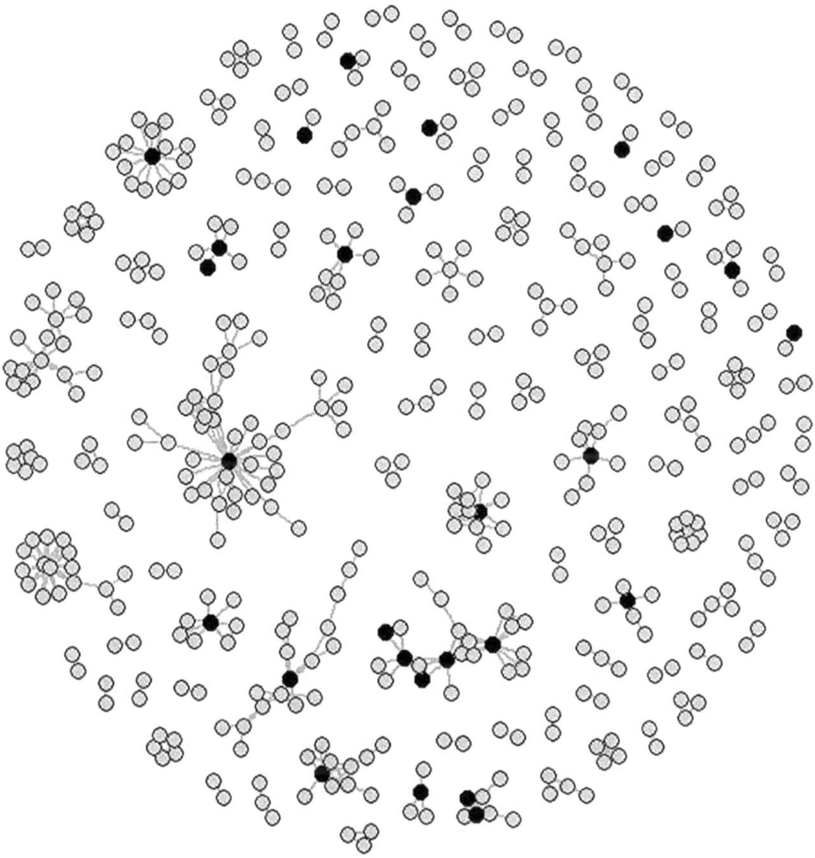

Figure 2.3.3 Cumulative merchant network, 1311–61. Dataset: Saβ (1953)

Source: Figure produced with the igraph[61] package (Fruchterman Reingold layout) in the programming language and software R. Node color black indicates "dominus" status

colors mark “dominus” status in black and other nodes in grey. A general visual impression is that the merchant network is fairly fragmented. Many isolated partnerships of size two (dyads) characterize the networks, but a few hubs stand out and connect relatively large components.

Besides fragmentation, the network is characterized by inequalities insofar as some traders have substantially more partnership ties compared to other traders.[62] At the center left of Figure 2.3.3, one large component includes H. Mornewech, the most prominent user of this register as described by Sprandel. Mornewech was a central, influential node in the trade network and had “dominus” status as marked by a black color. Most of the partners he traded with did not themselves trade with each other, which manifests in a structure resembling a network star. Most other active traders identified by Sprandel possess a similar ego-network structure. This trade structure resembles the feudal type of organization as described by van Doosselaere in his study of commerce in medieval Genoa.[63] Another large component that is located at the center bottom shows a different pattern and includes multiple traders with “dominus” status. This component includes H. Pape who acts like a broker and connects otherwise disconnected groups of traders. The interesting difference to the previous structure is that this councilor traded with other councilors as well indicating a potentially different type of partner selection among societates users. As H. Pape started to use the register in 1342, which is seven years after the last entry of H. Mornewech, this observation could indicate a potential change in trade organization over time such as a trend toward status “homophily” as observed by van Doosselaere in late medieval Genoa.[64]

In order to investigate (potential) homophilous tie patterns over time, a slightly more dynamic perspective is taken, and the data is subset into two periods, based on observations from the historian Cordes who distinguishes a main period (1311–1339) and a late period (1340–1361). Cordes uses this split because he finds a decline in societates use at this time, as well as a change in character of the entries at the beginning of the 1340s.[65] Moreover, while network depiction in Figure 2.3.3 leads to interesting first interpretations, these impressions are mere visual ones. As a step to quantify and generalize the finding of a potential change in partner selection, quantitative measures are calculated in order to assess whether nodes tend to connect to nodes with similar characteristics. To empirically measure “homophily,” assortativity measures[66] from social physics and S14[67] as used in van Doosselaere’s study of medieval Genoa are calculated. The corresponding values are correlations between binary attributes that trading partners might share. In assortative networks, nodes with high status (labeled “dominus”) are linked to other nodes of high status. Similarly, positive S14 values indicate a tendency for homogenous partnerships correcting for differential participation in trade by people with different attributes. Both measures vary between -1 (completely dissortative) and 1 (perfect assortative mixing).

Table 2.3.1 shows the assortativity coefficients and S14 values for the characteristics “dominus” and “council family background” in the main and late period

Table 2.3.1 Assortativity coefficients and S14 values for main period (1311–39) and late period (1340–61) of societates trade register. In each cell, first coefficients are based on Cordes at al. (2003) data followed by coefficients in parentheses based on Saβ (1953) data. All numbers were calculated in the programming language and software R. Assortativity was calculated with the igraph[68] package.

	Assortativity		*S14*	
	Main period 1311–1339	*Late period 1340–1361*	*Main period 1311–1339*	*Late period 1340–1361*
Dominus, city council member	−0.132 (−0.106)	0.021 (−0.023)	0.125 (0.125)	0.021 (0.021)
Council family background	0.002 (0.059)	0.525 (0.464)	0.020 (0.020)	0.528 (0.426)

of the societates register. The attribute "dominus" serves as an indicator of status, power and wealth. "Council family background" captures the quality of family background and whether a person stems from an established council family, e.g. old elites. In Table 2.3.1, it can be seen that assortativity coefficients for the attribute "dominus" are close to zero in all periods and across datasets. There is likely to be no assortativity regarding "dominus" status, e.g. current political and economic power. There is no evidence that "patricians" are more likely to trade with each other during this period, which contrasts with the findings for Genoa at roughly the same time. Whereas there is no correlation between the attribute "council family background" of connected nodes in the main period of the societates register (1311–1339), there is a positive association in the late period 1340–1361 (r=.53). This suggests that over time there may have been changes in partner selection related to old elite families, a potential similarity to what van Doosselaere found for Genoa. Correlation coefficients suggest that family status was not a prevalent selection criteria in the main period; however, the data show that similarity in family status often characterized partnerships in the late period. Note that the assortativity measures from "social physics"[69] are similar (if not identical) to the S14 values as used in the sociological study of Genoa by van Doosselaere, as well as other social scientists before. Furthermore, the results are stable across datasets, which provides some confidence. Status homophily in medieval Lübeck during the period 1340–1361 roughly corresponds to values found for the period 1269–1296 in medieval Genoa.[70]

In order to further look at homophily between traders listed in the societates register, Exponential Random Graph Modeling (ERGM) is used as a more rigorous test than the previously presented correlation measures.[71] "ERGM is a tool for examining patterns of relationships . . . and identifying how the characteristics of the network members and larger social forces can explain or predict the observed patterns of relationships."[72] This tool is similar to general linear models and logistic regression but more realistic because it does not require the assumption of independent observations and can capture various dependencies in social life. ERGMs can help answer the question whether ties between traders with council family background, as observed earlier in the late period, appear more likely than would be expected by chance. The hypothesis that traders born into council families in Lübeck are more likely than expected to trade can be tested for each time period. After various tests with model fit parameters like AIC and BIC, as well as goodness of fit tests, the following terms were selected that seemed to work best across time periods and dataset: edges, nodefactor, nodematch, gwdegree. Edges captures network density, nodefactor shows categorical variables, geometrically weighted degrees "account for the decreasing degree distribution in observed networks"[73] and nodematch captures homophily.[74]

Table 2.3.2 shows the estimates from exponential random graph models for main and late period and for each dataset separately. The "edges" coefficients

Table 2.3.2 Exponential Random Graph Models for main period (1311–39) and late period (1340–61) in societates trade register. In each cell, first numbers are coefficients and second numbers in parentheses are standard errors. All numbers were calculated in the programming language and software R using the statnet[75] package.

	Cordes at al. (2003) data		*Saß (1953) data*	
	Main period 1311–39 Coeff. (S.E.)	*Late period 1340–61 Coeff. (S.E.)*	*Main period 1311–39 Coeff. (S.E.)*	*Late period 1340–61 Coeff. (S.E.)*
Edges (constant)	−4.92***	−4.34***	−5.02***	−4.69***
	(0.61)	(0.35)	(0.54)	(0.34)
Homophily				
1. Kinship (all	3.57***	4.22***	3.76***	4.79***
names)	(0.19)	(0.27)	(0.17)	(0.27)
2. Council family	−0.38*	0.52**	−0.20	0.06
background	(0.19)	(0.20)	(0.17)	(0.22)
3. Dominus,	−0.94·	0.23	−0.85·	0.12
councilor	(0.52)	(0.26)	(0.49)	(0.26)
Structural terms				
• Gwdegree	0.79**	−0.64**	0.61**	0.18
	(0.26)	(0.24)	(0.23)	(0.27)
Categorical variables, Main effects				
A. Dominus/	0.22	0.41*	0.28	0.62**
councilor	(0.53)	(0.21)	(0.48)	(0.22)
B. Council family	−0.31·	0.05	0.01	−0.18
background	(0.17)	(0.13)	(0.15)	(0.16)
Fit: AIC, BIC	3586, 3648	1973, 2024	3764, 3826	1772, 1824

Signif. codes: '***' 0.001
'**' 0.01
'*' 0.05
'.' 0.1

reflect densities of the networks in each period. Negative signs indicate that densities are below 50% as observed in most social networks.[76] Among the coefficients for individual attributes, positive and significant results for "dominus" in the late period in both datasets ($p < 0.05$ and $p < 0.01$) were found. There is a higher likelihood for councilors and other powerful individuals (domini) to form ties compared to others during the late period of the register, but no significant pattern exists in the main period. In addition, coefficients for "homophily" depict log-odds of the presence of ties between actors with same characteristics. The results from the ERGMs mostly confirm the previous results in Table 2.3.1. There is evidence that domini or councilors do not tend to interact with each other or are even less likely to trade with each other than we would expect by chance. This is true for all datasets and across time periods. Results for traders with council family backgrounds cannot be fully repeated across datasets using ERGMs. While the Cordes et al. dataset shows the expected pattern that traders from old town elites did not trade with each other in the early period but were more likely to do so in the late period ($p < 0.05$ and $p < 0.01$), the same coefficients are not statistically in the Saβ dataset. Thus, using ERGMs, only partial evidence for homophily between Hansards from old council families in the late period was found.

In the first row of estimates capturing homophily, it can be seen that there is a strong propensity for "kinship" among traders with the same last name – broadly defined – to trade with each other compared to traders with different names. All coefficients of "kinship" are highly statistically significant ($p < 0.001$) and show in a positive direction. Thus, the results confirm the impressions of Burkhardt who observed that traders to Bergen were mostly related via family ties.[77] Hansa trade in the 14th century was to a large extent family trade and therefore based on close, personal relations including trust. Finally, if results based on the published Cordes et al. data are taken more seriously than the unpublished Saβ dataset, network statistics may indicate potential changes in partner selection, which could have contributed to increased capital and wealth accumulation among elites.[78] Linking Sprandel's finding of increasing capital size with the finding of increased "status homophily" may extend the perspective on how wealth may have been accumulated during the 14th century. While Sprandel treats the decline in the use of societates as simply the end of a particular practice, my analyses indicate that it was perhaps the breeding ground for the linking of political and economic domains and a model for the subsequently powerful medieval Hansa, which engaged in both politics and trade.[79]

Discussion: challenges of social network analysis applied to Hansa trade and a few guidelines for new research

In this chapter it is claimed that social network analysis is an innovation that should be taken up in history. Important features of organization could be missed (or gotten wrong) if scholars are not taking a network perspective. For instance, if studies are not taking a network perspective, they fail to reveal the underlying relational micro-structures of economic organization. Studies that are focusing on wealth and economic output only could miss crucial similarities between societies. The distinctness of economic arrangements that are potentially associated with economic development may be overstated. Traditional methods of economic history are unable to identify patterns of social and economic relations.

In the present case study of medieval Hansa trade it was shown that the Northern German town Lübeck had relatively similar micro-level structures compared to the richer Italian town Genoa. Thus, differences between northern Europe and southern Europe may have been smaller than previously assumed by historians. Hammel-Kiesow points out that northern and southern European regions had relatively similar legal structures in trade.[80] The present study extends this emphasis on similarities to the level of partner selection in trade.

To the extent that network structures and cultures overlap, van Doosselaere's study may give reason to believe that the warrior culture that characterized the Italian town during the times of the Crusades also persisted in Lübeck at the time when Genoa was forming a mercantile oligarchy. This interpretation, however, may be doubted as Hansards are known to have relied on violence as an ultima ratio only.[81] One might conclude that similar micro-level structures may prevail in different social and cultural contexts. A look at the specific timing of the micro-level patterns reveals that the economic organization of Lübeck in the first half of the 14th century resembles the structures in Genoa 100–200 years earlier.[82] While these results may be interpreted as micro-evidence for the backwardness of economic organization in the north, there is also evidence that types of organization changed starting about a decade before the Black Death. Long-term changes of economic organization in Genoa may have happened relatively quickly in the north. However, the data to make reliable statements in this regard are absent. Future investigations of societates records should focus more on the complex nature of partnership ties and potential changes during the 14th century.

This chapter showed that economic history is a particularly suitable field for social network analysis due to the relational topics like trade but also due to the large stocks of data available even for periods far back in history. The network approach brings new insights to economic history by potentially offering new answers to old questions or new perspective on historical data stimulating new questions and research. Network studies in sociology have already made substantial contributions to the study of economic history using the network approach. They were particularly successful when they pointed to mechanisms and processes driving social change. In my short application of the social network approach, I have investigated briefly how changes in partner selection as a distinct network process may have contributed to concentration of capital and wealth in medieval Lübeck,[83] an outcome that was noted by historians but has not been explained well.

While the network approach has many benefits, specific challenges of historical network research should not be ignored. One of the greatest challenges in historical studies is related to availability and quality of data. Historians can only work with documents that survived over time and can be read with sufficient clarity.[84]

Only careful interpretations of data limitations and bias help researchers to overcome these issues. Data limitations have consequences for the types of questions that can be answered, which means that questions are more source dependent in historical research compared to more contemporary research, which can even design a study to answer a question of interest. My own case study, including an analysis of one of the oldest systematic trade records in northern Europe, the societates register, has data limitations as well, particularly regarding the identity of individuals. Names were not standardized in the Late Middle Ages, which is problematic for the identification of unique individuals (nodes). I have tried to deal with this problem by using two different transcriptions of the original medieval documents in order to increase reliability. The two datasets used were well-suited because they were extracted by experts with a 50 year gap between recordings.

One of the biggest problems for social network analysis more generally is related to missing data. Results are misleading, for instance, if key bridging actors are not included in the data but a key feature of the "true" network. In that case, observed network fragmentation does not correspond to the actual historical constellations and network representations, and measures and conclusions become misleading. The extent of measurement bias depends on network measures and features of networks.[85] Missing data in the case study presented earlier refers to trade in the town of Lübeck that took place during the period of observation but was not recorded in the societates register. As the number of partnership entries in the register decreased over time, missing values become a substantially meaningful issue in relation to the use of the recording practice besides the methodological aspects already mentioned that related to the network structure. Historical studies covering long time spans need to reflect the meaning of missing values, and potential changes could indicate interesting research questions themselves.

Another issue in social network research is related to visualizations. My case study on Hansa trade indicates that network visualizations may be problematic when they are cumulative and cover several years or decades. Some individuals may die or exit a network within the period of observation while others join.[86] Dates of birth and the death of persons and their social relationships offer a challenge; researchers have to make decisions to either hide or show the dead.[87] Many practical considerations to deal with issues like duration of ties, e.g., decay functions or the continued importance of old ties, demand researchers to have sufficient information available. However, historians in particular face the problem that available sources do not always provide information about the start or end of relationships. "Yet assigning an infinite duration to ties because we lack precise information is just an arbitrary choice among others, and generally not the best: Do the oldest always have the highest social capital? It is unlikely."[88]

A final problem relates to aspects of validity and the meaning of network ties. Researchers should ask themselves about the meanings of ties, whether ties existed at all or if these are an artefact of coding. Network methods might be misleading if a hypothesized tie between two actors does not mean a tie for the relevant actors themselves but only to the researcher. Consider the example in Figure 2.3.1, the hypothetical illustration of Hansa traders connecting medieval markets. Here, connections between traders (right graph) are not directly observed in the data but simply inferred from their trade activities in Hansa towns/markets (left graph). In the historical case of the medieval Hansa, it is likely that traders who traded in the same, relatively small towns and markets knew each other, particularly if they traded similar or related goods. It is an established finding in social science research that spatial proximity and shared social foci serve as important predictors of tie formation.[89] This gives confidence to use the two-mode approach for the Hansa case without observing actual interactions or relationships among traders.

In addition, Gould noted that network ties should actually represent something meaningful, e.g. information flows between two nodes A and B.[90] Researchers should be explicit about what they think network ties do, depending on their research question and knowledge about the case. The meaning of network ties is the key to the understanding of social processes as drivers of social change. In my Hansa case study, network ties constitute trade partnerships and manifest a capital transfer, which usually involved risks for the parties involved. This aspect links the type of ties under investigation to topics such as economic development and the rise of the "efficient" West. North and Thomas considered societas and commenda as contractual arrangements "designed to provide capital and a working partner by a voluntary association, to spread and reduce risks, and to improve information flows."[91] Because long-distance trade meant various uncertainties, traders in medieval Lübeck used the societates register to gain more legal certainty. Network ties establishing partnerships in these official records may be characterized by a lack of trust among partners.

It is in these instances of interpretation of meanings in which traditionally working (economic) historians may make important contributions. The methods training historians have, e.g. critical assessment of sources and hermeneutics, sensitizes historians for questions of meaning. Similarly, historians are particularly trained in the handling of fragmentary written records, which is helpful in the context of missing network data. This expertise is very valuable, and historical sociologists may very well learn from historians in these regards. Nevertheless, recently, network experts in sociology have started to pick up tie meanings as a research topic in the context of Renaissance Florence. Gondal and McLean argue "that tie-meaning leaves traces in the structure of a network: particular meanings of a given type of tie are more likely to manifest in certain micro- and macrostructural network configurations than others."[92] The cultural turn in network research has produced a number of valuable studies on what networks are or how networks and culture are related.[93]

The more we know about the quality and content of social relationships the better and the more detailed methods can be applied, such as the application of weighted network measures.[94] It is also here where historians could apply their case-related knowledge for the appropriate use of tie weights, e.g. in the selection of relevant events in two-mode data. For the present case of the societates register, the application of monetary values as tie weights would be a logical next step due to the focus on capital distribution of preceding economic historians.[95] While monetary values could potentially explain distrust among partners, it is striking that this official recording practice attracted many traders from the same families. It seems plausible to understand this pattern not as a lack of trust among family members but rather due to the fact that these worked with non-family members in larger partnerships.[96] Triadic patterns of interactions or even larger constellations in the context of partner selection of nodal attributes offer a potential for new research. Most studies of partner selection from sociology focus on dyadic mechanisms.[97]

To conclude this chapter, a few simple steps for starting a new historical network project can be pointed out. Researches may begin with a detailed look at the data from the historical case of interest to see what is available, how systematic it is and then think about what can be done with it.[98] Scholars who are not tied to a particular case may be encouraged to search for data that is as complete as possible and includes various types of social relations as well as information about the attributes of individuals. Other scholars tied to a particular case may think about which types of relations can be identified in their historical documents and whether attributes of individuals are available. Alternatively, which different types of meanings can the ties in the data have and which structures can be expected for these in a given period? In an exploratory stage, scholars may look for interesting patterns on various levels of analysis (individual, group or total network). Do the observed network patterns support or contradict existent histories, or can these be specified? If the data is subset into multiple periods, are there changes in the observed patterns? Does the pattern historians have found stay the same if the split in periods is changed? Do the data allow an annual split rather than larger periods?

After data exploration and ideally the identification of an analytically or historically interesting pattern, researchers may continue by making sure findings are reliable. Network scholars who are interested in making generalizations may want to make sure that observed patterns deviate from what could be expected by chance. Visualizations may conceal or overemphasize patterns that could be supported by descriptive and multivariate statistics. Statistical models as those described in the selected best practice studies section are a good way to handle this and may account for alternative explanatory factors and confounding. A threat to most network arguments and scholars is the structure vs. agency question. Network arguments may address whether the observed structural patterns are caused by individual attributes. The mere partitioning of networks into periods may also have an impact on the results. Detailed knowledge about cases is mandatory for carrying out historical network research not only to make decisions about meanings of ties or their duration but also to take into account and interpret other exogenous factors. The aim of a network-based study, however, should be to identify interesting endogenous, relational structures and processes that matter in the economy. Social network analysis offers new opportunities to economic history to find these patterns.

Notes

1 The author wishes to thank Prof. Dr. Katherine Stovel, Dr. Martin Stark, Dr. Marten Duering, and Prof. Dr. Hammel-Kiesow.

2 Christian Marx, "Economic Networks," in *European History Online (EGO)*, ed. Leibniz Institute of European History (IEG) (Mainz, 2012), http://ieg-ego.eu/en/threads/european-networks/economic-networks (accessed 9 September 2016).

3 Social network analysis can be defined as the set of tools commonly used by network scholars to study the structure of social relationships among a set of actors. Standard introductions to methods of social network analysis are: Stanley Wasserman and Katherine Faust, *Social Network Analysis: Methods and Applications* (New York: Cambridge University Press, 1994); and much less technical John Scott, *Social Network Analysis: A Handbook* (Thousand Oaks: Sage Publications, 2000).

4 Peter Bearman, "Big Data and Historical Social Science," *Big Data & Society*, no. 2 (2015): pp. 1–5.

5 Ibid., p. 2.

6 Wasserman and Faust, *Social Network Analysis*.

7 Two examples that I will also summarize later are: Emily Erikson and Peter Bearman, "Malfeasance and the Foundations for Global Trade: The Structure of English Trade in the East Indies, 1601–1833," *American Journal of Sociology*, no. 112 (2006): pp. 195–230 and Quentin van Doosselaere, *Commercial Agreements and Social Dynamics in Medieval Genoa* (Cambridge: Cambridge University Press, 2009).

8 Mike Burkhardt, "Networks as Social Structures in Late Medieval and Early Modern Towns: A Theoretical Approach to Historical Network Analysis," in *Commercial Networks and European Cities, 1400–1800*, ed. Andrea Caracausi and Christof Jeggle, pp. 13–43 (Brookfield: Pickering & Chatto, 2014). Burkhardt sees an urgent need for a more precise network notion in the historical sciences: "It is not enough to have the feeling that there might be a network or a linkage. If I [Burkhardt] talk about a network, I need to be able to prove that it existed. And the base of proof for a network analysis is found in empirical data collection" (Ibid., p. 17). In sociology, Emirbayer and Goodwin (1994:1414) pointed out that network analysts carefully operationalize ideas such as social structure or cohesion.

9 Roger V. Gould, "Uses of Network Tools in Comparative Historical Research," in *Comparative Historical Analysis in the Social Sciences*, ed. James Mahoney and Dietrich Rueschemeyer, 7th ed., Cambridge Studies in Comparative Politics, pp. 241–69 (New York: Cambridge University Press, 2003).

10 Petri Ylikoski, "Social Mechanism," in *International Encyclopedia of the Social Sciences & Behavioral Sciences*, ed. James D. Wright, 2nd ed., pp. 415–20 (New York: Eslevier, 2015).

11 Wasserman and Faust, *Social Network Analysis*, pp. 169–219.

12 Ibid., pp. 249–90.

13 Ibid., pp. 164, 180–1.

14 Harrison C. White, Scott A. Boorman and Ronald L. Breiger, "Social Structure from Multiple Networks: Blockmodels of Roles and Positions," *American Journal of Sociology* 81 (1976): pp. 730–80.

15 Alan Agresti and Barbara Finlay, *Statistical Methods for the Social Sciences* (Upper Saddle River: Pearson Prentice Hall, 2009).

16 Matthias Bixler, "Historical Network Research: Taking Stock," in *Debtors, Creditors, and Their Networks: Social Dimensions of Monetary Dependence from the Seventeenth*

to the Twentieth Century, ed. Andreas Gestrich and Martin Stark, p. 65 (London: German Historical Institute London Bulletin Supplement, 2015).

17 Peter Hedström and Peter Bearman, eds., *The Oxford Handbook of Analytical Sociology* (Oxford: Oxford University Press, 2009).

18 Karen Barkey, "Historical Sociology," in *The Oxford Handbook of Analytical Sociology*, ed. Peter Hedström and Peter Bearman, p. 712 (Oxford: Oxford University Press, 2009).

19 Chris Lorenz, "History and Theory," in *The Oxford History of Historical Writing*, ed. Axel Schneider and Daniel R. Woolf, 5 vols., Historical Writing Since 1945, p. 21 (Oxford: Oxford University Press, 2011).

20 Fernand Braudel, *On History* (Chicago: University of Chicago Press, 1980), Translated by Sarah Matthews, pp. 64, 71.

21 John F. Padgett and Paul D. McLean, "Organizational Invention and Elite Transformation: The Birth of Partnership Systems in Renaissance Florence," *American Journal of Sociology*, no. 111 (2006): pp. 1463–568.

22 Gould, "Uses of Network Tools in Comparative Historical Research," p. 242.

23 Max Weber, *The Methodology of the Social Sciences* (Glencoe: Free Press, 1949), p. 92.

24 Mike Burkhardt, *Der Hansische Bergenhandel im Spätmittelalter: Handel- Kaufleute – Netzwerke* (Köln: Böhlau, 2009), pp. 189, 219, 365.

25 Gould, "Uses of Network Tools in Comparative Historical Research," p. 242.

26 Henning Hillmann and Brady L. Aven, "Fragmented Networks and Entrepreneurship in Late Imperial Russia," *American Journal of Sociology*, no. 117 (2011): pp. 484–538.

27 Agresti and Finlay, *Statistical Methods for the Social Sciences*.

28 Emily Erikson, *Between Monopoly and Free Trade: The English East India Company, 1600–1757* (Princeton: Princeton University Press, 2014).

29 Erikson and Bearman, "Malfeasance and the Foundations for Global Trade: The Structure of English Trade in the East Indies, 1601–1833," pp. 195–230.

30 van Doosselaere, *Commercial Agreements and Social Dynamics in Medieval Genoa*.

31 Ibid., pp. 63, 72.

32 Ibid., pp. 85, 94–9.

33 Ibid., p. 110.

34 Georg Simmel, *Soziologie: Untersuchungen über die Formen der Vergesellschaftung* (Frankfurt am Main: Suhrkamp, 1908/1992).

35 Katherine Stovel and Lynette Shaw, "Brokerage," *Annual Review of Sociology*, no. 38 (2012): pp. 139–58.

36 Tore Opsahl, "Triadic Closure in Two-Mode Networks: Redefining the Global and Local Clustering Coefficients," *Social Networks*, no. 35 (2013): pp. 159–67.

37 Gabor Csardi and Tamas Nepusz, "The Igraph Software Package for Complex Network Research," *InterJournal, Complex Systems 1695*, p. 2006.

38 Ronald L. Breiger, "The Duality of Persons and Groups," *Social Forces*, no. 53 (1974): pp. 181–90.

39 Scott, *Social Network Analysis*, p. 102.

40 Stephan Selzer, *Die mittelalterliche Hanse* (Darmstadt: WBG – Wissenschaftliche Buchgesellschaft, 2010).

41 Bernd Wurpts, Katie E. Corcoran and Steven Pfaff, "The Diffusion of Protestantism in Northern Europe: Historical Embeddedness and Complex Contagions in the Adoption of the Reformation," *Social Science History* 42 (2018): pp. 213–44, doi:10.1017/ssh.2017.49 (accessed 13 December 2018).

42 Lutz Kaelber, "Max Weber's Dissertation," *History of the Human Sciences*, no. 16 (2003): pp. 27–56.

43 van Doosselaere, *Commercial Agreements and Social Dynamics in Medieval Genoa*.

44 Rolf Hammel-Kiesow, "Schriftlichkeit und Handelsgesellschaften niederdeutsch-hansischer und oberdeutscher Kaufleute im späten 13. und im 14. Jahrhundert," in *Von Nowgorod bis London: Studien zu Handel, Wirtschaft und Gesellschaft im mittelalterlichen Europa Festschrift für Stuart Jenks zum 60. Geburtstag*, ed. Marie-Luise Heckmann and Jens Röhrkasten, pp. 213–41 (Göttingen: V & R Unipress, 2008b).
45 Ahasver von Brandt, "Lübeck in der deutschen Geistesgeschichte: Ein Versuch," *Zeitschrift des Vereins für Lübeckische Geschichte und Altertumskunde*, 31 (1949): pp. 149–88.
46 van Doosselaere, *Commercial Agreements and Social Dynamics in Medieval Genoa*, p. 30.
47 Rolf Hammel-Kiesow, *Die Hanse* (München: C.H. Beck, 2008a); Philippe Dollinger, *Die Hanse* (Stuttgart: Alfred Kröner, 1989).
48 Dollinger, *Die Hanse*, pp. 85–8.
49 Hammel-Kiesow, *Die Hanse*, p. 60.
50 Two data sources were used for the following analyses. Albrecht Cordes, Klaus Friedland and Rolf Sprandel, eds., *Societates: Das Verzeichnis der Handelsgesellschaften im Lübecker Niederstadtbuch 1311–1361*, Quellen und Darstellungen zur Hansischen Geschichte Neue Folge/Band LIV (Köln: Böhlau, 2003), pp. 59–122; Karl-Heinz Saß, *Societates-Register. Regesten zum Niederstadtbuch* 1 (Archiv der Hansestadt Lübeck: Unpublished Documents, 1953), pp. 53–94.
51 Albrecht Cordes, "Rechtshistorische Einführung," in Cordes; Friedland; Sprandel, *Societates*, pp. 11–43.
52 Albrecht Cordes, *Spätmittelalterlicher Gesellschaftshandel im Hanseraum*, Quellen und Darstellungen zur Hansischen Geschichte Neue Folge/Band XLV (Köln: Böhlau, 1998), p. 223.
53 Rolf Sprandel, "Wirtschaftsgeschichtliche Einführung," in Cordes; Friedland; Sprandel, *Societates*, p. 1 folgs.
54 Ibid., p. 7.
55 Hammel-Kiesow, "Schriftlichkeit und Handelsgesellschaften niederdeutsch-hansischer und oberdeutscher Kaufleute im späten 13. und im 14. Jahrhundert," p. 223; Cordes, *Spätmittelalterlicher Gesellschaftshandel im Hanseraum*, p. 111f.
56 van Doosselaere, *Commercial Agreements and Social Dynamics in Medieval Genoa*, p. 75.
57 Albrecht Cordes, "Rechtshistorische Einführung," in Cordes; Friedland; Sprandel, *Societates*, p. 13.
58 Ibid., p. 43.
59 Not coded as dominus here were two cases in which traders were marked as dominus and were identified as clerics plus one case of a one woman labeled domina. These individuals likely did not have much political power in regards to city and Hansa politics.
60 Emil Ferdinand Fehling, *Lübeckische Ratslinie von den Anfängen der Stadt bis auf die Gegenwart*, Veröffentlichungen zur Geschichte der Freien und Hansestadt Lübeck (Lübeck: Max Schmidt-Römhild, 1925).
61 Csardi and Nepusz, "The Igraph Software Package for Complex Network Research".
62 van Doosselaere, *Commercial Agreements and Social Dynamics in Medieval Genoa*, p. 104.
63 Ibid., pp. 104–7.
64 Ibid., pp. 104–7, 112. Sociologists use the term "homophily" in general when they describe the often-observed pattern that individuals tend to interact with similar others. Miller McPherson, Lynn Smith-Lovin and James M. Cook, "Birds of a Feather: Homophily in Social Networks," *Annual Review of Sociology*, no. 27 (2001): pp. 415–44. For a more recent review including literature from economic sociology and organization studies see: Mark T. Rivera, Sara B. Soderstrom and

Brian Uzzi, "Dynamics of Dyads in Social Networks: Assortative, Relational, and Proximity Mechanisms," *Annual Review of Sociology*, no. 36 (2010): pp. 91–115. Individuals have been found to prefer homophilous ties because these have expressive benefits and often include trust. Heterophilous ties to partners with different attributes and capabilities are frequently preferred in contexts of collaborations or economic production.

65 Cordes, *Spätmittelalterlicher Gesellschaftshandel im Hanseraum*, p. 111. The early period includes 322 vertices and 342 edges (density=0.007) whereas the late period contains 151 vertices and 255 edges (density=0.023) in the Cordes et al. dataset. The dataset based on Saβ (1953) includes 329 vertices and 373 edges in the main period and 153 vertices connected by 232 edges in the late period. There are 23 and nine traders with dominus status in the early and late period, respectively. The average number of connections per node (mean degree) is 2.12 in the early period and 3.43 in the late period. A look at the triad census shows 105 complete triangles in the main period and 242 in the late period of the societates register.

66 Mark E.J. Newman, "Mixing Patterns in Networks," *Physical Review*, E 67, 026126 (2003): p. 2. See also for how to calculate this in Eric D. Kolaczyk and Gabor Csardi, *Statistical Analysis of Network Data with R* (New York: Springer, 2014), pp. 66–7.

67 S14 was developed by Gower and Legendre (1987). In order to calculate S14, I created the two by two matrix including pairs of attributes of adjacent nodes. The attribute pairs consider whether a trader has the attribute (1) or does not have the attribute (0). The present binary attributes are council membership (1 = yes, 0 = no) and having a last name of a councilor (1 = yes, 0 = no). Each partnership tie between trader A and trader B is coded as one of these alternatives: A(1)/B(1), A(1)/B(0), A(0)/B(1), A(0)/B(0). Across the entire network this is summarized in a two by two table where the cell entries indicate the total number of ties having each type of configuration. This "mixing matrix" was produced in R using code developed by Gary Weissman. Unlike the original code the mixmat function used here had the parameter "use.density=FALSE" instead of the "use.density=TRUE" specified in the code by Gary, https://gist.github.com/gweissman/2402741(accessed 30 January 2019). Alternatively, the mixing matrix can also be calculated in statnet (see Jenine K. Harris, *An Introduction to Exponential Random Graph Models* (Thousand Oaks: Sage Publications, 2014), p. 43). S14 values were calculated in R based on the formula presented in van Doosselaere, *Commercial Agreements and Social Dynamics in Medieval Genoa*, p. 113, and David Krackhardt, "Assessing the Political Landscape: Structure, Cognition, and Power in Organizations," *Administrative Science Quarterly*, no. 35 (1990): p. 350. Assortativity coefficients were calculated using the assortativity function in the igraph package.

68 John Scott, "Social Physics and Social Networks," in *The Sage Handbook of Social Network Analysis*, ed. John Scott and Peter J. Carrington, pp. 55–66 (Thousand Oaks: Sage Publications, 2011).

69 Van Doosselaere, *Commercial Agreements and Social Dynamics in Medieval Genoa*, p. 113.

70 Csardi and Nepusz, "The Igraph Software Package for Complex Network Research".

71 An early introductory text for ERGMs is: Garry Robins, Pip K. Y. Pattison, and Dean Lusher, "An Introduction to Exponential Random Graph (p*) Models for Social Networks," *Social Networks*, no. 29 (2007): p. 1. The software package to estimate ERGMs in R, as used in this chapter, is statnet: Mark S. Handcock et al., "Statnet: Software Tools for the Statistical Modeling of Network Data," (2003), http://statnetproject.org (accessed 13 December 2018).

72 Harris, *An Introduction to Exponential Random Graph Models*, pp. 5–6.

73 Ibid., p. 27.

74 Ibid., p. 55.

75 Handcock et al., "Statnet: Software Tools for the Statistical Modeling of Network Data".
76 Harris, *An Introduction to Exponential Random Graph Models*, p. 46.
77 Burkhardt, *Der Hansische Bergenhandel im Spätmittelalter*, pp. 189, 219, 365.
78 Dollinger, *Die Hanse*, p. 88.
79 John F. Padgett and Walter W. Powell, *The Emergence of Organizations and Markets*, Core Textbook (Princeton: Princeton University Press, 2012). Bernd Wurpts, Networks into Institutions or Institutions into Networks? Evidence from the Medieval Hansa. (University of Washington, Ph.D. Dissertation, 2018).
80 Hammel-Kiesow, "Schriftlichkeit und Handelsgesellschaften niederdeutsch-hansischer und oberdeutscher Kaufleute im späten 13. und im 14. Jahrhundert," p. 223.
81 Helen Zimmern, *The Hansa Towns* (New York: G. P. Putnam's Sons, 1889), p. 34.
82 van Doosselaere, *Commercial Agreements and Social Dynamics in Medieval Genoa*, p. 107.
83 Dollinger, *Die Hanse*, p. 88.
84 Ulf C. Ewert and Stephan Selzer, "Social Networks," in *A Companion to the Hanseatic League*, ed. Donald J. Harreld, p. 164 (Boston: Brill, 2015).
85 Jeffrey A. Smith and James Moody, "Structural Effects of Network Sampling Coverage I: Nodes Missing at Random," *Social Networks* 35 (2013): pp. 652–68.
86 Burkhardt, "Networks as Social Structures in Late Medieval and Early Modern Towns: A Theoretical Approach to Historical Network Analysis," pp. 13–43.
87 Claire Lemercier, "Taking Time Seriously: How Do We Deal with Change in Historical Networks?," in *Knoten und Kanten III: Soziale Netzwerkanalyse in Geschichts- und Politikforschung*, ed. Markus Gamper et al., pp. 183–212 (Bielefeld: Transcript Verlag, 2015), p. 206.
88 Ibid., p. 195.
89 Rivera, Soderstrom and Uzzi, "Dynamics of Dyads in Social Networks: Assortative, Relational, and Proximity Mechanisms," pp. 105–7.
90 Gould, "Uses of Network Tools in Comparative Historical Research," pp. 241–69.
91 Douglass Cecil North and Robert Paul Thomas, *The Rise of the Western World: A New Economic History* (Cambridge: Cambridge University Press, 1973), p. 54.
92 Neha Gondal and McLean Paul D., "Linking Tie-Meaning with Network Structure: Variable Connotations of Personal Lending in a Multiple-Network Ecology," *Poetics*, no. 41 (2013): p. 123.
93 Harrison C. White, *Identity and Control: How Social Formations Emerge* (Princeton: Princeton University Press, 2008); Jan A. Fuhse, "The Meaning Structure of Social Networks," *Sociological Theory*, no. 27 (2009): pp. 51–73; Paul Douglas McLean, *Culture in Networks* (Malden: Polity, 2017).
94 Tore Opsahl, Filip Agneessens, and John Skvoretz, "Node Centrality in Weighted Networks: Generalizing Degree and Shortest Paths," *Social Networks* 32 (2010): pp. 245–51. For an application of weighted ties in the context of the medieval Hansa and the early diffusion of Protestantism see: Wurpts, Corcoran and Pfaff, "The Diffusion of Protestantism in Northern Europe: Historical Embeddedness and Complex Contagions in the Adoption of the Reformation" pp. 213–44.
95 Rolf Sprandel, "Wirtschaftsgeschichtliche Einführung," in Cordes; Friedland; Sprandel, *Societates*, pp. 1–9.
96 For a discussion of more reasons why medieval traders used official recording procedures see: van Doosselaere, *Commercial Agreements and Social Dynamics in Medieval Genoa*, p. 103.
97 Rivera, Soderstrom and Uzzi, "Dynamics of Dyads in Social Networks: Assortative, Relational, and Proximity Mechanisms" pp. 91–115.
98 Here I want to thank Prof. Henning Hillmann for some general remarks made in correspondence when I was developing one of my studies. Also, a helpful introduction

for the planning stage of network research is: Garry Robins, *Doing Social Network Research: Network Research Design for Social Scientists* (Los Angeles: Sage Publications, 2015).

Bibliography

Agresti, Alan, and Barbara Finlay. *Statistical Methods for the Social Sciences*. Upper Saddle River: Pearson Prentice Hall, 2009.

Barkey, Karen. "Historical Sociology." In *The Oxford Handbook of Analytical Sociology*. Edited by Peter Hedström and Peter Bearman, pp. 712–33. Oxford: Oxford University Press, 2009.

Bearman, Peter. "Big Data and Historical Social Science." *Big Data & Society*, no. 2 (2015): pp. 1–5.

Bixler, Matthias. "Historical Network Research: Taking Stock." In *Debtors, Creditors, and Their Networks: Social Dimensions of Monetary Dependence from the Seventeenth to the Twentieth Century*. Edited by Andreas Gestrich and Martin Stark, pp. 43–67, 3. London: German Historical Institute London Bulletin Supplement, 2015.

Brandt, Ahasver von. "Lübeck in der deutschen Geistesgeschichte: Ein Versuch." *Zeitschrift des Vereins für Lübeckische Geschichte und Altertumskunde* 31 (1949): pp. 149–88.

Braudel, Fernand. *On History*. Translated by Sarah Matthews. Chicago: University of Chicago Press, 1980.

Breiger, Ronald L. "The Duality of Persons and Groups." *Social Forces*, no. 53 (1974): pp. 181–90.

Burkhardt, Mike. *Der Hansische Bergenhandel im Spätmittelalter: Handel- Kaufleute – Netzwerke*. Köln: Böhlau, 2009.

———. "Networks as Social Structures in Late Medieval and Early Modern Towns: A Theoretical Approach to Historical Network Analysis." In *Commercial Networks and European Cities, 1400–1800*. Edited by Andrea Caracausi and Christof Jeggle, pp. 13–43. Brookfield: Pickering & Chatto, 2014.

Cordes, Albrecht. *Spätmittelalterlicher Gesellschaftshandel im Hanseraum*. Quellen und Darstellungen zur Hansischen Geschichte Neue Folge/Band XLV. Köln: Böhlau, 1998.

———. "Rechtshistorische Einführung." In *Societates: Das Verzeichnis der Handelsgesellschaften im Lübecker Niederstadtbuch 1311–1361*. Quellen und Darstellungen zur Hansischen Geschichte Neue Folge/Band LIV. Edited by Albrecht Cordes, Klaus Friedland, and Rolf Sprandel, pp. 11–43. Köln: Böhlau, 2003.

Cordes, Albrecht, Klaus Friedland, and Rolf Sprandel, eds. *Societates: Das Verzeichnis der Handelsgesellschaften im Lübecker Niederstadtbuch 1311–1361*. Quellen und Darstellungen zur Hansischen Geschichte Neue Folge/Band LIV. Köln: Böhlau, 2003.

Csardi, Gabor, and Tamas Nepusz. "The Igraph Software Package for Complex Network Research." *InterJournal, Complex Systems* 1695 (2006).

Dollinger, Philippe. *Die Hanse*. Stuttgart: Alfred Kröner, 1989.

Emirbayer, Mustafa, and Jeff Goodwin. "Network Analysis, Culture, and the Problem of Agency." American Journal of Sociology, no. 99 (1994): pp. 1411–1454.

Erikson, Emily. *Between Monopoly and Free Trade: The English East India Company, 1600–1757*. Princeton: Princeton University Press, 2014.

Erikson, Emily, and Peter Bearman. "Malfeasance and the Foundations for Global Trade: The Structure of English Trade in the East Indies, 1601–1833." *American Journal of Sociology*, no. 112 (2006): pp. 195–230.

Ewert, Ulf Christian, and Stephan Selzer. "Social Networks." In *A Companion to the Hanseatic League*. Edited by Donald J. Harreld, pp. 162–93. Boston: Brill, 2015.

Fehling, Emil Ferdinand. *Lübeckische Ratslinie von den Anfängen der Stadt bis auf die Gegenwart*. Veröffentlichungen zur Geschichte der Freien und Hansestadt Lübeck. Lübeck: Max Schmidt-Römhild, 1925.

Fuhse, Jan A. "The Meaning Structure of Social Networks." *Sociological Theory*, no. 27 (2009): pp. 51–73.

Gamper, Markus, Linda Reschke, Michael Schönhuth, and Marten Düring, eds. *Knoten und Kanten III: Soziale Netzwerkanalyse in Geschichts- und Politikforschung*. Bielefeld: Transcript Verlag, 2015.

Gondal, Neha, and McLean Paul D. "Linking Tie-Meaning with Network Structure: Variable Connotations of Personal Lending in a Multiple-Network Ecology." *Poetics*, no. 41 (2013): pp. 122–50.

Gould, Roger V. "Uses of Network Tools in Comparative Historical Research." In *Comparative Historical Analysis in the Social Sciences*. Edited by James Mahoney and Dietrich Rueschemeyer. 7th ed., pp. 241–69. Cambridge Studies in Comparative Politics. New York: Cambridge University Press, 2003.

Hammel-Kiesow, Rolf. *Die Hanse*. München: C.H. Beck, 2008a.

———. "Schriftlichkeit und Handelsgesellschaften niederdeutsch-hansischer und oberdeutscher Kaufleute im späten 13. und im 14. Jahrhundert." In *Von Nowgorod bis London: Studien zu Handel, Wirtschaft und Gesellschaft im mittelalterlichen Europa Festschrift für Stuart Jenks zum 60. Geburtstag*. Edited by Marie-Luise Heckmann and Jens Röhrkasten, pp. 213–41. Göttingen: V & R Unipress, 2008b.

Handcock, Mark S., David R. Hunter, Carter T. Butts, Steven M. Goodreau, and Martina Morris. "statnet: Software Tools for the Statistical Modeling of Network Data." (2003). http://statnetproject.org (accessed 13 December 2018).

Harris, Jenine K. *An Introduction to Exponential Random Graph Models*. Thousand Oaks: Sage Publications, 2014.

Hedström, Peter, and Peter Bearman, eds. *The Oxford Handbook of Analytical Sociology*. Oxford: Oxford University Press, 2009.

Hillmann, Henning, and Brady L. Aven. "Fragmented Networks and Entrepreneurship in Late Imperial Russia." *American Journal of Sociology*, no. 117 (2011): pp. 484–538.

Kaelber, Lutz. "Max Weber's Dissertation." *History of the Human Sciences*, no. 16 (2003): pp. 27–56.

Kolaczyk, Eric D., and Gabor Csardi. *Statistical Analysis of Network Data with R*. New York: Springer, 2014.

Krackhardt, David. "Assessing the Political Landscape: Structure, Cognition, and Power in Organizations." *Administrative Science Quarterly*, no. 35 (1990): pp. 342–69.

Lemercier, Claire. "Taking Time Seriously: How Do We Deal with Change in Historical Networks?" In *Knoten und Kanten III: Soziale Netzwerkanalyse in Geschichts- und Politikforschung*. Edited by Markus Gamper et al., pp. 183–212. Bielefeld: Transcript Verlag, 2015.

Lorenz, Chris. "History and Theory." In *The Oxford History of Historical Writing*. Edited by Axel Schneider and Daniel R. Woolf. 5 vols., pp. 13–35. Historical Writing Since 1945. Oxford: Oxford University Press, 2011.

Marx, Christian. "Economic Networks." In *European History Online (EGO)*. Edited by Leibniz Institute of European History (IEG). Mainz, 2012. http://ieg-ego.eu/en/threads/european-networks/economic-networks (accessed 9 September 2016).

McLean, Paul Douglas. *Culture in Networks*. Malden: Polity Press, 2017.

McPherson, Miller, Lynn Smith-Lovin, and James M. Cook. "Birds of a Feather: Homophily in Social Networks." *Annual Review of Sociology*, no. 27 (2001): pp. 415–44.

Newman, Mark E.J. "Mixing Patterns in Networks." *Physical Review* E 67, no. 026126 (2003): pp. 1–14.

North, Douglass Cecil, and Robert Paul Thomas. *The Rise of the Western World: A New Economic History*. Cambridge: Cambridge University Press, 1973.

Opsahl, Tore. "Triadic Closure in Two-Mode Networks: Redefining the Global and Local Clustering Coefficients." *Social Networks*, no. 35 (2013): pp. 159–67.

Opsahl, Tore, Filip Agneessens, and John Skvoretz. "Node Centrality in Weighted Networks: Generalizing Degree and Shortest Paths." *Social Networks* 32 (2010): pp. 245–51.

Padgett, John F., and Paul Douglas McLean. "Organizational Invention and Elite Transformation: The Birth of Partnership Systems in Renaissance Florence." *American Journal of Sociology*, no. 111 (2006): pp. 1463–568.

Padgett, John F., and Walter W. Powell. *The Emergence of Organizations and Markets*. Core Textbook. Princeton: Princeton University Press, 2012.

Rivera, Mark T., Sara B. Soderstrom, and Brian Uzzi. "Dynamics of Dyads in Social Networks: Assortative, Relational, and Proximity Mechanisms." *Annual Review of Sociology*, no. 36 (2010): pp. 91–115.

Robins, Garry. *Doing Social Network Research: Network Research Design for Social Scientists*. Los Angeles: Sage Publications, 2015.

Robins, Garry, Pip Kalish Yuval Pattison, and Dean Lusher. "An Introduction to Exponential Random Graph (p*) Models for Social Networks." *Social Networks*, no. 29 (2007): pp. 173–91.

Saß, Karl-Heinz. *Societates-Register. Regesten zum Niederstadtbuch* 1. Archiv der Hansestadt Lübeck: Unpublished Documents, 1953, pp. 53–94.

Scott, John. *Social Network Analysis: A Handbook*. Thousand Oaks: Sage Publications, 2000.

———. "Social Physics and Social Networks." In *The Sage Handbook of Social Network Analysis*. Edited by John Scott and Peter J. Carrington, pp. 55–66. Thousand Oaks: Sage Publications, 2011.

Selzer, Stephan. *Die mittelalterliche Hanse*. Darmstadt: WBG – Wissenschaftliche Buchgesellschaft, 2010.

Simmel, Georg. *Soziologie: Untersuchungen über die Formen der Vergesellschaftung*. Frankfurt am Main: Suhrkamp, 1908/1992.

Smith, Jeffrey A., and James Moody. "Structural Effects of Network Sampling Coverage I: Nodes Missing at Random." *Social Networks* 35 (2013): pp. 652–68.

Sprandel, Rolf. "Wirtschaftsgeschichtliche Einführung." In Cordes; Friedland; Sprandel, *Societates*, pp. 1–9.

Stovel, Katherine, and Lynette Shaw. "Brokerage." *Annual Review of Sociology*, no. 38 (2012): pp. 139–58.

van Doosselaere, Quentin. *Commercial Agreements and Social Dynamics in Medieval Genoa*. Cambridge: Cambridge University Press, 2009.

Wasserman, Stanley, and Katherine Faust. *Social Network Analysis: Methods and Applications*. New York: Cambridge University Press, 1994.

Weber, Max. *The Methodology of the Social Sciences*. Glencoe: Free Press, 1949.

White, Harrison C. *Identity and Control: How Social Formations Emerge*. Princeton: Princeton University Press, 2008.

White, Harrison C., Scott A. Boorman, and Ronald L. Breiger. "Social Structure from Multiple Networks: Blockmodels of Roles and Positions." *American Journal of Sociology* 81 (1976): pp. 730–80.

Wurpts, Bernd. Networks into Institutions or Institutions into Networks? Evidence from the Medieval Hansa. University of Washington, Ph.D. Dissertation, 2018.

Wurpts, Bernd, Katie E. Corcoran, and Steven Pfaff. "The Diffusion of Protestantism in Northern Europe: Historical Embeddedness and Complex Contagions in the Adoption of the Reformation." *Social Science History* 42 (2018): pp. 213–44. doi:10.1017/ssh.2017.49 (accessed 13 December 2018).

Ylikoski, Petri. "Social Mechanism." In *International Encyclopedia of the Social Sciences & Behavioral Sciences*. Edited by James D. Wright. 2nd ed., pp. 415–20. New York: Eslevier, 2015.

Zimmern, Helen. *The Hansa Towns*. New York: G. P. Putnam's Sons, 1889.

2.4 Flemish merchant networks in early modern Seville. Approaches, comparisons, and methodical considerations

Eberhard Crailsheim

For some time now, networks have been a dominant paradigm in economic history.[1] Investigating networks enables a multi-polar approach to an historical economic setting, which has many advantages over traditional perspectives that are, for example, based on bilateral or linear exchange relations between regions or states.[2] Networks seem to have the intriguing potential to show what "really" lies behind economic developments. The analysis of economic networks seems to shed light on the structures that lie underneath the obvious and lead to new insights into our commercial conduct. In my opinion, this is true, but one has to bear in mind a caveat: many historical studies dedicated to network research remain on the surface and analyse networks only by discussing or displaying the protagonists and their relations, without using the potential of specific network tools. Social Network Analysis (SNA) goes one step further: utilizing network and graph theories, SNA calculates and visualizes the structures of social relations and hence can indeed show different results from what seems obvious.[3]

Christian Marx has listed five areas of research that can be considered relevant for economic history: (1) (long-distance) trade networks, (2) credit networks, (3) networks of financial capital and economic elites, (4) internal company structures, and (5) innovations networks and regional clusters.[4] For the early modern period, it is especially the first two areas that are relevant. Following the traces of goods and credits, networks can be recreated, which helps us to understand merchants' lines of commerce, communication, and influence.[5]

Network studies in economic history are not exclusively dedicated to the economic sphere. In early modern times, they are strongly interrelated with studies of migration and Diaspora, focusing on cross-cultural connections, as well as on links between members of one group (e.g. family, origin, language, religion).[6] In this regard, one has to stress the value of "trust" in commercial relations,[7] as an integral element of commercial networks. Trust helps to reduce social complexity and, hence, diminishes transaction costs.[8] The study of trade networks in early modern times offers the opportunity to focus on the network structure of the commerce but, at the same time, to include the agency of the merchants.[9] Investigations on historical trade networks have often focused on research objects such as merchandise,[10] (port) cities,[11] merchant communities,[12] families,[13] trading companies,[14] or entire oceans.[15] The controversy about the beginning of globalization is one of the most appealing topics in this respect. Following the traces of merchants and their goods supports the argument that globalization started not in the nineteenth century but, at the latest, with the crossing of the Atlantic in 1492.[16] The incorporation of America into the European trade circuit accelerated the commercial processes in the "Old World". This so-called "European expansion" drew much attention in economic historiography, especially long-distance trade across the oceans. One of the hot spots of early globalization and a centre for long-distance trade was Seville.

At the beginning of the early modern period, the city of Seville was a bustling centre of trade and commerce that connected the New World with the Old. Almost all traffic between the continents was channelled through this city, which is why it attracted a large number of merchants from all over Europe. To control and monitor this traffic, the Spanish monarchy implemented thorough bureaucracy, and, in order to facilitate the commerce, 24 notary offices were at the service of the numerous Spanish and foreign merchants, eager to participate in the auspicious Indies[17] trade. This implied the generation of large amounts of data, which allows the reconstruction of networks of merchants who traded in Seville during that time. My investigations on this topic have focused on foreign merchant communities in Seville and explored the composition of the foreigners in the city, their relevance within the Indies trade and their networks – in particular those of the large Flemish community.[18] For that purpose, I have studied files of naturalization, which can be considered a necessary requisite for foreigners to participate directly in the Indies trade, as well as notary files of selected offices between the years 1580 and 1640. On the following pages, I will outline my sources and concerns that emerged during the recreation of the networks, display models of Flemish kinship and professional networks, elucidate the example of one outstanding merchant, and contemplate the specific value of social network analysis tools to gain additional insights into the characteristics of merchant communities' social structures.

Seville and the Indies trade

In 1492, the slow process of globalization took a huge leap forward. Two worlds formerly separated were suddenly connected, and both Europe and America transformed profoundly. In America, the political and social changes, along with the so called "Columbian Exchange" (the biological consequences of the connection)[19] triggered the collapse of the existing society, which also had intense repercussions in the economic sector. For Spain, the discoveries enabled a political, cultural, and economic Golden Age in the sixteenth century. Moreover, in the rest of Europe, the New World generated countless transformations. One result on the economic level was that commerce in general experienced a fundamental acceleration, which was in particular due to one product: silver bullion. Since the middle of the sixteenth century, the growing American production of silver compensated for the decline of the Old World silver mining industry, and plenty of silver entered Europe via Spain.[20] Besides contributing to a more dynamic European economy, this influx of silver also enabled the intensification of the European Asian trade, especially via the Cape of Good Hope; China in particular had a great demand for silver coins.[21] Hence, the Spanish American trade was an essential element of early globalization.

In the centre of the Spanish American trade was the city of Seville, which had a royal monopoly and was the only legal port of call for America. Between 1492 and 1600, the number of its inhabitants doubled from 60,000 to 120,000, by which time it had become one of the largest cities in Europe.[22] The number of large scale merchants rose from 26 in 1492 to 350 in 1533.[23] To guarantee the supply of its American territories and to control all traffic across the Atlantic, the Spanish Crown founded a number of institutions, many of which were located in Seville. The most important one was the *Casa de la Contratación de las Indias*, the House of Trade, a government agency that controlled all American traffic and watched over the Spanish trade monopoly with its territories.[24] Within this monopoly system, the rich Castilian[25] merchants in Seville created a Consulate (*Consulado de Cargadores a Indias*), and only the *Cargadores* were allowed to have traffic moving to and from America.[26] Consequently, a large number of merchants from Spain and all over Europe travelled to Seville to participate in the rich American trade. Many of them were agents of foreign merchant houses, some of which established their own business in Seville. An active participation in the Indies trade, however, was only allowed for subjects of the Crown of Castile. The Flemish, Portuguese, Italian, French, English, and German merchants, who established their houses in Seville,[27] consequently had to rely on Castilian intermediaries to gain access to trade with American goods, such as indigo, hides, cochineal, and, above all, silver. Therefore, many foreign merchants started selling European goods that were in great demand in America, for example textiles and metalware, to Castilian merchants in Seville in exchange for American goods and silver coins.[28] Thus, for a certain period, the Castilian merchants in Seville were in a very fortunate position between the American and the European merchants. The monopoly converted them into brokers, many of whom stopped engaging in the perilous trade themselves and became commissioners for more venturesome foreigners.[29] This situation in itself seems to offer the intriguing task of analysing and reconstructing the merchants' networks in Seville as well as asking for the economically most dominant and powerful groups of merchants that were involved in the American and the European trades.

Sources

To differentiate between the merchants, I focused on their places of origin. Secondary sources only revealed limited information,[30] but the files of the *Archivo General de Indias* (AGI) in Seville were very useful. The AGI contains the largest inventory of American-related records for the early modern period, including much data referring to the Spanish Indies trade. At some point of their lives, many eminent foreign merchants in Seville applied for naturalization in Spain. Overall, such an administrative step did not bring many advantages for foreign merchants in Spain, but in Seville it meant direct access to the Indies trade.[31] Once they had become "Spaniards" they were able to trade directly with America, without any Spanish intermediary. However, such naturalization could not be obtained easily. At the beginning of the seventeenth century, a merchant had to prove that he had been living in Spain for over 20 years, that he possessed more than 4,000 ducats,[32] and that he did not trade on credit, in order to obtain a letter of naturalization. Moreover, he had to be married to a Spanish woman.[33] The applications were collected and dealt with in the *Casa de la Contratación*.[34] The files, which are now at the AGI, contain much information regarding the merchants' family, godfathers and godmothers, witnesses to weddings and other legal processes, as well as some other connections. Frequently, the places of origin of these people were indicated, which enabled the creation of a list of foreigners in Seville. All the applications between 1570 and 1650 were included, summing up to 313.

With this list of merchants from many foreign "nations",[35] I could start the investigation of the notary archives of Seville so as to trace the course of the transatlantic and north western European commerce.[36] The records of the 24 notary offices, which existed simultaneously between 1580 and 1640, are collected in the *Archivo de Protocolos de Sevilla* (APS), being part of the *Archivo Histórico Provincial* of Seville. Each year, between 64 and 122 volumes,[37] with about a thousand folios each, were filled with information regarding selling, buying, credit, insurance, or real estate business. To get a reasonable sample and still gain insight into changes over time (at the peak of the Spanish *Siglo de Oro*), the research team[38] focused on four years with an interval of 20 years: 1580, 1600, 1620, and 1640. For each year, two to three months (between January and July)[39] were examined in two or three different offices.[40] The attention was given to the offices V, XII, XVI, and XXIV, which hold much information about foreigners. Within the respective volumes,[41] a total of 1,696 files were selected for the investigation, which included 3,488 individuals.

Reconstruction of the networks

After about two months of research in the AGI and six months in the APS, we started to summarize the data of each file in an Access database. We included all relations that the files contained (mainly family, friendship, business, and legal relations),[42] thereby recreating the networks of the merchants.[43] For usability, I decided to group the different relations into "private" and "commercial". The former included the AGI data, which involved family and friends ("strong ties"), while the later included the business connections ("weak ties") from the APS. However, the AGI files also revealed the names of witnesses who could vouch for the good behaviour of the applying merchants, an aspect that could neither be linked easily to the private nor to the purely commercial spheres. Therefore, I added a third network category of "semi-private" connections to complete the picture. Compared to biographical investigations, the amount of information on each merchant is relatively small. Yet, to know the most fundamental *private relations* of the merchants (parents, wives, children, godfathers, best men) provides the possibility for partly identifying "family companies" and understanding the entwinement of the private and the commercial spheres. The witnesses for the merchants applying for naturalizations had to be selected with care. In this regard, they reflect the chance for the applicant to rely on powerful partners, even though it is not clear whether they came from a private or commercial relation – hence *semi-private connections*. The commercial networks, based on *business relations*, can be considered a good statistical sample of the notary activity of the investigated period, as they cover about ten percent of the notary files (two to three notary offices of 24 were investigated; the relevant period covered two to three months). The commercial networks hence display a reasonable share of the documented commercial activity of the city of Seville.[44]

Another issue was the time gap of 20 years between the samples of the APS. Due to these large gaps, it is impossible to know if, for example, some of the encountered merchants gained importance for several months, only to disappear shortly after. However, the denominations "resident" or even "citizen" of Seville were clear evidence that many lived in the city for a large period of time. Moreover, a few nodes appeared in more than one of the selected years, indicating at least a continuity of 20 years (or a return after 20 years). However, to keep the networks of the selected years coherent, without the additional dimension of time, each of the years was analysed and assessed separately. Hence, the diachronic connections were omitted in the display and in the calculations – but included in the qualitative analysis. Consequently, I reconstructed one network for each year, which was sub-structured into networks for each nation (the relations within all documents that included members of one nation) and ego-networks for each merchant.

Flemish networks

The Flemish networks were at the heart of my investigation and formed an integral part of the commerce of Seville.[45] The Flemings were the largest foreign community in Seville for much of the first half of the seventeenth century. In addition, they strongly participated in one way or the other (directly or through intermediaries) in the Indies trade.[46]

Figure 2.4.1 shows the reconstruction of their private network on the basis of family and friendship ties, taken from the naturalization files of the AGI. It displays the (limited) connectedness within the Flemish private network that could be reconstructed through the sources. As the inquiries of the Seville officials were, above all, about the family in law of the naturalized merchant (who had to be married to a Spanish woman), it is not surprising that their wives were the most central nodes of the network. The most central of these wives have been highlighted (betweenness centrality, discussed later).

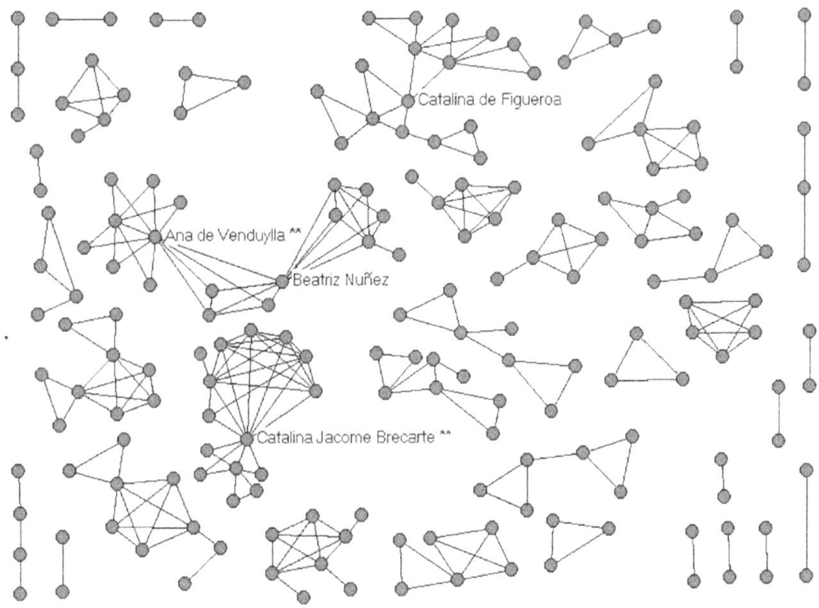

Figure 2.4.1 The private Flemish network 1580–1650

(Source: AGI, Contratación 50A, 50B, 51B, 596A and 596B) (** = Flemish or daughter of Flemings)[47]

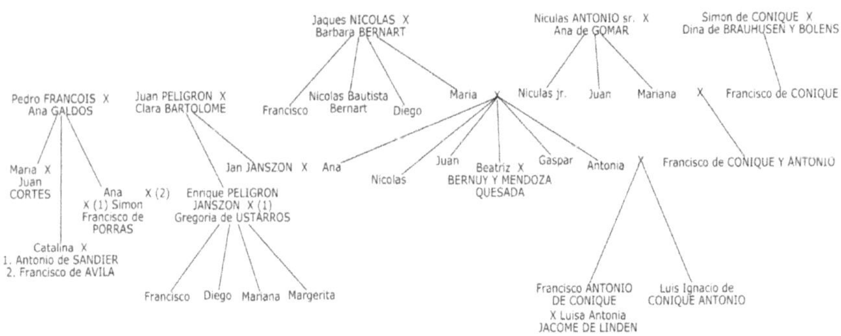

Figure 2.4.2 The connections between the families Nicolas, Antonio, De Conique, Peligron, and Francois (the numbers in front of some of the names indicate the first or second marriage)

(Source: Stols 1971, vol. 2, Stamboom 12; combined with data from the AGI Contratación 50B and 596A, s.f.)[48]

The information of the AGI, moreover, enables the recreation of family trees, evidencing family connections, as for example between the families Nicolas, Antonio, De Conique, Peligron and Francois (Figure 2.4.2), which show a strong cohesion between members of the Flemish nation in Seville.

The semi-private networks, made of ties between the merchants who sought naturalization and their witnesses in the course of the process of naturalization, consist of connections that probably ranged from friendship to business partnerships. Such hybrid semi-private networks are much larger than the private ones and enable the detection of the most central (betweenness) actors Roberto Marcelis and Juan Leclerque with a pronounced bottle-neck or broker function (Figure 2.4.3). Both actors were first-rate-merchants of Seville with Flemish origin who applied for naturalization in 1610 and participated in the Indies trade. Roberto Marcelis had come from Antwerp to Seville in 1588. There he became part of a colourful network, containing Flemings, Italians, Frenchmen, Portuguese, and of course Spaniards, enabling him to participate successfully in the Indies trade.[49] Juan Leclerque was born in 1567 in Lille, which belonged to Flanders at that time. He was connected with a number of compatriots in Seville as well as Spanish stock brokers, many of whom were amongst his witnesses. In the Flemish network of 1600, he was the most central of all nodes (betweenness centrality), in possession of a great deal of real estate, many governmental bonds and strongly related to the eminent Genoese merchant-banker Jacome Mortedo.[50] Both merchants were important as connectors between their places of origin and the merchants of Seville, which gave them commercial advantages and guaranteed their success.

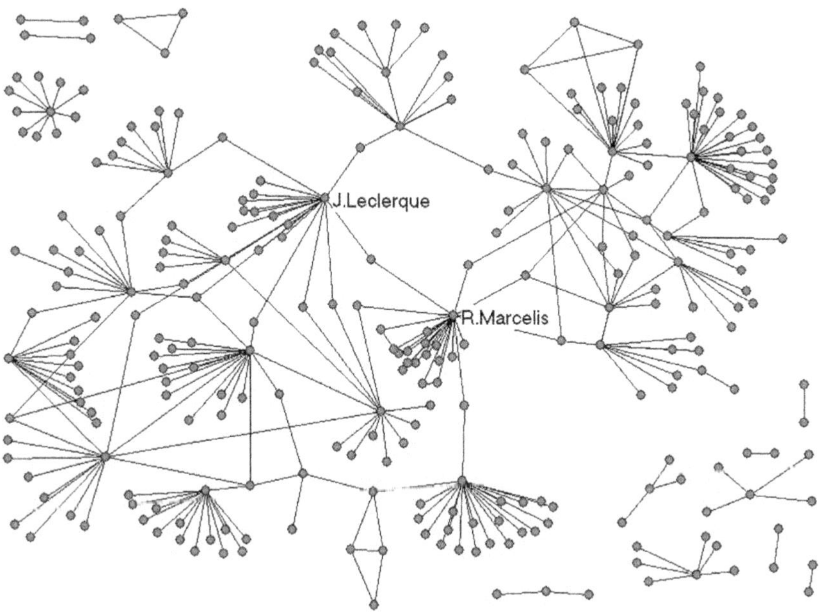

Figure 2.4.3 The semi-private Flemish network 1580–1650

(Source: AGI, Contratación 50A, 50B, 51B, 596A and 596B)

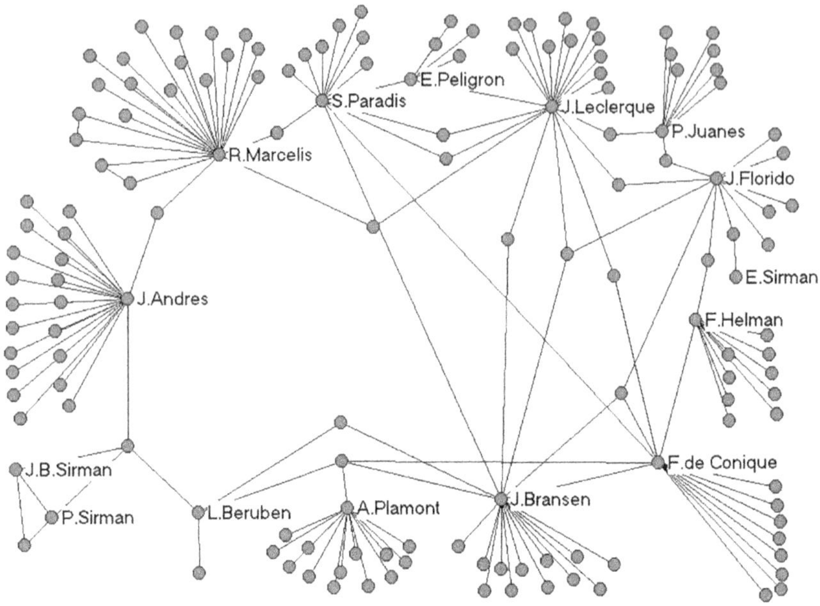

Figure 2.4.4 The First Circle of the Semi-Private Flemish Network
(Source: AGI, Contratación 50A, 50B, 51B, 596A and 596B)

Taking these two as separators, two groups of merchants emerged that had a certain internal cohesion. The left side of Figure 2.4.3 is displayed again in Figure 2.4.4 and visualizes, for example, a very powerful Flemish merchant group with interesting connections among each other. Within this group, I would like to point to the merchant Francisco de Conique (bottom right), whom I will take as an example in the next section because of his strong connectedness, his typical profile as a foreign merchant in Seville,[51] and his central position in the commerce of Seville in 1600 and 1620.[52]

Visualizing networks has the potential to make their structures and interrelations evident. In my investigation, it helped to provide a clearer picture of the commercial landscape of each period. In the selected year, 1580 (Figure 2.4.5),[53] only 13 Flemings appeared, and their network (total: 47 nodes) was relatively small and shattered, with various sub-networks. In 1600 (Figure 2.4.6), it had already grown to a considerable size (65 Flemings/229 nodes), and one large sub-network was identified. Furthermore, Francisco de Conique appears for the first time (at the bottom). In 1620, the Flemish network reached its extension maximum (153 Flemings/443 nodes). Figure 2.4.7 shows that, with over 400 nodes, it becomes increasingly difficult to make sense of such network representations. In that case, it can be extremely useful to calculate the (betweenness) centrality of the actors, to see who were the most central and influential actors

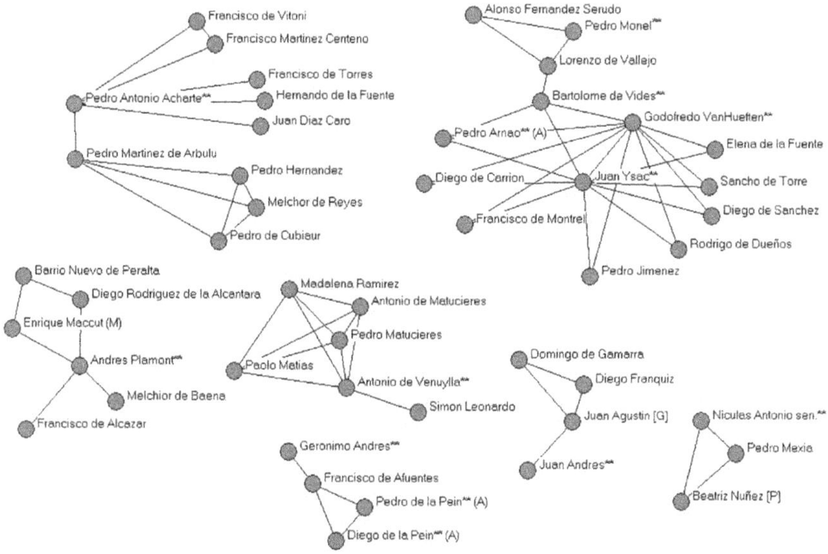

Figure 2.4.5 The Flemish network of 1580

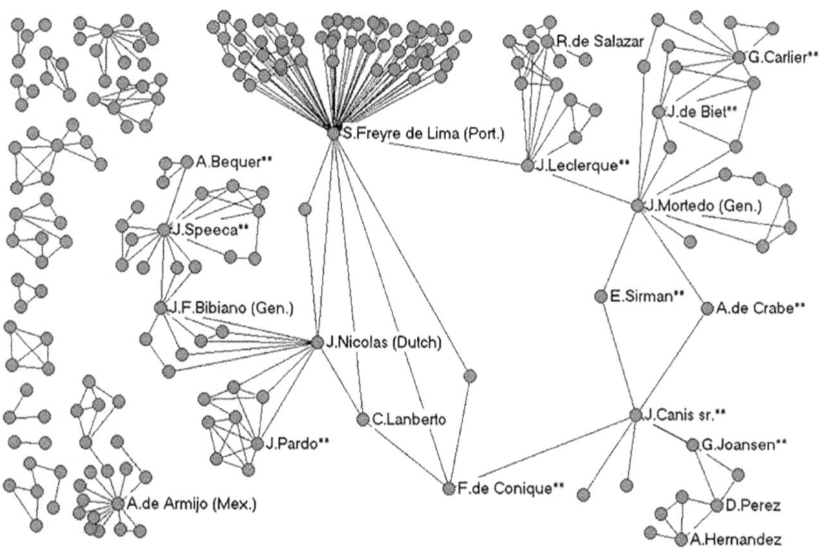

Figure 2.4.6 The Flemish network of 1600 (most prominent actors indicated)

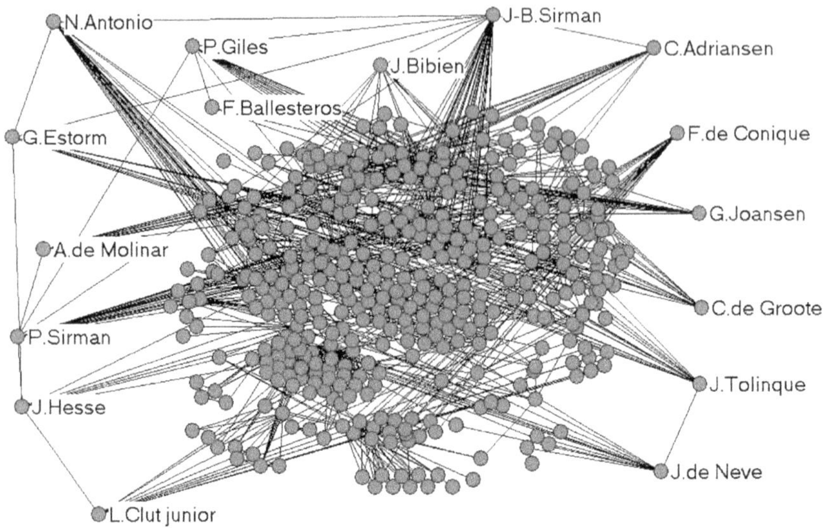

Figure 2.4.7 The Flemish network of 1620 (most prominent Flemish actors indicated)

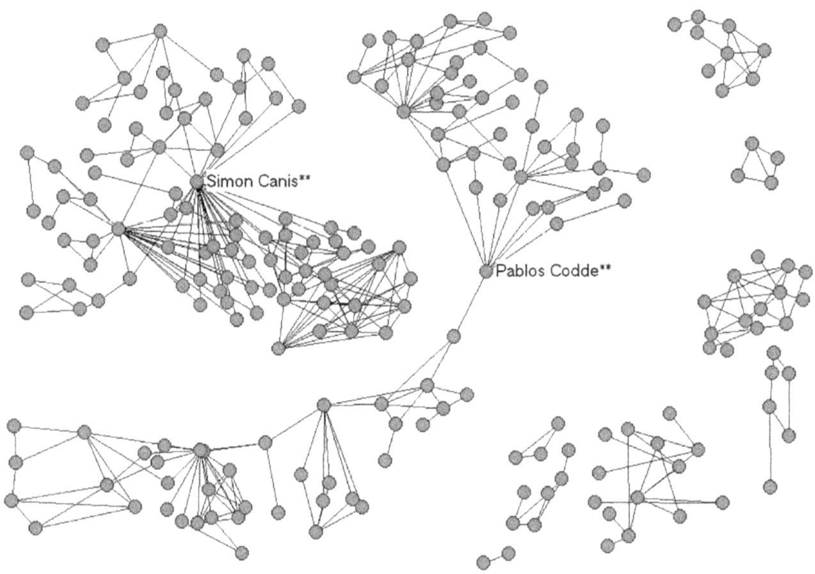

Figure 2.4.8 The Flemish network of 1640 (two most prominent actors indicated)

(Francisco de Conique on the right side). In 1640, the Flemish network had decreased again (62 Flemings/224 nodes), showing two dominant sub-networks (Figure 2.4.8).[54] The structural changes that occurred between the years can be explained through a series of factors, also taking into consideration that the investigated files of each year are samples, and there is a chance that they might not always be representative. Moreover, the character of notary documents must be taken into consideration, as notaries were not always employed at the same rate for the same kinds of transactions. However, some trends can be explained by changes in the commercial landscape. In 1600, for instance, the activities of Genoese banks in Seville as well as the bankruptcy of a wealthy Portuguese slave trade influenced the structure of the network. In addition, the political setting was important: for example, in 1620 Spain was at truce with the Northern Netherlands (1609–1621), and trade with the Dutch was permitted. This might be part of the explanation for the peak of Flemish and Dutch activity during that time.

Francisco de Conique

By way of example, I would like to focus on Francisco de Conique, a typical Flemish businessman of Seville, the eighth most central node of the Flemish network from 1600 and seventh most central node in 1620.[55] He was part of a much ramified family network, including the rich Flemish merchant families of the Antonio, Peligron, Sandier, and Janszon (see Figure 2.4.2).[56] These connections indicate that marriages within the group of Flemish merchants in Seville were common, and the Flemish nation was rather endogamous, displaying a strong homophily, i.e., cohesion within the Flemings in Seville.[57]

Francisco de Conique can be considered an exemplary and successful foreign merchant of his time. At the age of about 15 (in or shortly before 1580), coming from Antwerp, he arrived in Seville, probably sent by his family to work as an apprentice for a business partner, common practice at that time. In 1595, at 30, he founded a company with his compatriot Pedro Lemaire. Together, they conducted their business with the help of different agents in various European cities. Frequently, these agents were compatriots or even family, for example, Isaac Lemaire was stationed in Holland, Abraham Lemaire in Seeland, and David Lemaire in London. Thereby, the proximity between the private and the business world becomes evident, indicating the difficulty in distinguishing between both types of networks.

At the age of 32, Francisco de Conique married Mariana Antonio Gomar, the daughter of a compatriot, Niculas Antonio from Brabant, who had been a nobleman, merchant, and resident of Seville since he was very young. At the same time, Conique was granted citizenship in Seville, a status which implied a certain social prestige for the foreign merchant. He purchased a considerable amount of land in Andalusia in about 1600 (when he was 35); was knighted by the order of Santiago, which included numerous privileges and obligations; and became an alderman of Seville. From that time on, he played an important role in the political and social life of the city. Twice Francisco de Conique acted as a witness (semi-private connections) in the process of naturalizations for Flemish merchants: in 1594 for Francisco Helmann and in 1607 for Salomon Paradis. Moreover, in 1611, Francisco de Conique was best man at the wedding of the German Andres Labermeyr.

Among the first details regarding the trade of Francisco de Conique stands his illicit trade with political enemies of the Spanish Crown. During the prohibitive laws, when the commerce with some northern countries had been greatly restricted since 1585, Francisco de Conique was able to send olive oil, wine, figs, and cochineal (a dye from America) to England, Holland, and New Zealand and receive cloth in return. By investigating his correspondence in 1596, the Spanish authorities discovered his contraband trade, impeached him, and finally confiscated all his possessions. Nevertheless, in 1600, he received his letter of naturalization and was again (or still) in possession of much wealth and real estate in Seville. Presuming that the confiscation was carried out thoroughly, it remains unclear why his properties were returned to him and/or how he was able to recover so quickly. In any case, in 1600, Francisco de Conique was back in business and participated, for example, in the financing of a Portuguese slave trader. In 1605, he sold 80 barrels of Polish tar from Gdansk, and in 1608 he delivered Hungarian copper to the royal officials for the production of Spanish cannons.

Yet, the most important commercial sector for Francisco de Conique was the Indies trade. Even though he had received a letter of naturalization in 1600, he additionally received royal permission in 1609 to trade to the Indies. Between 1618 and 1649, he appears to be a member of the corporation of Indies traders, the *Consulado de Cargadores a Indias*.[58] But already in 1614, he sent goods to Mexico worth 4,545 ducats. Since then, Francisco de Conique became more and more active as a financer of the Indies trade. From 1619, he issued sea loans amounting to several thousand ducats to other Indies traders. In 1620 (the year that reveals most information about him), he gave credits and insured ships for more than 4,000 ducats, authorized partners to collect money from his debtors in Peru, Mexico, and today's Columbia, bought tobacco from the Indies in exchange for European merchandise, purchased Flemish wax in Seville for 1,700 ducats and French linen for 800 ducats (most likely for the American market), and sold even more French linen (*ruanes*) in Cartagena de Indias (today's Columbia).

Francisco de Conique's commercial conduct also displays the close relation between family and business. Together with his brother in law, Niculas Antonio junior, he chartered two thirds of the cargo hold of a galleon, sailing for the Indies in 1620. Captain of the ship and owner of the last third of the hold was Francisco Nicolas, who also appears to be the brother in law of Niculas Antonio junior (see Figure 2.4.2). A direct descendant of Francisco de Conique also participated in the commerce of Andalusia: Simon de Conique, one of Francisco's sons, who lived in Antwerp for many years but came to Seville later on. There, he received a letter of naturalization in 1635, 35 years after his father.[59] Simon de Conique was actively involved in the flourishing wool production of the town of Écija,[60] in the vicinity of Seville. Father and son were in business contact, and it is very likely that they worked closely together because Francisco had well-established connections within the European and especially the American textile markets.

The value of the betweenness centrality

How representative is the example of Francisco de Conique, and which position did the Flemish community in Seville have? The composition of the important Indies merchants can be exemplified by taking a look at an extraordinary "tax collection" from 1640.[61] About four million ducats were levied from 670 merchants in Spain; about one quarter of that amount was collected from just a few of the wealthiest Indies merchants of Seville, and 41 of them paid almost half a million ducats. Of this amount, 45 percent came from Spaniards while 55 percent came from foreigners (or their direct descendants). Among the foreigners, the Flemings paid the lion's share with 19 percent, ahead of the Portuguese with 16 percent.[62] The investigated four years confirm the eminent position of the Flemings: altogether, the Flemings (and some Dutch) accounted for 37 percent of all detected foreigners (305 nodes), well ahead of the second, the Portuguese, with 22 percent (187 nodes). Moreover, in each of the years separately, the Flemings were more numerous than the Portuguese.[63]

One could deduce by these numbers that the Flemings were the most dominant and important group of foreigners in the city. Yet, the power of a merchant community does not reside necessarily in the sheer size. To get the merchants or merchant groups with most influence in the distribution of information and resources, the "node-based betweenness-centrality" can be taken as a better indicator. The basic principle of the node-based betweenness-centrality states that nodes occurring on many shortest paths between other nodes have a higher betweenness than those that do not.[64] The merchants with the highest betweenness centrality have as much power as brokers and as much influence as merchants. Calculating the most central nodes of the years, the result was unexpected: the Flemings were never in first position. For all four years, it was the Portuguese who were the most central foreign merchants. This is particularly surprising in the year 1620, when the Flemings represented 40 percent of all foreigners in the notary records, while the Portuguese only represented 20 percent. In that year, the most influential foreign merchants were Miguel Fernandez Pereyra, Antonio Martinez Dorta, and Agustín Perez, all Portuguese. Only then, with Niculas Antonio and Juan Bautista Sirman, came the first Flemings (Table 2.4.1).[65]

Hence, in spite of being a smaller community, it was the Portuguese who had the more influential positions in the investigated commercial network of Seville of that year. They probably had more influence on the flow of information and made better informed business choices than the Flemings. These results of the betweenness centrality can be supported by the analysis of the letters of naturalization. Overall, only wealthy and influential merchants applied for such a letter. Between 1570 and 1650, 313 foreigners handed in such a request: 38 percent of them were Portuguese (116 applications), and only 28 percent were Flemings or Dutch (86 applications).[66] In conclusion, even though the Flemings were the most numerous foreign merchant group in Seville, it was the Portuguese who played the dominant role in the commercial networks.

Table 2.4.1 The 20 most central nodes of the network of the year 1620 (Crailsheim, *The Spanish Connection*, 269)

Name	*Residence*	*Origin*	*Naturalized*	*Cargador*
Miguel Fernandez Pereyra	Seville	Portugal		
Antonio Martinez Dorta	Seville	Portugal		
Lorenzo Bernal	Seville			
Agustin Perez	Seville	Portugal	1624	
Niculas Antonio	Seville	Flanders	1613	1635–37
Juan Bautista Sirman	Seville	Flanders	1617	
Pedro de la Farxa	Seville	French	1623	1635–39
Tomas de Mañara	Seville	Genoa	1607	1611–48
Simon Lopez de Granada	Seville	Portugal	1587	
Simon Fernandez de la Fuente	Seville			
Enrique de Andrada	Seville	Portugal	1618	1627–48
Gaspar de Rojas	Seville	Peru		
Francisco de Herrera Hurtado	Toledo			1623–28
Paolo Geronimo Semino	Seville	Genoa		
Manuel Gomez de Acosta	Seville	Portugal	1641	1637–59
Luis Lopez de Molina	Seville	Portugal		
Juan Lozano	Seville	Peru		
Fernando Carrillo	Seville	Flanders		
Antonio Maria Bucarelli	Seville	Florence	1616	1629–38
Francisco Lopez Talavan	Seville			1635–37

Conclusion

The methods of historical network research can contribute significantly to economic history. Fundamental source material for economic history, such as notary files, include documentation of financial and commercial transactions, proxies, and other types of dealings that show the connections between merchants and enable the quantitative reconstruction of commercial networks, from at least early modern times. One of the most dynamic places of that era was the city of Seville, which was the hub for the Indies trade and attracted numerous merchants from abroad. In Seville, Spanish and foreign businessmen spun their social and commercial webs all the way to the distant commercial centres of Europe and America. Based on the manifold source material – which resulted from the strict bureaucratic requirements of Spain – and the mercantile activities of the notaries in Seville, the reconstruction of these merchant networks was feasible.

In this chapter, the focus was laid on the Flemish networks in Seville between the years 1570 and 1650. The visualization of the networks emphasizes the structural aspect of the social and commercial relations and enables the determination of neuralgic nodes (for example the women within the private networks) as well as those who can be considered brokers. The network approach, moreover, forces the historian to reconsider categories of analysis and relations and to think about new ones, such as a semi-private layer of relations in this case (in the ambiguous area between private and commercial relations). Additionally, changes in the network structure over the years compel the researcher to think about the external and internal factors that might have caused these changes. Moreover, following the traces of single merchants' networks, like the Fleming, Francisco de Conique, offers the chance to adequately assess his position within a group or within the commercial setting of his time.

The last argument of this chapter pointed to the fact that it is not always the size of a group that defines its importance. Unarguably, the Flemings were a crucial and strong group within the commerce of Seville. But when it came to evaluating their importance as brokers as well as their access to potential partners, the calculation of the betweenness centrality showed that they only came second to the Portuguese community in Seville, in spite of being the largest in numbers.

Hence, in my view, the power of networks resides in two essential points. First, by visualizing networks, complex social structures become more evident, which enable the researcher to better understand social and commercial relations and consequently to go beyond initial questions. Second, by calculating centrality, social network analysis offers a way to transcend apparent findings and acquire more accurate results, refining or even correcting previous ones. In other words, social network analysis helps to differentiate the substantial from the superficial.

Notes

1 Christian Marx, "Economic Networks," in *European History Online (EGO)*, ed. Leibniz Institute of European History (IEG) (Mainz, 2012), www.ieg-ego.eu/marxc-2012-en (accessed 13 December 2018).

2 Manuel Herrero Sánchez and Klemens Kaps, "Connectors, Networks and Commercial Systems: Approaches to the Study of Early Modern Maritime Commercial History," in *Merchants and Trade Networks in the Atlantic and the Mediterranean, 1550–1800: Connectors of Commercial Systems*, ed. Manuel Herrero Sánchez and Klemens Kaps, p. 11 (London: Routledge, 2016).

3 Claire Lemercier, "Formale Methoden der Netzwerkanalyse in den Geschichtswissenschaften: Warum und Wie?," *Österreichische Zeitschrift für Geschichtswissenschaften ÖZG* 23, no. 1 (2012): pp. 16–41. The different contributions in *Merchants and Trade Networks in the Atlantic and the Mediterranean, 1550–1800: Connectors of Commercial Maritime Systems*, ed. Manuel Herrero Sánchez and Klemens Kaps show a possible spectrum of network analysis in early modern trade relations. Besides, applied SNA can be found e.g. in Kalus, Lesiak, Crespo Solana.

4 See Marx, "Economic Networks".

5 One of the earliest to set the focus on commercial networks was Sanjay Subrahmanyam, *Merchant Networks in the Early Modern World, 1450–1800* (London, New York: Routledge, 1996).

6 Jaime Contreras, ed., *Familia, religión y negocio. El sefardismo en las relaciones entre el mundo ibérico y los Países Bajos en la Edad Moderna* (Madrid: Fundación Carlos de Amberes, 2003) [... actas del tercer Seminario Internacional de Historia (Madrid-Alcalá de Hernares, 27–29 de junio de 2002) ...]. Seminario Internacional de Historia.; See Xabier Lamikiz, *Trade and Trust in the Eighteenth-Century Atlantic World: Spanish Merchants and Their Overseas Networks* (Rochester: Boydell & Brewer, 2010); Tamara Ganjalyan, "Armenische Handelsnetzwerke," www.ieg-ego.eu/ganjalyant-2016-de (accessed 13 December 2018).

7 See Niklas Luhmann, *Vertrauen* (Stuttgart: UTB, 2000); Martin Hartmann and Claus Offe, *Vertrauen: Die Grundlage des sozialen Zusammenhalts* (Frankfurt, New York: Campus Verlag, 2001).

8 See Lamikiz, *Trade and Trust in the Eighteenth-Century Atlantic World*; Sheilagh C. Ogilvie, *Merchant Guilds, Social Capital and the Commercial Revolution: Institutions and Economic Development in Medieval and Early Modern Europe*, Cambridge Studies in Economic History 2 (Cambridge: Cambridge University Press, 2011).

9 See Mustafa Emirbayer and Jeff Goodwin, "Network Analysis, Culture, and the Problem of Agency," *American Journal of Sociology* 99, no. 6 (1994): pp. 1411–54.

10 See Maximilian Kalus, *Pfeffer – Kupfer – Nachrichten: Kaufmannsnetzwerke und Handelsstrukturen im europäisch-asiatischen Handel am Ende des 16. Jahrhunderts* (Augsburg: Wißner, 2010); Manuel Suárez Rivera, "Caballero, vasco y mercader de libros: Tomás Domingo de Acha, sus redes mercantiles y de distribución (1771–1814)," *Estudios de Historia Novohispana* 50 (1 June 2014): pp. 125–73; Emilio B. Gimeno, "Las redes emigratorias auvernesas y el desarrollo de la metalurgía del cobre en el sur de Aragón," in *Circulation des marchandises et résaux commerciaux dans les Pyrénées (XIIIe-XIXe siècles): 7e Curs d'Historia d'Andorra col-loqui d'Andorra*, ed. Jean-Michel Minovez and Patrice Poujade, pp. 245–74 (Toulouse: CNRS-Université de Toulouse-Le Mirail, 2005); Rafael M. Girón Pascual, "Redes mercantiles en la Castilla del siglo XVI a través de las 'licencias de saca de lana con destino a Italia'," in *De la tierra al cielo: Líneas recientes de la investigación en historia moderna*, ed. Eliseo Serrano Martín, Publicación número 3239 de la Institución "Fernando el Católico", pp. 757–71 (Zaragoza: Institución "Fernando el Católico", 2013).

11 Klaus Weber, *Deutsche Kaufleute im Atlantikhandel, 1680–1830: Unternehmen und Familien in Hamburg, Cádiz und Bordeaux*, Schriftenreihe zur Zeitschrift für Unternehmensgeschichte 12 (München: C.H. Beck, 2004); Antonio García de León, "La malla inconclusa: Veracruz y los circuitos comerciales lusitanos en la primera mitad del siglo XVII," in Ibarra; Valle Pavón, *Redes sociales e instituciones comerciales en el imperio español, siglos XVII a XIX*, pp. 41–84; Montserrat Cachero Vinuesa, "Redes mercantiles en los inicios del comercio atlántico: Sevilla entre Europa y América, 1520–1525," in *Redes y negocios globales en el mundo ibérico, siglos XVI – XVIII*, ed. Nikolaus Böttcher, Bernd Hausberger and Antonio Ibarra, Bibliotheca ibero – americana 137, pp. 25–52 (Madrid: Iberoamericana, 2011).

12 Subrahmanyam, *Merchant Networks in the Early Modern World, 1450–1800*; Oñate Lario and Marial Carmen, *La colonia mercantil británica e irlandesa en Cádiz a finales del siglo XVIII* (Cádiz: Servicio de Publicaciones, Universidad de Cádiz, 2001), the 2003 edition oft he Annales (58.3); Guillermo Lohmann Villena, *Plata del Perú, riqueza de Europa: Los mercaderes peruanos y el comercio con la metrópoli en el siglo XVII* (Lima: Fondo Editorial del Congreso del Perú, 2004); Ana Crespo Solana, *Mercaderes atlánticos: Redes del comercio flamenco y holandés entre Europa y el Caribe* (Córdoba: Universidad de Córdoba, Servicio de Publicaciones CajaSur, 2009), Estudios de historia moderna Colección maior 33.; Germán Jiménez Montes, "La comunidad flamenca en Sevilla durante el reinado de Felipe II y su papel en las redes mercantiles antuerpienses," in *Comercio y cultura en la edad moderna: Comunicaciones de la XIII Reunión Científica de la Fundación Española de Historia Moderna*, ed. Juan J. Iglesias Rodríguez et al., pp. 43–56 (Seville: Universidad de Sevilla, 2015).

13 Michiel Baud, "Families and Migration: Towards an Historical Analysis of Family Networks," in *Economic and Social History in the Netherlands: Family Strategies and Changing Labour Relations*, ed. Michiel Baud et al., pp. 83–107 (Amsterdam: The Netherlands Economic History Archives, 1994); Contreras, ed., *Familia, religión y negocio. El sefardismo en las relaciones entre el mundo ibérico y los Países Bajos*

en la Edad Moderna; Actas del tercer Seminario Internacional de Historia (Madrid-Alcalá de Hernares, 27–29 de junio de 2002).

14 Ruth Pike, "Partnership Companies in Sixteenth-Century Transatlantic Trade: The De la Fuente Family of Seville," *The Journal of European Economic History* 24, no. 1 (2005): pp. 245–62; Amélia Polonia, Sara Pinto and Ana Ribeiro, "Trade Networks in the First Global Age: The Case Study of Simón Ruiz Company," in *Spatio-Temporal Narratives: Historical GIS and the Study of Global Trading Networks (1500–1800)*, ed. Ana Crespo Solana, pp. 140–77 (Newcastle upon Tyne, UK: Cambridge Scholars Pub., 2014).

15 Renate Pieper and Philipp Lesiak, "Redes mercantiles entre el Atlántico y el Mediterráneo en los inicios de la Guerra de los Treinta Años," in Ibarra; Valle Pavón, *Redes sociales e instituciones comerciales en el imperio español, siglos XVII a XIX*, pp. 19–39; Gert Oostindie and Jessica V. Roitman, *Dutch Atlantic connections, 1680–1800: Linking empires, bridging borders* (Leiden: Brill, 2014); Álvaro Aragón Ruano, "The Mediterranean Connections of Basque Ports (1700–1841): Trade, Trust and Networks," *The Journal of European Economic History* 44, no. 3 (2015): pp. 51–90.

16 Bernd Hausberger, *Die Verknüpfung der Welt: Geschichte der frühen Globalisierung vom 16. bis zum 18. Jahrhundert*, Expansion – Interaktion – Akkulturation 27 (Wien: Mandelbaum, 2015). See also Nikolaus Böttcher, Bernd Hausberger and Antonio Ibarra, eds., *Redes y negocios globales en el mundo ibérico, siglos XVI – XVIII*, Bibliotheca ibero – americana 137 (Madrid: Iberoamericana, 2011); Ana Crespo Solana, "The Iberian Peninsula in the First Global Trade: Geostrategy and Mercantile Network Interests (XV to XVIII Centuries)," in *Global Trade before Globalization (VIII-XVIII): Symposium London, 27–28 September, Brunei Gallery*, ed. Frederico Mayor Zaragoza, pp. 1–25 (Madrid: Fundación Cultura de Paz, 2006); Ana Crespo Solana, ed., *Spatio-Temporal Narratives: Historical GIS and the Study of Global Trading Networks (1500–1800)* (Newcastle upon Tyne, UK: Cambridge Scholars Pub., 2014); Polonia, Pinto and Ribeiro, "Trade Networks in the First Global Age: The Case Study of Simón Ruiz Company," pp. 140–77; Visualization Methods and Spatial Projections.

17 "Indies trade" refers to the trade between Europe and the West Indies, i.e., America.

18 More detailed results of this investigation can be consulted in Eberhard Crailsheim, *The Spanish Connection: French and Flemish Merchant Networks in Seville (1570–1650)*, Wirtschafs- und Sozialhistorische Studien 19 (Köln, Weimar, Wien: Böhlau-Verlag, 2016).

19 Cf. Alfred Crosby W., *The Columbian Exchange: Biological and Cultural Consequences of 1492, Contributions in American Studies 2* (Wesport, CT: Greenwood, 1972); Renate Pieper, "Raw Materials from Overseas and Their Impact on European Economies and Societies (XVI–XVIII Centuries)," in *Prodotti e Tecniche d'Oltremare nelle Economie Europee: Secc. XIII–XVIII; atti della ventinovesima settima di studie, 14–19 aprile 1997*, ed. Simonetta Cavaciocchi, 2nd ed. 29, pp. 359–83 (Florence: Le Monnier, 1998).

20 Cf. Stanley J. Stein and Barbara H. Stein, *Silver, Trade, and War Spain and America in the Making of Early Modern Europe* (Baltimore, MD, London: Johns Hopkins University Press, 2000); Renate Pieper, Bernd Hausberger and Omar Velasco, "Las repercusiones de los metales preciosos americanos en Europa, siglos XVI y XVIII," in *Oro y plata en los inicios de la economía global: De las minas a la moneda*, ed. Bernd Hausberger and Antonio Ibarra, pp. 273–97 (México, D.F.: El Colegio de México Centro de Estudios Históricos, 2014).

21 Cf. Artur Attman, *The Bullion Flow between Europe and the East 1000–1750* (Göteborg: Kungl. Vetenskaps- och Vitterhets – Samhället, 1981); Dennis Owen Flynn, *World Silver and Monetary History in the 16th and 17th Centuries*, Collected Studies Series C537 (Aldershot, UK, Brookfield, VT, USA: Variorum, 1996).

22 Francisco Morales Padrón, *Historia de Sevilla: La ciudad del quinientos*, 3rd ed. (Sevilla: Universidad de Sevilla, 1989), p. 65; Antonio Domínguez Ortiz, "La población de Sevilla a mediados del siglo XVII," in *Los extranjeros en la vida española durante el siglo XVII y otros artículos*, ed. Antonio Domínguez Ortiz and León C. Alvarez Santaló, pp. 243–62 (Seville: Diputación de Sevilla Area de Cultura y Ecología, 1996), p. 253.

23 Frédéric Mauro, "Merchant Communities, 1350–1750," in *The Rise of Merchant Empires: Long Distance Trade in the Early Modern World 1350–1750*, ed. James D. Tracy, p. 280 (Cambridge: Cambridge University Press, 1990).

24 Cf. Acosta Rodríguez, Antonio, Rodríguez González, Luis Adolfo and Enriqueta Vila Vilar, eds., *La Casa de la Contratación y la navegación entre España y las Indias* (Sevilla: Universidad de Sevilla; Consejo superior de investigaciones científicas; Fundación El Monte, 2003).

25 The Crown of Castile included most of today's Spain (except for the kingdom of Aragon), the American territories, and the Canary Islands.

26 Julián B. Ruiz Rivera and Manuela Cristina García Bernal, *Cargadores a Indias* (Madrid: Mapfre, 1992).

27 Cf. Eberhard Crailsheim, "Extranjeros entre dos mundos: Una aproximación proporcional a las colonias de mercaderes extranjeros en Sevilla, 1570–1650," *Jahrbuch für Geschichte Lateinamerikas*, no. 48 (2011): pp. 179–202.

28 Cf. Antonio García-Baquero González, *La carrera de Indias: Suma de la contratación y océano de negocios* (Sevilla: Algaida, 1992); Lorenzo Sanz, *Comercio de España con América en la época de Felipe II*, 1 vols. (Valladolid: Diputación Provincial de Valladolid, 1986).

29 Cf. Antonio Domínguez Ortiz and León Carlos Alvarez Santaló, eds., *Los extranjeros en la vida española durante el siglo XVII y otros artículos* (Seville: Diputación de Sevilla Area de Cultura y Ecología, 1996) León Carlos Álvarez Santaló; cf. Jacob van Klaveren, *Europäische Wirtschaftsgeschichte Spaniens im 16. und 17. Jahrhundert: Forschung zur Sozial- und Wirtschaftsgeschichte* (Stuttgart: Fischer, 1960).

30 E.g. Eddy Stols, *De Spaanse Brabanders of de handelsbetrekkingen der Zuidelijke Nederlanden met de Iberische Wereld 1598–1648*, 2 vols. (Brussel: Paleis der Academiën, 1971); Roland Baetens, *De Nazomer van Antwerpens welvaart: De diaspora en het handelshuis: De Groote tijdens de eerste helft der 17 de eeuw*, 2 vols. (Brussels: Gemeentekrediet van België, 1976).

31 Cf. José M. Díaz Blanco and Natalia Maillard Àlvarerz, "¿Una intimidad supeditada a la ley? Las estrategias matrimoniales de los cargadores a indias extranjeros en Sevilla (siglos XVI-XVII)," *Nuevo Mundo Mundos Nuevos, Coloquios*, https://journals.openedition.org/nuevomundo/28453 (accessed 13 December 2018).

32 The Spanish ducat was a golden coin of about 3.5 grams, worth about 55 litres of wheat. Jean Pierre Amalric et al., *Léxico Histórico de España: Siglos XVI a XX*, Textos Auxiliares 3 (Madrid: Taurus Ediciones, 1990), pp. 149–51; Enrique Martínez Ruiz, *Diccionario de historia moderna de España*, Fundamentos 136, p. 229 (Madrid: Istmo, 1998); Martínez Ruiz, *Diccionario de historia moderna de España*, pp. 265–6; APS 9984, fols. pp. 409–10, 431–2.

33 AGI Contratacion 50B, s.f. Cf. Recopilacion de las Leyes de Indias, ley 31–32, tiulo 27, libro 9; Díaz Blanco and Maillard Àlvarerz, "¿Una intimidad supeditada a la ley? Las estrategias matrimoniales de los cargadores a indias extranjeros en Sevilla (siglos XVI-XVII)," pp. 4, 8.

34 Most of them can be found in the AGI, Contratación 50A, 50B, 51B, 51A, 596A and 596B.

35 A "nation" is a corporation of foreign merchants abroad. Cf. Ana Crespo Solana, "El concepto de ciudadanía y la idea de nación según la comunidad flamenca de la Monarquía Hispánica," in *Las corporaciones de nacion en la monarquia hispanica (1580–1750): Identidad, patronazgo y redes de sociabilidad*, ed. Bernardo J. García García and Óscar Recio Morales, Serie Leo Belgicus, pp. 389–412 (Madrid: Fundacion Carlos de Amberes, 2014).
36 Notary files mostly do not indicate the origin of the mentioned individuals.
37 Thanks to Anna-Lena Glesinski for the evaluation of the data.
38 Eberhard Crailsheim and Philipp Lesiak.
39 The Indies fleets were supposed to leave Seville between April and August of each year, and it was hoped to include relevant data on the Indies trade by selecting these months for the investigation.
40 If few foreign individuals were encountered, the search was extended to one more office and/or month.
41 For the year 1580: office number V (file number 3494) and XXIV (16714, 16715); 1600: XII (7421), XVI (9983, 9984) and XXIV (16766); 1620: V (3607), XVI (10060) and XXIV (16869 and 16870); and 1640: XII (7497) and XXIV (16979). Besides, some complementing information was drawn from the files APS 1607, 2607, 3697, 6979, 7420, 7496, 9390, 10996, 16867, 16969, and 18484. Out of these volumes, all files were selected that were relevant in regard to foreign activities or Indies trade.
42 We created a special field for relations in the Access data base in which we marked all apparent relations between the individuals in each document. For the recreation of the network of a certain period of time, the content of each one of these fields in the relevant documents represented the basis.
43 Useful introductory literature to social network analysis: David Knoke and James H. Kuklinski, "Network Analysis: Basic Concepts," in *Markets, Hierachies and Networks: The Coordination of Social Life*, ed. Grahame Thompson et al. (London, Newbury Park, New Delhi: Sage Publications, 1991); Robert A. Hanneman and Mark Riddle, "Introduction to Social Network Methods," http://faculty.ucr.edu/~hanneman/nettext/ (accessed 13 December 2018); Dorothea Jansen, *Einführung in die Netzwerkanalyse: Grundlagen, Methoden, Forschungsbeispiele*, 2nd ed. (Opladen: Leske + Budrich, 2003); a historical orientation: Linton C. Freeman, *The Development of Social Network Analysis: A Study in the Sociology of Science* (Vancouver, BC: Empirical Press, 2004); more recently: Markus Gamper, Linda Rescke and Düring, Marten, eds., *Knoten und Kanten III: Soziale Netzwerkanalyse in Geschichts- und Politikforschung* (Bielefeld: Transcript Verlag, 2015). Applied programs were UCINET, https://sites.google.com/site/ucinetsoftware/ (accessed 13 December 2018), NetDraw, https://sites.google.com/site/netdrawsoftware/home (accessed 13 December 2018) and Inkscape https://inkscape.org/ (accessed 13 December 2018).
44 It has to be taken into account that a share of the city's trade was carried out without the involvement of the notary offices.
45 Cf. Eddy Stols, "La colonía flamenca de Sevilla y el comercio de los Paises Bajos españoles en la primera mitad del siglo XVII," in *Anuario de Historia Económica y Social 2*, 2nd ed., pp. 363–81 (1969); Carolina Abadía Flores, "La comunidad flamenca en Sevilla en el siglo XVI," in *Archivo Hispalense*, pp. 173–92 (2010).
46 Cf. Eberhard Crailsheim, "Behind the Atlantic Expansion: Flemish Trade Connections of Seville in 1620," *Research in Maritime History*, no. 42 (2010): pp. 21–46.
47 Figures 2.4.1 and 2.4.3–2.4.8 have been created with UCINET and NetDraw, based on a Spring-Embedder-Algorithm.
48 This figure has been created with Inkscape.
49 Crailsheim, *The Spanish Connection*, pp. 212–13.
50 Ibid., pp. 201–2, 257–9.

51 Enriqueta Vila Vilar, "Los europeos en el comercio americano: Sevilla como plataforma," in *Latin America in the Atlantic World: (1500–1850); Essays in Honor of Horst Pietschmann = El mundo atlántico y América Latina*, ed. Renate Pieper, Lateinamerikanische Forschungen 33, p. 291 (Köln, Weimar, Wien: Böhlau, 2005).

52 His betweenness centrality put him in both years among the top 5 Flemings. Crailsheim, *The Spanish Connection*, pp. 254, 298.

53 The following figures are based on the previously indicated sources from the APS.

54 Explanation for figures 5–8: ** = Flemish, A = Antwerp, G/Gen. = Genua, M. = Madrid, Mex. = Mexico, P/Port. = Portugal.

55 If not indicated differently, the information about Francisco de Conique comes from AGI Contratación 50B and 596A, s.f.; AGI Indiferente 428, L. 33, fols. 108–109; in 1600 from the file APS 9984, fols. 409–410, 431v – 432; and in 1620 from APS 10060, fols. 143r – 145r, 146r – 147v, 178r – 182v, 197r – 199r, 207r – 208v, 231r – 232r, 305r – 306v, 307r – 308r, 310r – 314v; APS 16869, fols. 168r – 171r, 242, 286, 291r – 293v, 338r – 340v, 404r – 405v. 569r – 572v, 679r – 682r; APS 16870, fols. 135r – 137r, 192; for deliveries of copper: AGI Contratacón 3893, fols. 12r-15v; AGI Recaudos 1618, fols. 29r-31v. Also from Antonio-Miguel Bernal Rodríguez, *La financiación de la Carrera de Indias (1492–1824): Dinero y crédito en el comercio colonial español con América* (Sevilla, Madrid: Fundación El Monte, 1992), pp. 246, 249, 578; Pierre Chaunu and Huguette Chaunu, *Séville et l'Atlantique, 1504–1650*, 8 vols. (Paris: A. Colin, 1955), vol. 5, pp. 362–3; Carlos Gómez-Centurión Jimenez, *Felipe II, la empresa de Inglaterra e il comercio septentrional (1566–1609)* (Madrid: Editorial Naval, 1988), p. 293.; Lorenzo Sanz, *Comercio de España con América en la época de Felipe II*, pp. 74–88.; Michèle Moret, *Aspects de la société marchande de Séville au début du XVIIe siècle* (Paris: Marcel Rivière et Cie, 1967), pp. 47, 77.; Stols, *De Spaanse Brabanders of de handelsbetrekkingen der Zuidelijke Nederlanden met de Iberische Wereld 1598–1648*, pp. 160–1, Stamboom 12; Valentín de Vázquez Prada, *Lettres marchandes d'Anvers,* (Paris: S.E.V.P.E.N, 1961), pp. 73–4; and Vila Vilar, "Los europeos en el comercio americano: Sevilla como plataforma," in *Latin America in the Atlantic World*, pp. 166, 291–2.

56 The family Nicolas, which also appears in the figure, was actually Dutch.

57 For the concept of homophily, cf. Miller McPherson, Lynn Smith-Lovin and James M. Cook, "Birds of a Feather: Homophily in Social Networks," *Annual Review of Sociology* 27 (2001): p. 415.

58 Enriqueta Vila Vilar, "Una amplia nómina de los hombres de comercio Sevillano del S. XVII," in *Minervae baeticae: Boletín de la Real academia Sevillana de Buenas Letras* 30, 139–91, p. 153 (2002).

59 The fact that he had to apply for another letter indicates that he probably had been born in Flanders.

60 Cf. Antonio Vidal Ortega and Enriqueta Vila Vilar, "El comercio lanero y el comercio trasatlántico: Écija en la encrucijada," in *Écija y el nuevo mundo: Actas del VI Congreso de Historia (celebrado en Écija, del 15 al 17 de noviembre de 2001)*, pp. 57–67 (Écija, Sevilla: Ayuntamiento de Écija; Diputación Provincial deSevilla, 2002).

61 In 1640, the Spanish Crown was in need of much money for its imperial projects and collected four million ducats from its wealthiest merchants. Over half of it was collected in Seville. Juana Gil-Bermejo García, "Mercaderes sevillanos II: Una relación de 1640," *Archivo Hispalense*, no. 188 (1978): pp. 25–52.

62 Nine percent came from Frenchmen, eight percent from Genoese, and three percent from the Britons.

63 Crailsheim, *The Spanish Connection*, pp. 81–2, 91.

64 Calculated through UCINET. Cf. Linton C. Freeman, "Centrality in Social Networks Conceptual Clarification," *Social Networks Social Networks*, no. 1 (1978/1979): pp. 215–39.

65 Among the most central 20 merchants, seven were Portuguese and only three were Flemings.
66 Crailsheim, *The Spanish Connection*, p. 83 (sample total 305).

Bibliography

Abadía Flores, Carolina. "La comunidad flamenca en Sevilla en el siglo XVI." In *Archivo Hispalense*, pp. 173–92, 2010.

Acosta Rodríguez, Antonio, Rodríguez González, Luis Adolfo, and Enriqueta Vila Vilar, eds. *La Casa de la Contratación y la navegación entre España y las Indias*. Sevilla: Universidad de Sevilla; Consejo superior de investigaciones científicas; Fundación El Monte, 2003.

Amalric, Jean Pierre et al. *Léxico Histórico de España: Siglos XVI a XX*. Textos auxiliares 3. Madrid: Taurus Ediciones, 1990.

Aragón Ruano, Álvaro. "The Mediterranean Connections of Basque Ports (1700–1841): Trade, Trust and Networks." *The Journal of European Economic History* 44, no. 3 (2015): pp. 51–90.

Attman, Artur. *The Bullion Flow between Europe and the East 1000–1750*. Göteborg: Kungl. Vetenskaps- och Vitterhets – Samhället, 1981.

Baetens, Roland. *De Nazomer van Antwerpens welvaart: De diaspora en het handelshuis: De Groote tijdens de eerste helft der 17 de eeuw*. 2 vols. Brussels: Gemeentekrediet van België, 1976.

Baud, Michiel. "Families and Migration: Towards an Historical Analysis of Family Networks." In *Economic and Social History in the Netherlands: Family Strategies and Changing Labour Relations*. Edited by Michiel Baud et al., pp. 83–107. Amsterdam: The Netherlands Economic History Archives, 1994.

Bernal Rodríguez, Antonio-Miguel. *La financiación de la Carrera de Indias (1492–1824): Dinero y crédito en el comercio colonial español con América*. Sevilla, Madrid: Fundación El Monte, 1992.

Böttcher, Nikolaus, Bernd Hausberger, and Antonio Ibarra, eds. *Redes y negocios globales en el mundo ibérico, siglos XVI – XVIII*. Bibliotheca ibero – americana 137. Madrid: Iberoamericana, 2011.

Cachero Vinuesa, Montserrat. "Redes mercantiles en los inicios del comercio atlántico: Sevilla entre Europa y América, 1520–1525." In *Redes y negocios globales en el mundo ibérico, siglos XVI – XVIII*. Edited by Nikolaus Böttcher, Bernd Hausberger and Antonio Ibarra, pp. 25–52. Bibliotheca ibero – americana 137. Madrid: Iberoamericana, 2011.

Chaunu, Pierre, and Huguette Chaunu. *Séville et l'Atlantique, 1504–1650*. 8 vols. Paris: A. Colin, 1955.

Contreras, Jaime, ed. *Familia, religión y negocio. El sefardismo en las relaciones entre el mundo ibérico y los Países Bajos en la Edad Moderna*. Madrid: Fundación Carlos de Amberes, 2003.

Crailsheim, Eberhard. "Behind the Atlantic Expansion: Flemish Trade Connections of Seville in 1620." *Research in Maritime History*, no. 42 (2010): pp. 21–46.

———. "Extranjeros entre dos mundos: Una aproximación proporcional a las colonias de mercaderes extranjeros en Sevilla, 1570–1650." *Jahrbuch für Geschichte Lateinamerikas*, no. 48 (2011): pp. 179–202.

———. *The Spanish Connection: French and Flemish Merchant Networks in Seville (1570–1650)*. Wirtschafs- und Sozialhistorische Studien 19. Köln, Weimar, Wien: Böhlau-Verlag, 2016.
Crespo Solana, Ana. "The Iberian Peninsula in the First Global Trade: Geostrategy and Mercantile Network Interests (XV to XVIII Centuries)." In *Global Trade before Globalization (VIII–XVIII): Symposium London, 27–28 September, Brunei Gallery*. Edited by Frederico Mayor Zaragoza, pp. 1–25. Madrid: Fundación Cultura de Paz, 2006.
———. *Mercaderes atlánticos: Redes del comercio flamenco y holandés entre Europa y el Caribe*. Estudios de historia moderna Colección maior 33. Córdoba: Universidad de Córdoba, Servicio de Publicaciones CajaSur, 2009.
———. "El concepto de ciudadanía y la idea de nación según la comunidad flamenca de la Monarquía Hispánica." In *Las corporaciones de nacion en la monarquia hispanica (1580–1750): Identidad, patronazgo y redes de sociabilidad*. Edited by Bernardo J. García García and Óscar Recio Morales, pp. 389–412. Serie Leo Belgicus. Madrid: Fundacion Carlos de Amberes, 2014.
———, ed. *Spatio-Temporal Narratives: Historical GIS and the Study of Global Trading Networks (1500–1800)*. Newcastle upon Tyne, United Kingdom: Cambridge Scholars Pub., 2014.
Crosby W., Alfred. *The Columbian Exchange: Biological and Cultural Consequences of 1492, Contributions in American studies 2*. Wesport, CT: Greenwood, 1972.
Díaz Blanco, José M., and Natalia Maillard Àlvarerz. "¿Una intimidad supeditada a la ley? Las estrategias matrimoniales de los cargadores a indias extranjeros en Sevilla (siglos XVI-XVII)." *Nuevo Mundo Mundos Nuevos, Coloquios*. https://journals.openedition.org/nuevomundo/28453 (accessed 13 December 2018).
Domínguez Ortiz, Antonio. "La población de Sevilla a mediados del siglo XVII." In *Los extranjeros en la vida española durante el siglo XVII y otros artículos*. Edited by Antonio Domínguez Ortiz and León C. Alvarez Santaló, pp. 243–62. Seville: Diputación de Sevilla Area de Cultura y Ecología, 1996.
Domínguez Ortiz, Antonio, and León Carlos Alvarez Santaló, eds. *Los extranjeros en la vida española durante el siglo XVII y otros artículos*. Seville: Diputación de Sevilla Area de Cultura y Ecología, 1996.
Écija y el nuevo mundo: Actas del VI Congreso de Historia (celebrado en Écija, del 15 al 17 de noviembre de 2001). Écija, Sevilla: Ayuntamiento de Écija; Diputación Provincial de Sevilla, 2002.
Emirbayer, Mustafa, and Jeff Goodwin. "Network Analysis, Culture, and the Problem of Agency." *American Journal of Sociology* 99, no. 6 (1994): pp. 1411–54.
Flynn, Dennis Owen. *World Silver and Monetary History in the 16th and 17th Centuries*. Collected Studies Series C537. Aldershot [UK], Brookfield, Vt., USA: Variorum, 1996.
Freeman, Linton C. "Centrality in Social Networks Conceptual Clarification." *Social Networks Social Networks*, no. 1 (1978/1979): pp. 215–39.
———. *The Development of Social Network Analysis: A Study in the Sociology of Science*. Vancouver, BC: Empirical Press, 2004.
Gamper, Markus, Linda Rescke, and Marten Düring, eds. *Knoten und Kanten III: Soziale Netzwerkanalyse in Geschichts- und Politikforschung*. Bielefeld: Transcript Verlag, 2015.
Ganjalyan, Tamara. "Armenische Handelsnetzwerke." www.ieg-ego.eu/ganjalyant-2016-de (accessed 13 December 2018).
García-Baquero González, Antonio. *La carrera de Indias: Suma de la contratación y océano de negocios*. Sevilla: Algaida, 1992.

García de León, Antonio. "La malla inconclusa: Veracruz y los circuitos comerciales lusitanos en la primera mitad del siglo XVII." In Ibarra; Valle Pavón, *Redes sociales e instituciones comerciales en el imperio español, siglos XVII a XIX*, pp. 41–84.

Gil-Bermejo García, Juana. "Mercaderes sevillanos II: Una relación de 1640." *Archivo Hispalense*, no. 188 (1978): pp. 25–52.

Gimeno, Emilio B. "Las redes emigratorias auvernesas y el desarrollo de la metalurgía del cobre en el sur de Aragón." In *Circulation des marchandises et résaux commerciaux dans les Pyrénées (XIIIe-XIXe siècles): 7e Curs d'Historia d'Andorra col-loqui d'Andorra*. Edited by Jean-Michel Minovez and Patrice Poujade, pp. 245–74. Toulouse: CNRS-Université de Toulouse-Le Mirail, 2005.

Girón Pascual, Rafael M. "Redes mercantiles en la Castilla del siglo XVI a través de las 'licencias de saca de lana con destino a Italia'." In *De la tierra al cielo: Líneas recientes de la investigación en historia moderna*. Edited by Eliseo Serrano Martín, pp. 757–71. Publicación número 3239 de la Institución "Fernando el Católico". Zaragoza: Institución "Fernando el Católico", 2013.

Gómez-Centurión Jimenez, Carlos. *Felipe II, la empresa de Inglaterra e il comercio septentrional (1566–1609)*. Madrid: Editorial Naval, 1988.

Hanneman, Robert A., and Mark Riddle. "Introduction to Social Network Methods." http://faculty.ucr.edu/~hanneman/nettext/ (accessed 13 December 2018).

Hartmann, Martin, and Claus Offe. *Vertrauen: Die Grundlage des sozialen Zusammenhalts*. Frankfurt, New York: Campus Verlag, 2001.

Hausberger, Bernd. *Die Verknüpfung der Welt: Geschichte der frühen Globalisierung vom 16. bis zum 18. Jahrhundert*. Expansion – Interaktion – Akkulturation 27. Wien: Mandelbaum, 2015.

Herrero Sánchez, Manuel, and Klemens Kaps. "Connectors, Networks and Commercial Systems: Approaches to the Study of Early Modern Maritime Commercial History." In *Merchants and Trade Networks in the Atlantic and the Mediterranean, 1550–1800: Connectors of Commercial Systems*. Edited by Manuel Herrero Sánchez and Klemens Kaps, pp. 1–36. London: Routledge, 2016.

———, eds. *Merchants and Trade Networks in the Atlantic and the Mediterranean, 1550–1800: Connectors of Commercial Systems*. London: Routledge, 2016.

Jansen, Dorothea. *Einführung in die Netzwerkanalyse: Grundlagen, Methoden, Forschungsbeispiele*. 2nd ed. Opladen: Leske + Budrich, 2003.

Jiménez Montes, Germán. "La comunidad flamenca en Sevilla durante el reinado de Felipe II y su papel en las redes mercantiles antuerpienses." In *Comercio y cultura en la edad moderna: Comunicaciones de la XIII Reunión Científica de la Fundación Española de Historia Moderna*. Edited by Juan J. Iglesias Rodríguez et al., pp. 43–56. Seville: Universidad de Sevilla, 2015.

Kalus, Maximilian. *Pfeffer – Kupfer – Nachrichten: Kaufmannsnetzwerke und Handelsstrukturen im europäisch-asiatischen Handel am Ende des 16. Jahrhunderts*. Augsburg: Wißner, 2010.

Klaveren, Jacob van. *Europäische Wirtschaftsgeschichte Spaniens im 16. und 17. Jahrhundert: Forschung zur Sozial- und Wirtschaftsgeschichte*. Stuttgart: Fischer, 1960.

Knoke, David, and James H. Kuklinski. "Network Analysis: Basic Concepts." In *Markets, Hierachies and Networks: The Coordination of Social Life*. Edited by Grahame Thompson et al. London, Newbury Park, New Delhi: Sage Publications, 1991.

Lamikiz, Xabier. *Trade and Trust in the Eighteenth-Century Atlantic World: Spanish Merchants and Their Overseas Networks*. Rochester: Boydell & Brewer, 2010.

Lario, Oñate, and Marial Carmen. *La colonia mercantil británica e irlandesa en Cádiz a finales del siglo XVIII*. Cádiz: Servicio de Publicaciones, Universidad de Cádiz, 2001.

Lemercier, Claire. "Formale Methoden der Netzwerkanalyse in den Geschichtswissenschaften: Warum und Wie?" *Österreichische Zeitschrift für Geschichtswissenschaften ÖZG* 23, no. 1 (2012): pp. 16–41.

Lohmann Villena, Guillermo. *Plata del Perú, riqueza de Europa: Los mercaderes peruanos y el comercio con la metrópoli en el siglo XVII*. Lima: Fondo Editorial del Congreso del Perú, 2004.

Lorenzo, Sanz. *Comercio de España con América en la época de Felipe II*. 1 vol. Valladolid: Diputación Provincial de Valladolid, 1986.

Luhmann, Niklas. *Vertrauen*. Stuttgart: UTB, 2000.

Martínez Ruiz, Enrique. *Diccionario de historia moderna de España*. Fundamentos 136, 229. Madrid: Istmo, 1998.

Marx, Christian. "Economic Networks." In *European History Online (EGO)*. Edited by Leibniz Institute of European History (IEG). Mainz, 2012. www.ieg-ego.eu/marxc-2012-en (accessed 13 December 2018).

Mauro, Frédéric. "Merchant Communities, 1350–1750." In *The Rise of Merchant Empires: Long Distance Trade in the Early Modern World 1350–1750*. Edited by James D. Tracy. Cambridge: Cambridge University Press, 1990.

McPherson, Miller, Lynn Smith-Lovin, and James M. Cook. "Birds of a Feather: Homophily in Social Networks." *Annual Review of Sociology* 27 (2001): p. 415.

Morales Padrón, Francisco. *Historia de Sevilla: La ciudad del quinientos*. 3rd ed. Sevilla: Universidad de Sevilla, 1989.

Moret, Michèle. *Aspects de la société marchande de Séville au début du XVIIe siècle*. Paris: Marcel Rivière et Cie, 1967.

Ogilvie, Sheilagh C. *Merchant Guilds, Social Capital and the Commercial Revolution: Institutions and Economic Development in Medieval and Early Modern Europe*. Cambridge Studies in Economic History 2. Cambridge: Cambridge University Press, 2011.

Oostindie, Gert, and Jessica V. Roitman. *Dutch Atlantic Connections, 1680–1800: Linking Empires, Bridging Borders*. Leiden: Brill, 2014.

Pieper, Renate. "Raw Materials from Overseas and Their Impact on European Economies and Societies (XVI–XVIII Centuries)." In *Prodotti e Tecniche d'Oltremare nelle Economie Europee: Secc. XIII-XVIII; atti della ventinovesima settima di studie, 14–19 aprile 1997*. Edited by Simonetta Cavaciocchi. 2nd ed., pp. 359–83, 29. Florence: Le Monnier, 1998.

Pieper, Renate, Bernd Hausberger, and Omar Velasco. "Las repercusiones de los metales preciosos americanos en Europa, siglos XVI y XVIII." In *Oro y plata en los inicios de la economía global: De las minas a la moneda*. Edited by Bernd Hausberger and Antonio Ibarra, pp. 273–97. México, DF: El Colegio de México Centro de Estudios Históricos, 2014.

Pieper, Renate, and Philipp Lesiak. "Redes mercantiles entre el Atlántico y el Mediterráneo en los inicios de la Guerra de los Treinta Años." In Ibarra; Valle Pavón, *Redes sociales e instituciones comerciales en el imperio español, siglos XVII a XIX*, pp. 19–39.

Pietschmann, Horst. *Latin America in the Atlantic World: (1500–1850): Essays in Honor of Horst Pietschmann = El mundo atlántico y América Latina*. Edited by Renate Pieper. Lateinamerikanische Forschungen 33. Köln, Weimar, Wien: Böhlau, 2005.

Pike, Ruth. "Partnership Companies in Sixteenth-Century Transatlantic Trade: The De la Fuente Family of Seville." *The Journal of European Economic History 24*, no. 1 (2005): pp. 245–62.

Polonia, Amélia, Sara Pinto, and Ana Ribeiro. "Trade Networks in the First Global Age: The Case Study of Simón Ruiz Company." In *Spatio-Temporal Narratives: Historical GIS and the Study of Global Trading Networks (1500–1800)*. Edited by Ana Crespo Solana, pp. 140–77. Newcastle upon Tyne, United Kingdom: Cambridge Scholars Pub., 2014.

Ruiz Rivera, Julián B., and Manuela Cristina García Bernal. *Cargadores a Indias*. Madrid: Mapfre, 1992.

Stein, Stanley J., and Barbara H. Stein. *Silver, Trade, and War Spain and America in the Making of Early Modern Europe*. Baltimore, MD, London: Johns Hopkins University Press, 2000.

Stols, Eddy. "La colonía flamenca de Sevilla y el comercio de los Paises Bajos españoles en la primera mitad del siglo XVII." In *Anuario de Historia Económica y Social 2*. 2nd ed., pp. 363–81. 1969.

———. *De Spaanse Brabanders of de handelsbetrekkingen der Zuidelijke Nederlanden met de Iberische Wereld 1598–1648*. 2 vols. Brussels: Paleis der Academiën, 1971.

Suárez Rivera, Manuel. "Caballero, vasco y mercader de libros: Tomás Domingo de Acha, sus redes mercantiles y de distribución (1771–1814)." *Estudios de Historia Novohispana* 50 (1 June 2014): pp. 125–73.

Subrahmanyam, Sanjay. *Merchant Networks in the Early Modern World, 1450–1800*. London, New York: Routledge, 1996.

Vázquez Prada, Valentín de. *Lettres marchandes d'Anvers,* Paris: S.E.V.P.E.N, 1961.

Vidal Ortega, Antonio, and Enriqueta Vila Vilar. "El comercio lanero y el comercio trasatlántico: Écija en la encrucijada." In *Écija y el nuevo mundo: Actas del VI Congreso de Historia (celebrado en Écija, del 15 al 17 de noviembre de 2001)*, pp. 57–67. Écija, Sevilla: Ayuntamiento de Écija; Diputación Provincial deSevilla, 2002.

Vila Vilar, Enriqueta. "Una amplia nómina de los hombres de comercio Sevillano del S. XVII." In *Minervae baeticae: Boletín de la Real academia Sevillana de Buenas Letras*, pp. 139–91, 30, 2002.

———. "Los europeos en el comercio americano: Sevilla como plataforma." In *Latin America in the Atlantic World: (1500–1850): Essays in Honor of Horst Pietschmann = El mundo atlántico y América Latina*. Edited by Renate Pieper, pp. 279–96. Lateinamerikanische Forschungen 33. Köln, Weimar, Wien: Böhlau, 2005.

Weber, Klaus. *Deutsche Kaufleute im Atlantikhandel, 1680–1830: Unternehmen und Familien in Hamburg, Cádiz und Bordeaux*. Schriftenreihe zur Zeitschrift für Unternehmensgeschichte 12. München: C.H. Beck, 2004.

2.5 Kinship networks in North Western German rural society (18th/19th centuries)

Christine Fertig

Introduction: kinship and social network analysis

In the late 19th century, sociology dismissed kinship as an archaic principle, deeply rooted in European foretime but disappearing with the ascent of modernity. Talcott Parsons' influential image of the modern nuclear family was drawn against a background of premodern embeddedness in kinship groups.[1] However, this idea has been challenged from two sides. Peter Laslett has shown that premodern families were not always extended families. Depending on social class, life phase and demographic coincidence, many families consisted of very few members.[2] On the other hand, historians have pointed to the difference between co-residence and kinship relationships beyond the household. Although often neglected, kinship networks had considerable influence on social, economic and political life. Kinship has turned out to be an important feature of human socialisation in early modern and modern times. Historians have stressed how kinship created affiliation, provided for support and social status, opened up access to resources and shaped political organisation.[3]

The systematic exploration of kinship is actually a rather difficult task. The reason for this lies in the cognatic character of the European kinship system. In European societies, the assignment of individuals to clearly contoured kinship groups was unknown since the Middle Ages. This means that relatives from both parents, mothers and fathers were regarded as equally significant. Therefore, only full siblings share the same set of relatives whilst their parents each have their own kindred. Couples and nuclear families were the central unit of Christian European societies; kinship in contrast was at first hand only a field of potential relationships that had to be activated and cultivated by individuals and families. The central position of the nuclear family was therefore not at all an outcome of modern life but dates back to medieval Europe.[4]

The indefinite character of European kinship impedes analysis of kinship networks. This might have contributed to the fact that kinship has remained in the background for a long time. Even with social network analysis, there are several hindrances that hamper the straightforward research of kinship. In addition to the lack of clear group membership, the shape of genealogical networks complicates social network analysis. Each person descends from exactly two parents, unless there is a lack of information about one or both of them. However, nobody can have more than two parents, so the discovery of notable differences between actors with regard to direct relations cannot be expected. On the other hand, people do not have children alone; they are usually the result of a marriage or relationship between two people. These features of genealogical networks provide for some difficulties when analysing this kind of data with the tools of formal social network analysis.

There are generally speaking three feasible solutions. (1) Simple descriptive statistics, such as cross tables, where kinship relations are treated as only one dimension among others, can be used. (2) Although there are no clearly defined groups within the European kinship system, groups can be defined and the kinship relations between them examined. The most sensible solution is the household unit, as a fundamental entity of social organisation. In this case, households, houses or farms can be defined as actors and the nature of relations between these actors, using social network analysis, looked at. (3) The P-graph procedure, introduced by Douglas R. White, can be used to reorganise and analyse genealogical networks. By defining couples, not persons, as actors and distinguishing between male and female descent, a network can be formed that allows for searching to find kinship relations.

In the following sections, all three solutions will be demonstrated on the basis of a study of two rural communities in 19th century North Western Germany. Both belonged to the Western Prussian province of Westphalia, with similar institutional and political conditions. In both cases, impartible inheritance and increasing population had led to strong social inequality, with few wealthy peasants, many smallholders and a large class of landless families. In Eastern Westphalia, many people lived by producing yarn and textiles, as did many in Löhne (district of Herford). In central Westphalia, agricultural conditions were very good, and, apart from some craftsmen, almost everybody lived from agriculture, as farm owners, servants or day labourers. There were no signs of protoindustrial activities in Borgeln (district of Soest). For both communities' information on persons and their life courses, family and godparent relations, as well as property and credit, have been collected and linked. Family reconstitutions form the basis of comprehensive databases, supplemented with all available data from parish registers, land registers, land title registers and court documents.[5]

The main question of this study concerns the contribution of kinship relations to the cohesion of these rural communities. Kinship relations structure the integration of societies in two respects: the construction of marriage and godparent networks can contribute to the integration of people of social distance – or not – as the case may be. Furthermore, in general, people can restrict their fields of action to their kindred, or they can be open to strangers. If the network perspective is taken seriously and social relations between people and families of different social classes are analysed, can we find what historians like Josef Moser and David W. Sabean called a 'rural class society'? In his great study on a southern German community, Sabean showed that the use people make of their kinship relations can shape local societies in very different ways. Families can open up to new people, integrating them through marriage and also godparenthood. Both variants can be seen as complementing blood kinship and can be used to enlarge the kinship network. However, if people chose relatives as marriage partners or as godparents for their children, they decided to restrict their social relations to partners who had been close before. It is important to remember that this choice always affected larger sets of people: marriages always connected families, not only marriage partners.[6]

The study of kinship relations can contribute to a better understanding in two ways. First, the existence of kinship relations within and between certain social groups provides information on the cohesion of any society. Second, the active construction of social networks, including marriage and godparenthood relations, indicate social closure or a general openness towards strangers. The following sections will first show how, with rather simple tools, a lot about the openness of wealthy families towards poorer members of their local society can be learned. Cross tabulations allow for some insight into the choices people make when they decide on their children's godparents. Second, formal social networks analysis will show how circles of marriage excluded or integrated the lower classes. The last section will introduce the P-Graph procedure, a rather specific tool of social network analysis that is able to discover structures in genealogical networks.

Relatives and godparents: construction of social networks

Whenever Christian parents were blessed with a child, they had to find godparents to celebrate a valid baptism. Beyond the spiritual role of a godparent, this relation was charged with multiple cultural and social connotations. They might differ between societies, but in general godparenthood comes along with behavioural expectations, supposing special attachment and benevolence between parents and godparents.[7] Hence, the choice of a godparent is not a trivial matter, and the acceptance of such a proposal includes the readiness to maintain the relationship and support each other. Therefore, godparents are important members of personal networks. How people in the past constructed and refined their personal networks can be seen by studying the choice of godparents. Two dimensions are of particular interest: did parents intend to broaden their social networks, or did they decide to stay for the most part within their social close range? Kinship relations with godparents are a good indicator as to whether families had the tendency to close or open up. The second dimension concerns the surpassing of social borders: were people eager or willing to cross social borders and establish long-lasting and compulsory relations with people of different social standing?

Anthropological research has shown how godparenthood was able to structure social relations in local societies.[8] Godparenthood was used to stabilise existing social relations, but also to expand social networks in new social fields. Martine Segalen has shown how godparenthood was used to re-integrate distant relatives in order to counteract growing distance and oblivion over time.[9] Historians have also studied the inner relatedness of societies, comparing godparenthood practices and social classes. Sandro Guzzi-Heeb pointed out that families in Val de Bagnes (Switzerland) avoided clientelistic configurations and spread their godparent relations, building strong social environments in the 19th century.[10] Often people had a clear preference for godparents of equal or higher range, but there were also attempts to create links to ordinary people or families.[11] For the Northern Swedish town of Umeå, Tom Ericsson could prove that research in godparent networks can heavily affect our understanding of social stratification. On the level of godparenthood relations he did not find any signs of segregation into different social classes, as deduced from usual occupational categories. Even without applying formal social network analysis, this is a good example of the epistemological value of a network perspective.[12]

In Table 2.5.1, both dimensions are brought together in a three-dimensional cross tabulation. The table compares for the two communities under research and again for peasants and non-peasants, whether people chose relatives as godparents or crossed social borders due to that choice. Without discussing every number in detail, it can be seen that there were obvious differences in how people constructed their godparent networks. (1) Peasants had a stronger tendency to choose other peasants, thereby excluding non-peasants from their personal networks. They were often ready to accept the proposal of a non-peasant, but they were hesitant to initiate it. (2) Efforts to restrict personal networks to well-known persons and to adhere to relatives were stronger with peasants than with non-peasants. Again, peasants tended to close their social circles. Only in very few cases did they ask a non-peasant who was not related to stand as a godparent. (3) There were remarkable differences between the two communities. Both effects were much stronger in Borgeln than in Löhne, pointing to a much better integration of social classes in Löhne than in Borgeln.

This section aimed to demonstrate that, even with simple descriptive statistics, a lot can be found out about social networks and the social fabric of society. The basic idea is that parents made a decision about the shape of their personal

Table 2.5.1 Networks of relatives and godparents, 1800–56

Parents	*Peasants*				*Non-peasants*			
Godparents	*Peasants*		*Non-peasants*		*Peasants*		*Non-peasants*	
... in Löhne	*N*	*%*	*N*	*%*	*N*	*%*	*N*	*%*
Close relatives	255	*62.2*	72	*50.7*	137	*38.3*	97	*46.6*
Distant relatives	108	*26.3*	50	*35.2*	134	*37.4*	57	*27.4*
Non-relatives	47	*11.5*	20	*14.1*	87	*24.3*	54	*26.0*
Total	**410**	***100.0***	**142**	***100.0***	**358**	***100.0***	**208**	***100.0***
... in Borgeln	*N*	*%*	*N*	*%*	*N*	*%*	*N*	*%*
Close relatives	721	*63.0*	98	*74.8*	286	*30.7*	157	*48.9*
Distant relatives	235	*20.5*	14	*10.7*	203	*21.8*	22	*6.9*
Non-relatives	188	*16.4*	19	*14.5*	444	*47.6*	142	*44.2*
Total	**1,144**	***100.0***	**131**	***100.0***	**933**	***100.0***	**321**	***100.0***

networks each time they chose a godparent for their newborn. They could either expand their network and integrate new people, or they could restrict their choices to people close in social range. To ask the wife of a peasant for whom one of the parents had worked several years before but without any other kind of enduring relation would be an example of the first strategy. For day-labourers, this could be a wise decision, because if the peasant's wife agreed, a lasting relation with a certain kind of mutual obligation that could help in difficult situations like unemployment or harvest failures would be established. However, if people wanted to avoid large, comprehensive networks, they adhered to already existing relationships. To ask a cousin to stand as godparent emphasised the value of this relationship, but it also prevented the creation of new external contacts. Apparently, peasants had a propensity to limit their commitments to outsiders. As a result, it can be noted that peasants and non-peasants obviously pursued different strategies. Non-peasants aimed at enlarging their personal networks, especially in the direction of peasants as potential employers and benefactors. Peasants, on the other hand, were often reluctant to embrace outsiders, particularly of lower social standing.

Families and kinship: relatedness of societies

In most cases, there is an important event before children are born and baptized, namely the parents' marriage. This is another pivotal moment for making decisions about the shape of one's personal network. In Christian Europe, marriage is essentially an agreement between two individuals. Without the consent of the two marriage partners, marriage will not attain legal validity. For this reason, it is hardly possible for families to force their offspring into unwanted marriages. However, this does not mean that families did not try to exert influence on marriage decisions. If parents disliked prospective marriage partners, they attempted to foil the relationship. But they were not the only ones who took an interest in marriage relationships: relatives, peer groups and local communities usually had an opinion on romantic relationships and marriage plans, causing them to frequently intervene if the marriage plans seemed to be inappropriate or threatening to social order.[13] One important reason for this interest is that marriages have a heavy impact on the overall network of a local community. Even if marriage is basically about the bond between husband and wife, it also connects families with each other – and sometimes the other members of one or the other family are not too happy about the partner choice. Similar to decisions concerning godparents, the choice of a marriage partner modifies personal networks. The impact is even stronger because a number of people are involved, such as marriage partners, fathers-in-law, sisters-in-law, aunts-in-law's husbands and so on.

Marriage circles have been a subject of discussion in many contexts. Mostly the tendency to marry within social class has been emphasized. However, David Sabean has shown in his study on Neckarhausen that in the early 18th century marriage mostly brought together people of very different social standing. In most cases, the fortune of one marriage partner was about twice as much as that of the other. These marriages connected rich and poor families, with benefits for both sides. Poorer families found support and job opportunities, whereas richer families built up an extended client base and had access to manpower, a crucial factor in premodern agriculture. With increasing population and resources becoming scarcer, the upper layer of local society started to shut out people who were worse off, thereby securing position and wealth. At first, they started to prefer kin as marriage partners, keeping things in the family, and then they avoided poorer families in general. In Neckarhausen, Sabean observed a clear shift from integrating marriage behaviour to a strategy that separated social classes. In the beginning, only the wealthiest families pursued such a strategy, but the middle classes soon followed.

The basic idea behind endogamous marriages – namely marriage within kinship or a social group – is that families avoid spreading resources outside their social circles. In the early 18th century, the wealth of Neckarhausen's upper class benefited the families of their poorer marriage partners. In the middle of the 19th century, however, lower classes were rigorously excluded from the social circles of the rich. They no longer found wealthy marriage partners and, as a consequence, didn't have affinal kin who could provide support and resources in times of crisis. The local society of Neckarhausen had therefore developed from a well-integrated and networked society into a separated class society.[14]

It took David Sabean months and years to analyse an abundance of source material on interpersonal connections in Neckarhausen. With social network analysis and a family reconstitution database, it is much easier to detect structures of connectedness and exclusion in social networks. In this section houses and farms are taken as a proxy for family groups, and marriages between these corporate actors are analysed. In societies with impartible inheritance, as common in North Western Germany, the farm and the house can be taken as a social and economic family anchor point.[15]

The basic question is whether marriages integrated social classes such as wealthy peasants and small holders and whether marriage partners and the resources they brought circulated throughout local societies. Farms and houses are taken as actors in these two networks, marriages are regarded as directed relations (arcs) and the procedure is the search for weak and strong components. A weak component is a part of the networks where all actors are connected to each other directly or indirectly by at least one path. In strong components, all actors have to be connected by at least two paths.

Figure 2.5.1 displays all marriages between farms and houses in Borgeln from 1750 to 1874. Actors are coloured according to their social class: larger farms in dark grey, small holders in medium grey and houses with very little land in light grey. Three results stand out. (1) About half of all the houses are connected, but some farms/houses did not have any marriage relation within the local context. This means that marriage partners either came from the outside community, from another parish or from a family without landed property. (2) The overall network is rather loosely connected. Very few houses have more than one or two marriage

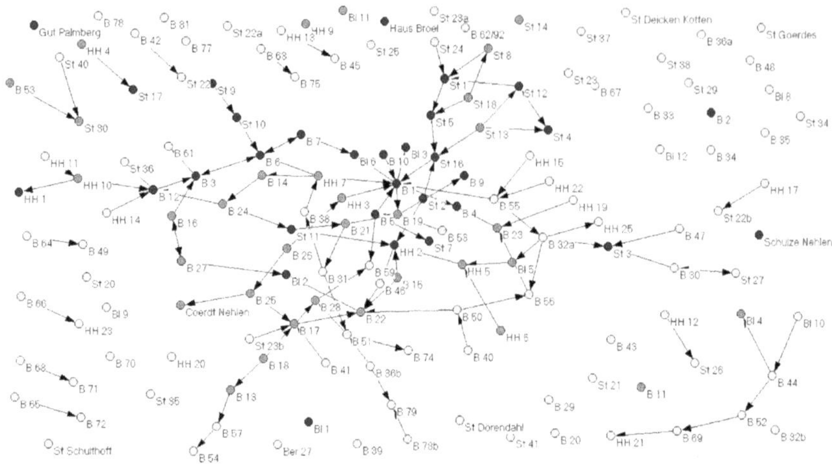

Figure 2.5.1 Marriages between farms and houses in Borgeln, 1750–1874

Note: Large farms (dark grey) with more than 160 Taler taxable net yield (ca. 15 ha), small farms (medium grey) with 18 to 160 Taler taxable net yield (ca. 2 to 15 ha) and houses with very little land below 18 Taler or 2 ha. Labels contain addresses if available, otherwise names of peasants. Formal network analysis has been carried out with PAJEK.

relations during this period of 125 years. (3) Although there are marriages across social borders it is not difficult to detect areas with concentrations of larger farms or smaller farms. Larger farms are in the centre of the core, and they tend to be connected to each other. Smaller farms are also well-integrated, but the houses are fairly frequently isolated.

In Löhne (Figure 2.5.2) about half of houses have marriage connections, but there are also remarkable differences from the results in Borgeln. Obviously, the network core is much more strongly connected, and there is less segregation between social classes within the core. In Borgeln there was one large weak component and several other much smaller components. In Löhne there is a large weak component that comprises of almost all the farms. The visual exploration of the network in Figure 2.5.2 can be supported by another analysis. The marriage network in Löhne contains a strong component of remarkable size. Every third farm was part of this network core. These farms were, at the same time, givers and recipients in the marriage network: they gave one of the family's children and his or her inheritance payment to another farm within the parish, and sooner or later they also received a marriage partner and resources from another farm. Only in Löhne was there a network core that was marked by reciprocity and the distribution of resources between larger and smaller peasants but still without much integration of families with few resources.

The search for the networks cores and components in particular has proved conducive for an analysis of a local marriage market. Results point to the

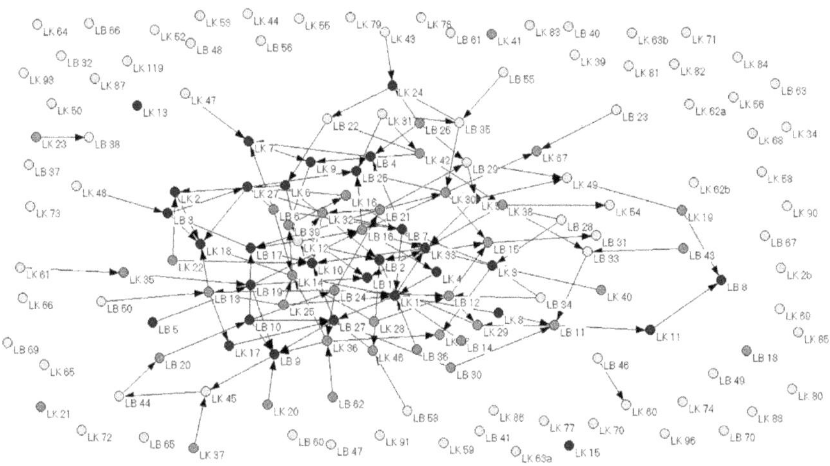

Figure 2.5.2 Marriages between farms and houses in Löhne, 1750–1874

Note: Large farms (dark grey) with more than 83 Taler taxable net yield (ca. 12 ha), small farms (medium grey) with 18 to 83 Taler taxable net yield (ca. 3 to 12 ha) and houses with very little land below 18 Taler or 3 ha. Labels contain addresses of farms and houses. Formal network analysis has been carried out with PAJEK.

same direction as the godparent networks. The two communities show different degrees of connectedness not only between but also among social classes. Lower classes are in general better integrated in godparenthood than through marriage, but that is not a very astonishing result. Farm heirs had to pay considerable amounts of inheritance compensations to their siblings, and they were strongly dependent on their marriage partners' dowry. There was a strong economic logic that forced farm heirs to choose partners with sustainable wealth, inherent in an impartible inheritance system where farm land was never divided. In such a system, children of poorer families had bad chances for marrying wealthy farm heirs. Nonetheless, the comparison between Borgeln and Löhne shows a much better integration of peasants in Löhne, irrespective of differences in farm size and wealth. In Borgeln, marriage partners rather crossed geographical borders and married into other parishes than crossing class borders: better to marry someone from a large farm further away than the poorer neighbour's daughter or son.

Social network analysis can help with the understanding of the social order of past societies and relational behaviour of people in the past. It is helpful to think about different perspectives and various methods for analysing past societies. The two approaches presented so far generate results that are compatible with each other, but, taken together, they provide a refined picture of similarities and differences. The next section will add to these results and also introduce an elaborate approach for the analysis of genealogical networks.

P-graph analysis: sophisticated tool for the analysis of genealogical networks

The previous section demonstrated how marriages connected farms and houses, leading to flows of persons and resources between them. This section will analyse the local kinship network as a whole and submit deeper insights not only into the network structure but also into possibilities for network analyses. The basic idea behind this approach is that of relinking. Every marriage with a relative leads back to the group of people to whom one is already related, creating a circle, as seen earlier in the densely linked network core in Löhne. The P-graph procedure uses this quality and identifies fields of condensed kinship relations in genealogical networks.[16] As a first step, the data have to be converted into a P-graph shape. This unconventional way of organising genealogical networks helps to solve the problem of having individual and corporate actors at the same time, and it concedes to searching for kinship circles. This procedure was developed by Douglas R. White and has been tested on an Austrian community.[17]

Figure 2.5.3 shows two representations of the same small network. On the left-hand side, 11 individuals and five couples are shown in a standard kinship notation. Women (triangles) and men (circles) are connected by horizontal lines when married. Descending from these horizontal lines, children are added by using vertical lines drawn below their parents. Couple 1, for example, has two sons, each of them married (couples 2 and 3). Couple 3 also has two sons, who are grandchildren of couple 1. One of these sons is married to the daughter of couple 2 – his own cousin. Even in the left notation, it can be seen that this kinship marriage forms a circle in the genealogical network. On the right-hand side, the same persons and couples are represented by a P-graph. The underlying principle is simple: nodes represent couples or unmarried individuals, and descent is encoded as arcs pointing towards the parents' node. In contrast to the standard notation, there is only one kind of actor (not individuals and

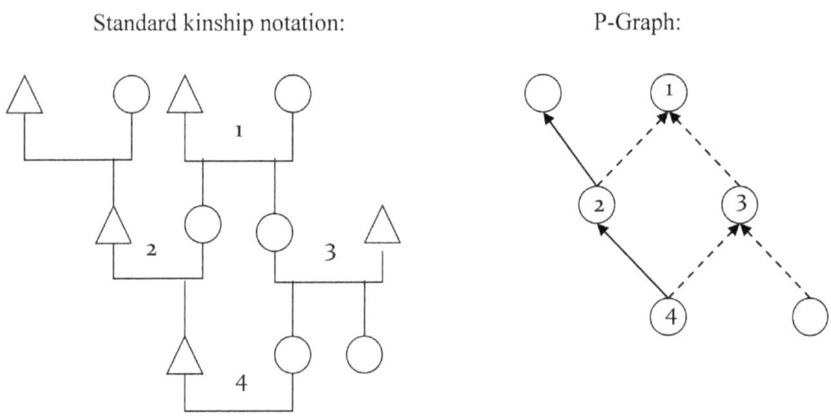

Figure 2.5.3 Standard kinship notation and the P-graph[18]

parents) and one kind of relationship (descent). However, relationships can have two manifestations, as male or female descent. The benefit of the P-graph representation is not necessarily apparent in visual representation – standard notation might be easier to understand. But it is much easier to perform network analysis on data transformed into P-graphs.

In order to find a kinship core in a local genealogical network, bi-components need to be discovered. Bi-components are subgraphs in which all members are connected at least twice.[19] These subgraphs do not contain any cut points or bridges – even if one actor was removed, all others would still be connected from the other side. What can be gained from such an analysis is the identification of condensed relatedness areas. All couples within such a component belong to an area of condensed kinship relations, although not everybody is married to a relative. In the godparent network, as well as the marriage network, kinship and class are not independent of each other. In this section, the first step is to identify the kinship core of both communities and then to compare its structure to the overall class structure. In a third step, the two communities will be compared to Feistritz, a small community in Austria that was the subject of Douglas R. White's and Lilyan Brudner's study. The main question is whether there was a correlation between kinship and social class.

Table 2.5.2 presents affiliation to the kinship core and social class for all three communities. Three results stand out: (1) Kinship cores only include minor parts of the communities. However, there is a substantial difference between Löhne on the one hand and Borgeln and Feistritz on the other. In Löhne, more couples (42.5%) were part of the kinship core whereas in the other two communities only about one third. (2) Everywhere, roughly half the peasants were part of the kinship core although in Borgeln and Feistritz this figure was slightly higher. However, in the Westphalian communities, peasants made up about one third of the kinship core whilst in the Austrian community almost 90% of

Table 2.5.2 Peasants and non-peasants in the kinship cores, in Löhne, Borgeln (1750–1874) and Feistritz (1860–1960)

		Peasants[1]		*Non-peasants*[2]		*Total*	
		N	%	*N*	%	*N*	%
Löhne	Kinship core	211	*49.9*	361	*39.2*	572	*42.5*
	Outside	212	*50.1*	561	*60.8*	773	*57.5*
	Total	423	*100.0*	922	*100.0*	1,345	*100.0*
Borgeln	Kinship core	188	*57.7*	428	*29.6*	616	*34.7*
	Outside	138	*42.3*	1,019	*70.4*	1157	*65.3*
	Total	326	*100.0*	1,447	*100.0*	1,773	*100.0*
Feistritz	Kinship core	182	*57.1*	25	*8.2*	207	*33.1*
	Outside	137	*42.9*	281	*91.8*	418	*66.9*
	Total	319	*100.0*	306	*100.0*	625	*100.0*

[1]in Feistritz: Heirs/Buyer

[2]in Feistritz: Residents; in Löhne and Borgeln land-poor and landless households.

Löhne: Pearson's r = 0.16; Borgeln: Pearson's r = 0.23; Feistritz: Pearson's r = 0.54

Sources: Databases Löhne and Borgeln; Brudner/White 1997, p. 193 (Table 2.5.2).

the core members were peasants. To some extent, this difference can be traced back to the varying ratios of peasants and non-peasants in the communities. (3) There are major differences with regard to the integration of non-peasants into the kinship core. In Löhne, 39.2% of the non-peasant couples were part of the kinship core; in Borgeln only 29.6%. However, the number is much smaller in Feistritz where only 25 couples, 8.2% of the non-peasants, have been integrated into the kinship core. This is an effective exclusion of non-peasants from social circles of the more wealthy peasants, an aspect that is not observable with so much rigour in Westphalia. Nonetheless, Löhne is once again the community with the most integration of non-peasants, who were more strongly left aside in Borgeln.

Conclusion

Social network analysis is an important tool for analysing kinship in past societies. Although the analysis is not straightforward due to the special character of kinship systems in Europe, there are nonetheless several options for approaching kinship networks. A first step can be simple descriptive statistics that allow the making of statements about social relations in any kind of society. Depending on the kinds of relations that are observed, a great deal can be learned about the integration or segregation of society. In this chapter, the relationship between kinship, godparenthood and social class has been taken as a means to get a better view of the social fabric in two Westphalian rural communities. Cross tables revealed that there were considerable differences between peasants' and non-peasants' networks on the one hand and the two communities on the other. In this rather simple analysis, it is in the looming what was supported by formal social network analysis: Borgeln in central Westphalia had a strong tendency towards pronounced class segregations, whereas Löhne in Eastern Westphalia could rather be regarded as a well-connected networks society where peasants and non-peasants had manifold relations across social borders.

The analysis of kinship networks is troubled by the question concerning how to make a workable dataset. Kinship is not only a relationship between individuals but also between families. Anthropological research has frequently shown that kinship groups have been corporate actors in kinship networks. In European societies, kinship relations also touch the interests of all family members, especially when it comes to marriage decisions. However, the unascertained character of European kinship structures causes problems for analysis, since clans cannot easily be used as the basis for analysis. In societies with impartible inheritance, farms and houses can be taken as the socio-economic basis of family groups, and the structure of kinship links between them can in turn reveal a lot about the degree of integration. Here the search for weak and strong components has confirmed the first results. In Löhne, all the peasants were well-connected, regardless of their farm size or wealth. In Borgeln, peasants rather looked for wealthy marriage partners from outside the community than to marry someone with a smaller dowry. In both communities, the land-poor had bad chances for marrying into a peasant farm – the economic logic that the farm heir needed a partner who brought resources in order to pay the inheritance compensations for his or her siblings was a powerful hindrance for poor marriage partners.

Another way to analyse kinship networks is the P-graph analysis. The genealogical network that is difficult to investigate in its standard notation is transformed into a network that is suitable for the search of bi-components. Thus, it is possible to find sectors of condensed kinship relations or kinship cores. The comparison of peasants' and non-peasants' affiliation to kinship cores reveals whether kinship is correlated strongly with social class or not. Results here again point towards a better integrated society in Löhne and a stronger segregated society in Borgeln. Both communities are better integrated than Feistritz, the Austrian community that was studied by the developers of this quite special and very useful procedure.

The examination of social networks has proved to be a useful tool in the science of history. Regarding the history of family and kinship network analysis, – it is still in its infancy, – but with fairly promising first results. Sources often allow for the reconstruction of social, economic and political conditions of local societies, and these data can be used for formal social network analysis. Depending on the character of sources and data, researchers have to consider different kinds of analysis, starting from rather simple and well-introduced statistics to special forms of network analysis. Social network analysis is a promising field in social history, and we can expect to see more application and results in the future.

Notes

1 Talcott Parsons, "The Kinship System of the Contemporary United States," *American Anthropologist* 45, no. 1 (1943), doi:10.1525/aa.1943.45.1.02a00030: pp. 22–38.
2 Peter Laslett, *The World We Have Lost: England before the Industrial Age* (New York: Scribner, 1965).
3 Sabean David W., Simon Teuscher and Jon Mathieu, *Kinship in Europe: Approaches to Long-Term Development (1300–1900)* (New York, Oxford: Berghahn Books, 2007).
4 Michael Mitterauer, "European Kinship Systems and Household Structures: Medieval Origins," in *Distinct Inheritances: Property, Family and Community in a Changing Europe*, ed. Hannes Grandits and Patrick Heady, pp. 35–51 (Münster: LIT Verlag, 2003); Michael Mitterauer, "A 'European Family' in the Nineteenth and Twentieth Centuries?," in *The European Way*, ed. Hartmut Kaelble, pp. 140–60 (New York, Oxford: Berghahn Books, 2004).
5 Christine Fertig, *Familie, verwandtschaftliche Netzwerke und Klassenbildung im ländlichen Westfalen (1750–1874)*, Quellen und Forschungen zur Agrargeschichte 54 (Stuttgart: Lucius & Lucius, 2012); Christine Fertig, "Rural Society and Social Networks in Nineteenth-Century Westphalia: The Role of Godparenting in Social Mobility," *Journal of Interdisciplinary History* 39, no. 4 (2009): pp. 497–522; Ulrich Pfister et al., "Life Course Strategies, Social Networks, and Market Participation in Nineteenth-Century Rural Westphalia: An Interpretative Essay," in *Social Networks, Political Institutions, and Rural Societies*, ed. Georg Fertig, Rural History in Europe 11, pp. 89–124 (Turnhout: Brepols Publishers, 2015).

6 Josef Mooser, *Ländliche Klassengesellschaft, 1770–1848: Bauern und Unterschichten, Landwirtschaft und Gewerbe im östlichen Westfalen*, Kritische Studien zur Geschichtswissenschaft 86 (Göttingen: Vandenhoek & Ruprecht, 1984); Sabean David W., *Property, Production, and Family in Neckarhausen, 1700–1870*, Cambridge Studies in Social and Cultural Anthropology 73 (Cambridge: Cambridge University Press, 1990); Sabean David W., *Kinship in Neckarhausen, 1700–1870* (Cambridge: Cambridge University Press, 1998).

7 Guido Alfani and Vincent Gourdon, eds., *Spiritual Kinship in Europe, 1500–1900* (London: Palgrave Macmillan UK, 2012); Guido Alfani, Philippe Castagnetti, and Vincent Gourdon, eds. *Baptiser. Pratique sacramentelle, pratique sociale, XVIe–XXe siècles*. (Saint-Étienne: Publ. de l'Univ. de Saint-Etienne, 2009).

8 Georg M. Foster, "Godparents and Social Networks in Tzintzuntzan," *Southwestern Journal of Anthropology* 25 (1969): pp. 261–78; Michael Schnegg and Douglas R. White, "Getting Connected: Kinship and Compadrazgo in Rural Tlaxcala, Mexiko," in *Networks, Resources and Economic Action*, ed. Clemens Greiner and Waltraud Kokot, Kulturanalysen 9, pp. 37–52 (Berlin: Dietrich Reimer Verlag, 2009); Sidney W. Mintz and Eric R. Wolf, "An Analysis of Ritual Co-Parenthood (Compadrazgo)," in *Friends, Followers and Factions: A Reader in Political Clientelism*, ed. Steffen W. Schmidt et al., pp. 1–15 (Berkeley: University of California Press, 1977).

9 Martine Segalen, *Fifteen Generations of Bretons: Kinship and Society in Lower Brittany 1720–1980* (Cambridge: Cambridge University Press, 1991).

10 Sandro Guzzi-Heeb, "Kinship, Ritual Kinship and Political Milieus in an Alpine Valley in 19th Century," *The History of the Family* 14, no. 1 (2009): pp. 107–23; Sandro Guzzi-Heeb, "Spiritual Kinship, Political Mobilisation and Social Cooperation: A Swiss Alpine Valley in the Eighteenth and Nineteenth Centuries," in *Spiritual Kinship in Europe, 1500–1900*, ed. Guido Alfani and Vincent Gourdon, pp. 183–203 (London: Palgrave Macmillan UK, 2012).

11 Solveig Fagerlund, "Women and Men as Godparents in an Early Modern Swedish Town," *The History of the Family* 5 (2000): pp. 347–57; Vincent Gourdon, "Aux Cœurs De La Sociabilité Villageoise Une Analyse De Réseau À Partir Du Choix Des Conjoints Et Des Témoins Au Mariage Dans Un Village D'Île-De-France Au XIXe Siècle," *Annales de Démographie Historique* 109 (2005): pp. 61–94; Louis Haas, "Mi Buno Compadre: Choosing Godparents and the Use of Baptismal Kinship in Renaissance Florence," *Journal of Social History* (1995): pp. 341–56; Christina Munno, "Prestige, Intégration, Parentèle: Les Réseaux De Parrainage Dans Une Communauté De Vénétie (1834–1854)," *Annales de Démographie Historique* 109, no. 1 (2005): pp. 95–130.

12 Tom Ericsson, "Godparents, Witnesses and Social Class in Mid-Nineteenth Century Sweden," *The History of the Family* 5 (2000): pp. 273–86.

13 Marion Lischka, *Liebe als Ritual. Eheanbahnung und Brautwerbung in der frühneuzeitlichen Grafschaft Lippe* (Paderborn: Ferdinand Schöningh, 2006); Jacques Le Goff and Jean-Claude Schmitt, eds., *Le Charivari* (Paris: École des Hautes Études en Sciences Sociales, 1981).

14 Sabean, *Property, Production, and Family in Neckarhausen, 1700–1870*; Sabean, *Kinship in Neckarhausen, 1700–1870*.

15 Christine Fertig, "Stem Families in Rural Northwestern Germany? Family Systems, Intergenerational Relations and Family Contracts," *The History of the Family* 23, no. 2 (2017), doi:10.1080/1081602X.2016.1265571: pp. 196–217.

16 Douglas R. White, "Structural Endogamy and the Network, Graphe De Parenté," *Mathématiques et sciences humaines* 137 (1997): pp. 101–25.

17 Lilyan A. Brudner and Douglas R. White, "Class, Property, and Structural Endogamy: Visualizing Networked Histories," *Theory and Society* 26 (1997): pp. 161–208.

18 See Thomas Schweizer, *Muster sozialer Ordnung. Netzwerkanalyse als Fundament der Sozialethnologie* (Berlin: Reimer Dietrich, 1996), p. 222.

19 Wouter de Nooy, Andrej Mrvar and Vladimir Batagelj, *Exploratory Social Network Analysis with Pajek* (Cambridge: Cambridge University Press, 2005), p. 140ff. and p. 237; Douglas R. White and Ulla Johansen, *Network Analysis and Ethnographic Problems: Process Models of a Turkish Nomad Clan* (Lanham u.a.: Lexington Books, 2005).

Bibliography

Alfani, Guido, and Vincent Gourdon, eds. *Spiritual Kinship in Europe, 1500–1900.* London, UK: Palgrave Macmillan, 2012.

Brudner, Lilyan A., and Douglas R. White. "Class, Property, and Structural Endogamy: Visualizing Networked Histories." *Theory and Society* 26 (1997): pp. 161–208.

Ericsson, Tom. "Godparents, Witnesses and Social Class in Mid-Nineteenth Century Sweden." *The History of the Family* 5 (2000): pp. 273–86.

Fagerlund, Solveig. "Women and Men as Godparents in an Early Modern Swedish Town." *The History of the Family* 5 (2000): pp. 347–57.

Fertig, Christine. "Rural Society and Social Networks in Nineteenth-Century Westphalia: The Role of Godparenting in Social Mobility." *Journal of Interdisciplinary History* 39, no. 4 (2009): pp. 497–522.

———. *Familie, verwandtschaftliche Netzwerke und Klassenbildung im ländlichen Westfalen (1750–1874).* Quellen und Forschungen zur Agrargeschichte 54. Stuttgart: Lucius & Lucius, 2012.

———. "Stem Families in Rural Northwestern Germany? Family Systems, Intergenerational Relations and Family Contracts." *The History of the Family* 23, no. 2 (2017): pp. 196–217. doi:10.1080/1081602X.2016.1265571.

Foster, Georg M. "Godparents and Social Networks in Tzintzuntzan." *Southwestern Journal of Anthropology* 25 (1969): pp. 261–78.

Gourdon, Vincent. "Aux Cœurs De La Sociabilité Villageoise Une Analyse De Réseau À Partir Du Choix Des Conjoints Et Des Témoins Au Mariage Dans Un Village D'Île-De-France Au XIXe Siècle." *Annales de Démographie Historique* (2005): pp. 61–94.

Guido Alfani, Philippe Castagnetti, and Vincent Gourdon, eds. Baptiser. *Pratique sacramentelle, pratique sociale, XVIe-XXe siècles.* Saint-Étienne: Publ. de l'Univ. de Saint-Etienne, 2009.

Guzzi-Heeb, Sandro. "Kinship, Ritual Kinship and Political Milieus in an Alpine Valley in 19th Century." *The History of the Family* 14, no. 1 (2009): pp. 107–23.

———. "Spiritual Kinship, Political Mobilisation and Social Cooperation: A Swiss Alpine Valley in the Eighteenth and Nineteenth Centuries." In *Spiritual Kinship in Europe, 1500–1900.* Edited by Guido Alfani and Vincent Gourdon, pp. 183–203. London, UK: Palgrave Macmillan, 2012.

Haas, Louis. "Mi Buno Compadre: Choosing Godparents and the Use of Baptismal Kinship in Renaissance Florence." *Journal of Social History* (1995): pp. 341–56.

Laslett, Peter. *The World We Have Lost: England before the Industrial Age.* New York: Scribner, 1965.

Le Goff, Jacques, and Jean-Claude Schmitt, eds. *Le Charivari.* Paris: École des Hautes Études en Sciences Sociales, 1981.

Lischka, Marion. *Liebe als Ritual. Eheanbahnung und Brautwerbung in der frühneuzeitlichen Grafschaft Lippe.* Paderborn: Ferdinand Schöningh, 2006.

Mintz, Sidney W., and Eric R. Wolf. "An Analysis of Ritual Co-Parenthood (Compadrazgo)." In *Friends, Followers and Factions: A Reader in Political Clientelism*. Edited by Steffen W. Schmidt et al., pp. 1–15. Berkeley: University of California Press, 1977.

Mitterauer, Michael. "European Kinship Systems and Household Structures: Medieval Origins." In *Distinct Inheritances: Property, Family and Community in a Changing Europe*. Edited by Hannes Grandits and Patrick Heady, pp. 35–51. Münster: LIT Verlag, 2003.

———. "A 'European Family' in the Nineteenth and Twentieth Centuries?" In *The European Way*. Edited by Hartmut Kaelble, pp. 140–60. New York, Oxford: Berghahn Books, 2004.

Mooser, Josef. *Ländliche Klassengesellschaft, 1770–1848: Bauern Und Unterschichten, Landwirtschaft Und Gewerbe im oÖstlichen Westfalen*. Kritische Studien zur Geschichtswissenschaft 86. Göttingen: Vandenhoek & Ruprecht, 1984.

Munno, Christina. "Prestige, Intégration, Parentèle: Les Réseaux De Parrainage Dans Une Communauté De Vénétie (1834–1854)." *Annales de Démographie Historique* 109, no. 1 (2005): pp. 95–130.

Nooy, Wouter de, Andrej Mrvar, and Vladimir Batagelj. *Exploratory Social Network Analysis with Pajek*. Cambridge: Cambridge University Press, 2005.

Parsons, Talcott. "The Kinship System of the Contemporary United States." *American Anthropologist* 45, no. 1 (1943): pp. 22–38. doi:10.1525/aa.1943.45.1.02a00030.

Pfister, Ulrich, Johannes Bracht, Christine Fertig, and Georg Fertig. "Life Course Strategies, Social Networks, and Market Participation in Nineteenth-Century Rural Westphalia: An Interpretative Essay." In *Social Networks, Political Institutions, and Rural Societies*. Edited by Georg Fertig, pp. 89–124. Rural History in Europe 11. Turnhout: Brepols Publishers, 2015.

Sabean, David W. *Property, Production, and Family in Neckarhausen, 1700–1870*. Cambridge Studies in Social and Cultural Anthropology 73. Cambridge: Cambridge University Press, 1990.

———. *Kinship in Neckarhausen, 1700–1870*. Cambridge: Cambridge University Press, 1998.

Sabean, David W., Simon Teuscher, and Jon Mathieu. *Kinship in Europe: Approaches to Long-Term Development (1300–1900)*. New York, Oxford: Berghahn Books, 2007.

Schnegg, Michael, and Douglas R. White. "Getting Connected: Kinship and Compadrazgo in Rural Tlaxcala, Mexiko." In *Networks, Resources and Economic Action*. Edited by Clemens Greiner and Waltraud Kokot, pp. 37–52. Kulturanalysen 9. Berlin: Dietrich Reimer Verlag, 2009.

Schweizer, Thomas. *Muster sozialer Ordnung. Netzwerkanalyse als Fundament der Sozialethnologie*. Berlin: Reimer Dietrich, 1996.

Segalen, Martine. *Fifteen Generations of Bretons: Kinship and Society in Lower Brittany 1720–1980*. Cambridge: Cambridge University Press, 1991.

White, Douglas R. "Structural Endogamy and the Network, Graphe De Parenté." *Mathématiques et sciences humaines* 137 (1997): pp. 101–25.

White, Douglas R., and Ulla Johansen. *Network Analysis and Ethnographic Problems: Process Models of a Turkish Nomad Clan*. Lanham u.a.: Lexington Books, 2005.

2.6 Mobility and movements in intellectual history

A social network approach

Christophe Verbruggen, Hans Blomme and Thomas D'haeninck

An actor-centred approach in intellectual history

It appears to be a great moment to be a scholar of intellectual history. The digitisation of documents, the availability of digital bibliographies, new digital tools and an abundance of data available online offer exciting new prospects.[1] With the increasing availability of both structured and unstructured digital data and the dissemination of relatively user-friendly tools and applications, it has become quite easy to analyse and visualize complex phenomena in a perceptive way. The loss of heuristic barriers facilitates the formal use of particular scientific methodologies, such as social network analysis (SNA), which is no longer predominantly used as a metaphor in historical research. Its concepts and methods have increasingly found their way into the historian's practice. However, as often in Digital Humanities, scholars sometimes tend to apply SNA methods without fully realising their theoretical implications or without starting out from research questions. Understanding whether and how networks can advance the understanding of data, without creating artificial complexity, is crucial.[2]

In this chapter we will illustrate (and temper) certain expectations with regard to network analysis and spatial methods by applying them to a historical case from our ongoing digital research project TIC Collaborative. Our global objective in this project is to understand the multiple meanings behind and effects of temporary (cultural) mobility, in particular the visits, exchanges, journeys and congress participation of students, lecturers, experts, scientists, activists et. in the long 19th century.[3] We focus on the involvement of social, legal and educational reformers and other kinds of socio-political activists from the Low Countries in (temporary) transnational intellectual networks on the one hand and their activities at home on the other hand. Mobility patterns are related with the capability and opportunity structures (like family, education, social status, gender, religion, ethnicity or capital) on which individuals or groups depend for their spatial and social mobility. In this chapter we will show how social network analysis can help us to interpret our complex (historical) information in a variety of ways, such as pattern identification (e.g. tracing clusters) or as a data reduction technique.[4] Next to our relational (network) approach, we cannot overlook the clear spatial component of transnational (intellectual) mobility. Spatial history has become an important means of doing research.[5] Mapping and plotting networks geographically provides scholars a way of exploring data in order to come up with new questions and unexpected findings. Many projects in the arts and humanities include a spatial and/or visual component, often making use of web-based research platforms and graphical interfaces.

As we stressed, it's critical for scholars to know how networks can support and enhance their research. Before we can develop some concrete network perspectives in the third section of this chapter, we will expand in the second section more broadly on how we use SNA to study intellectual mobility and (the dynamics of) intellectual and social movements. We propose an actor-oriented approach and focus on the cultural processes of communicative interaction that constitute the relevant networks, combining insights from the sociology of ideas, the history of science and the literature on (social) movements, in order to explain the dynamics of scientific and intellectual movements.[6]

One way to get a grip on cultural (ex-)change is the analysis of the contact zones where cultural goods (ideas, experiences, publications etc.) are exchanged and how these "intercultural spaces" are used.[7] Mobilities can be seen as cross-border movements of persons, objects, texts and ideas, both hidden as well as conspicuous. But, how can we trace the mobility of people and ideas in the spheres of politics and science? The social network approach is indeed highly applicable to the study of intellectuals, activists and their cognitive horizons because they are almost by definition active and "audacious" within network structures.[8] Promising is the identification of intermediary persons, a selected group of "mobilisers" (using the word recently coined by Stephen Greenblatt[9]) who are creators of social or political change in society through cultural means.

By interconnecting social units through social relationships, networks offer a flexible way to deal with cultural and social transfers, which are difficult to contain within specific boundaries, whether those boundaries are local or the boundaries of nation-states.[10] Whereas comparison is essentially synchronic, a focus upon transfer is diachronic. As regards the particular case of Transfergeschichte, however, such studies have often favoured the analysis of bilateral transfers. Critical of these shortcomings and its relationship with the nation or national history, Michael Werner and Benedicte Zimmermann[11] have introduced the perspective of "histoire croisée". They propose a reflexivity that "asks that historians understand their categories of analysis as well as their objects of study, as 'entangled products' of national crossings". Certainly, relational approaches cannot always capture the complex entities that are the product or the root of transnational processes – but formal network analysis certainly has the potential to foster empirical research and to make sense of the wide array of possibilities and source material that historians work with.[12]

Although transnational history traverses other fields such as political-institutional history (with a focus on political entities, organisations and movements), economic history (with a focus on businesses, commodity chains etc.), cultural, literary and artistic history and many other (sub)disciplines, it focuses on the people who forge connections and entanglements.[13] The network concept and the transnational paradigm were jointly elaborated in the work of

Margaret Keck and Kathryn Sikkink,[14] who launched their discussion of contemporary "transnational advocacy networks" by considering anti-slavery movements in the late 18th and early 19th centuries. They studied actors with shared values or purposes that were conceived or portrayed as universal: human rights, environmental problems, educational and social reform. From a long-term perspective, these are recurring and sometimes interrelated objectives highlighted in the "history of transnational issue networks".[15] Such networks were carriers for the import, transformation and export of ideas and practices to a new context. Although Keck and Sikkink did not use formal network analysis, their approach offers a good theoretical starting point for a deeper exploration of 19th-century intellectual mobility and movements.

Movement perspectives and collective action

In the 19th and 20th century, scientific, intellectual and artistic movements coincided with the rise of social movements and shared their underlying goal of fundamental social and cultural change. Both social and intellectual movements can be approached from a relational point of view. In this we endorse Frickel and Gross,[16] who suggested (see also Cuyala[17]). Interrelated objectives, movement dynamics and mobilisation structures can for instance be related to the so called "framing" of meanings, ideas and issues in different settings.[18] This approach ties in with the so-called new sociology of ideas,[19] in that the ideas circulating will be linked to and analysed in the context of the micro-, meso- and macrosocial and -historical contexts in which they are embedded. As Crossley[20] argued, the point of sociological analysis in general and SNA in particular is to get beneath the appearance of randomness to reveal the pattern and posit its explanation. Therefore, scholars aim to identify "mechanisms" that appear to constrain actors, afford them opportunities and/or exert a steering effect on the course of interactions.[21]

A longitudinal, relational approach towards the dynamics of intellectual movements may start by analysing multiple memberships, thus showing the evolution of networks and organisational exchanges.[22] They have been used several times as an indicator for the study of cultural transfers such as knowledge exchange, for instance by Rosenthal et al. who managed to create a genealogy of causes in the 19th century for New York State, focusing on the multiple memberships of women active in social reform movements.[23] In this pioneering study, biographical dictionaries were used to map the affiliations between women and organisations between 1840 and 1914. Not the interconnections between the women – but the interconnections between the organisations – were the primary subject of their study. The number of mutual members or joint ties allowed the authors to make clusters of women's reform organisations. Their analysis not only revealed a genealogy of causes but also allowed them to identify central and intermediary actors or "brokers", core/periphery structures and ultimately also differences between the organisational structure and cultures of 1848 and 1900.

Rosenthal et al. gave their visualizations a longitudinal dimension and plotted their results on a timeline. Integrating time and longitudinal change has always been crucial in historical network analysis. Strongly influenced by Bourdieus'[24] notion of "trajectory" (the series of positions successively occupied by an actor in the field), Giuffre argues that the status of the actors is defined by the relative positions of other actors to whom they are (indirectly) tied. Changes over time in a constantly shifting web of relationships indicate changes in status but from our perspective also changes in personal interests and, on a different level, also organisational change. Another good way to include and study network dynamics is the use of sequences or snapshots in time.[25, 26] If we want to capture change, we must take time seriously. We can do this by comparing and connecting snapshots. Snapshots implicitly imply that the actors and their ties have a start date and an end date. Bearing in mind that this is always somewhat arbitrary – even when we start from the actors' self-definitions of groups or networks – we add a clear temporal dimension to the development of networks and the interactions that form them.[27]

An excellent example of mapping the evolution of networks over time through the multiple memberships of activist cohorts (snapshots in time) is the book *Partisan Publics: Communication and Contention across Brazilian Youth Activist Networks*.[28] Drawing upon ethnography, as well as formal cluster methods, Mische combined information about events, organisations and individuals in a rare longitudinal network analysis of partisanship and civic associations in Brazilian youth politics in the eighties and nineties of the 20th century. Mische tracked the trajectories of five consecutive cohorts of youth activists through intersecting organisational clusters, such as the overlapping student movements and church-based activism. The underlying assumption is that overlapping relationships constitute a setting in which communication and meaningful interaction take place.

Co-presence at events is a related type of movement tie, in the sense that it builds on the same notion of the "duality of groups and persons". Diani and Mische stress the importance of the social settings in which discourses and alliances of movements are articulated.[29] Furthermore, they point out that events provide opportunities for previously disconnected groups to intermingle in the broad movement "public", which may lead to new connections, exchange of discourse and integration (or segmentation) of the field (see also Della Porta[30]). In fact, neither networks nor the settings in which they are performed are purely of a social nature. They are cultural as well. McLean[31] highlights three cultural elements that constitute networks: first he points to cultural tastes and worldviews, which are typically both influencing and reinforcing personal social networks in both directions. Closely related to this are the culturally informed competencies of social actors, which affect network formation, including both the use of cultural knowledge as well as the adoption of forms of interaction (such as etiquettes). Finally, McLean argues that cultural schemes (norms) can also affect how and with whom actors are willing to interact. Furthermore, spaces in which social events take place, themselves performative and relational constructions, can be bearers of culturally valuable resources. They have their own regularities, which cover the interaction and define what kind of cultural practices can be articulated (and which cannot).[32]

Movements are formed through multiple types of ties. Next to co-membership and co-presence, also co-authorship (which can be understood as a specific form of a shared project) and obviously direct relationships between actors can be perceived as such. To study movement dynamics, ideally, the interplay of multiple types should be taken into account.[33] Moreover, von Bülow argues that focussing on relations as perceived by the actors themselves can in certain cases be a more fruitful scientific approach than using formally existing relations.[34] It is true that, for example, co-presence or co-membership are not necessarily detectors of network ties, as perceived by actors themselves, as they do not typically say anything about an actor's actual behaviour in a movement, nor about his role or importance in it.[35] Taking this into account, several empirical studies have shown that these types of ties can prove to be meaningful relations for constituting networks. Organisations, events and publications can all be addressed as social aggregates or types of sociability. If the same person contributed to or participated in two organisations, events or publications, we can assume that they have something to do with each other (albeit not necessarily directly).

Empirical research has proven that network structure cannot be underestimated as a determining factor of the dynamics of movements.[36] In a formal network analysis we can subsequently measure network cohesion on the subgroup or individual level (centrality). The location of individual actors and centrality in networks can also be described in terms of cliques or subgroups. A given actor can connect different groups, while others can be isolated. In a next step, basic properties of whole networks, such as size and density, can also apply to the cohesive groups in a network. Since the definition of a clique insists that every member of a subgroup has a direct tie with each and every other member, this strict definition is often not suitable for historical research. Cluster analysis is probably the most accurate technique for finding subgroups within networks, as it takes into account the strength of the relationship between actors. However, when we look at it closely, most clustering techniques, such as hierarchical clustering, are not based on a cohesive approach. The cohesive group approach differs fundamentally from the "structural equivalent" approach. Actors who are structurally equivalent (not equivalent in general) can be connected to the same actors or can have similar ties without being connected themselves. Members of two different cliques or groups can be structurally equivalent. Examples from historical SNA in which structural equivalence has been used are rare. A good example is the content analysis of letters written by Cicero.[37]

Hierarchical clustering analysis has a lot in common with multidimensional scaling (MDS), which is composed of methods for plotting the proximities between network actors and positions and is based on the use of both similarities and dissimilarities. The use of both methods results in an overlapping hierarchical pattern of groups and structures that can be graphically represented and have to be interpreted by the researcher. The use of visually representing social networks has always had an important role in social network research. Network visualization improves the communication and potential significance of relational data and helps to explore network properties.[38] From the beginning of SNA, drawings of networks have been used both to discover insights in network structures and to communicate those insights to others.[39] SNA reflects the complexity of social structures and limits the risk of reductionism. As a result of visual persuasiveness, MDS can be very useful for the representation of relatively small networks. The procedure is based on the calculation of the shortest path from actor to actor. Actors in Figure 2.6.1 (symbolist journals) with stronger ties are placed closer to each other and vice versa. This does not imply that the visible strength is used as a basis for the calculation. The visible, direct strength of the relations between the actors, i.e., the number of shared authors, is indicated by the size (thickness) of the line.

This graphical exploration allowed us to distinguish social groups and clusters. The research questions regarding the existence of regionalist literary subfields within the symbolist literary movement could be partially answered by reading the patterns of relations (publication patterns and clusters of journals). Although it does not correspond with actual geographical locations, it provides a sense of spatiality, with a concentration of periodicals in Brussels (around *La Jeune Belgique*) and Paris (around *La Plume* and *Revue Blanche*) but also a

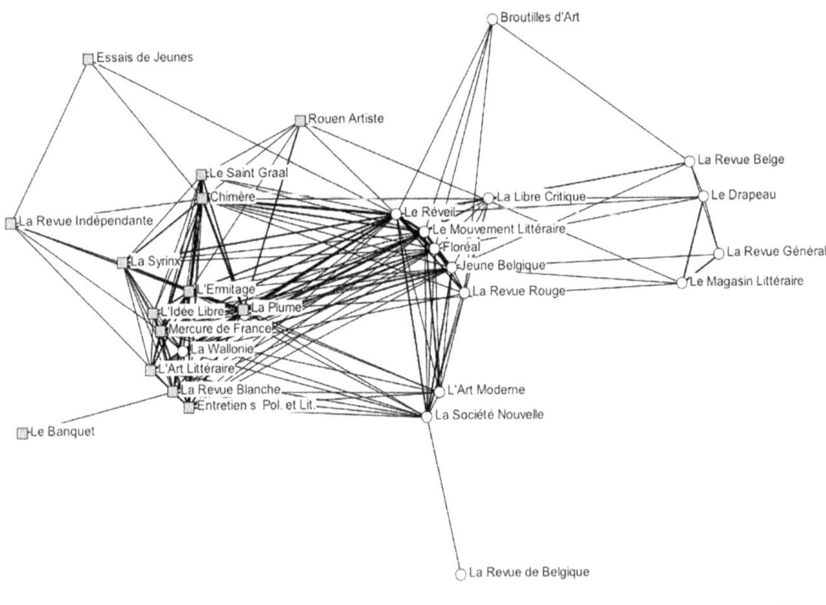

Figure 2.6.1 MDS-sociogram of Belgian and French literary journals 1892–93 (with at least three shared collaborators). The thickness of the line indicates the strength of the relationship (shared authors). Belgian journals are represented by white circles, French journals by grey squares. For a detailed interpretation, see: Laqua and Verbruggen.[40]

number of journals that were published in provincial towns and cities (notably *Chimère* and *Saint Graal*). Strikingly, the sociogram only partially reflects Belgium's socio-political divisions. However, the analysis also revealed new research issues, such as the close cooperation between Belgians (Le Réveil) and authors from the South of France (Chimère). These findings allow us to draw several conclusions regarding the French and Belgian literary field. For instance, it is evident that three Belgian Symbolist publications assumed a central position in literary exchanges, partly reflecting the aesthetic dominance of the Symbolist movement. The introduction of prosopography in combination with qualitative in-depth research would have been impossible without network analysis and graphical exploration. By reducing the complexity of the data, it was possible to identify relatively small social groups and central actors that are interesting for further research. At this stage, we have taken into account individuals with shared social attributes in a prosopography or group biography. In addition, empirical research based on personal documents such as memoirs, letters and diaries remained imperative for a more profound insight into the genesis of network structures. Thus, we advocate for an integrated approach that combines the use of qualitative methodologies, acknowledging the specificity and complexity of the multi-layered context, with formal network techniques that help detect patterns beyond the viewpoint of a given actor.[41]

The analysis of ego-networks or personal networks may also result in more in-depth knowledge about an object of study. The analysis of personal networks differs from a biography because it is a more systematic approach. For instance, in a co-authorship network, each author can be assigned as an ego, while co-authors or other contributors to the journals are the so called alters. Entire journals can also be assigned as an ego. The systematic analysis of structural properties of their network positions might reveal similarity or "homophily" between the ego and certain alters.[42] Also different from a more traditional biographical approach are the possibilities for graphical explorations. It is possible to measure structure within a personal network, but most analyses of personal network data "summarise the composition of the network as a set of variables" or social attributes.[43] Focussing on the relations around particular nodes not only offers a vast analytical added-value for research focussing on specific actors, but it also helps to overcome heuristic barriers, such as lack of (available) data and sources or, at the other end of the scale, an abundance of data on, for instance, network mobilisation in social movements.[44]

Science, expertise and activism: Belgian and Dutch 19th-century social reformers

In the following section we will illustrate our relational and actor-centred approach with a brief case study. Science, expertise and activism are domains of human activity that became increasingly international in the 19th and 20th centuries. This was, however, not a recent development but had been a feature of knowledge exchange since the early modern period (and before). The many recent and ongoing research projects studying knowledge circulation and the dynamics of the Republic of Letters illustrate this very well.[45] What was new in international exchange was not so much exchange across borders but the establishment of these (national) borders from the 19th century. The emergence of the period of the "modern" nation state, national identities and cultures coincided with a rising mobility and increasing economic integration.[46] However, the 19th-century nation state is not the most relevant category for investigating science as a social construct. This is because people labelled as "experts" set in motion processes of knowledge exchange and transformation that ultimately fuelled social and cultural reform efforts. These were based on relationships of mutual recognition, support and mobility across boundaries with regard to the 19th- and 20th-century colonial empires, even on an inter- and intra-imperial scale.

Tensions between national interests and universal aspirations also affected people engaged in social causes and advocacy, which makes it all the more relevant to address questions about activists and their networks within a transnational dimension. For example, to what extent, and in which domains, have women entered the transnational sphere? Debates about the theory and practice of social reform were not only confined to the context of nation states (in the making), but there were also many transnational connections between reform-minded citizens. A shared sense of urgency and a belief in the possibility that society could be changed for the better made engaged citizens travel beyond borders and search for new ideas and best practices that could solve social tensions. During the 19th century, ideas, policies and practices circulated in numerous ways (visits, journals, correspondence, international gatherings, learned associations and the like) and thus gave rise to a social and discursive field related to social reform.[47] Issues engendered networks, also across national borders, which have been compared to a "nébuleuse réformatrice".[48]

We are primarily looking at participation in international reform congresses, which we assume to be a strong indicator of transnational international engagement and social activism. Jamison et al.[49] see social movements as temporary public spaces, as moments of collective creation that provide societies with ideas, identities and even ideals. International congresses do meet the criteria of such definitions. These gatherings of reform-minded elites, originating from many different countries, can be seen as laboratories of new expert knowledge.[50] In the absence of international (non)governmental organisations, international congresses were the most important form of "scientific internationalisation".[51] They were – par excellence – the sites where scientists, administrators, politicians, artists and others met and exchanged ideas. They were places, in other words, where "rooted cosmopolitans"[52] connected the local, the national and the global. As such we consider social reformers and experts as part of a global field of discourse and practice, without making rigid a priori distinctions between science, knowledge and expertise. Our main research interest is the internationalisation of the social question and the emergence and development of institutional ties, generated by multiple memberships of social reformers. Above all, we are looking for different and changing patterns of attending international congresses. The changing web of relationships indicates changes in status but, from our perspective, also changes in personal interests and, as mentioned before, on a different level, also organisational change.

The central dataset we are using comes from TIC Collaborative, which is a Virtual Research Environment (VRE) for the study of 19th- and early 20th-century international organisations and (scientific) congresses.[53] The database contains biographical information of 20.000+ social reformers, activists and experts and their affiliations with international congresses and 400+ non-governmental international and transregional organisations founded before 1914. The VRE is powered by Nodegoat,[54] a web-based database management platform with a graphical interface. It is first and foremost well-suited for the spatial exploration of data in order to come up with new questions and unexpected findings. Nodegoat is primarily concerned with the creation and contextualisation of single objects that move through time and space, but queries and selections can also be made for network analysis outside Nodegoat or a multivariate analysis in the context of a prosopography.

In this case study, we focus on Belgian and Dutch participants in a large selection of thematically related international congresses between the first international congress on statistics held in Brussels in 1853 and the beginning of WWI in 1914. Nico Randeraad and Chris Leonards[55] have used the TIC VRE to conduct research on the congress visits of a wider selection of nationalities. We selected over 300 congresses with a direct or indirect focus on education, women's rights, moral and cultural reform. In total, 7,202 reformers originating from the Low Countries, who made 9,559 congress visits, are included in the dataset. 19th-century congresses can both be perceived as events as well as organisations. They were often a first step towards institutionalisation or functioned more or less as organisations by frequently providing a forum for experts to exchange their experiences and ideas. Co-presence or co-membership (as participants were often referred to as members) can be valuable ties to constitute a meaningful network, even knowing that they do not necessarily say something about an actor's actual behaviour within these networks as we argued earlier. The latent patterns in the transnational social reform network we want to visualize refer to Belgian and Dutch reformers clustered by shared congress visits.

We used a hierarchical clustering technique, which is a way to re-evaluate an entire network and to group actors together who share similar positions with regard to the totality of positions in the network. Our activists from the Low Countries visiting socio-cultural reform congresses were plotted in a two-step approach. First, our data was entered and pre-processed in Gephi. A projection technique, via the Jaroslav Kuchar's plugin, was used to convert the two-mode network (persons and congresses) to a hierarchically clustered one-mode network of congresses. Second, we calculated the network properties (degree centrality and modularity) in Gephi, which are visualized in Figure 2.6.2 via the size and colour of the nodes and vertices.

Figure 2.6.2 shows which congresses were visited by reformers from the Low Countries (central graph), which were not visited (49 congresses, black nodes,

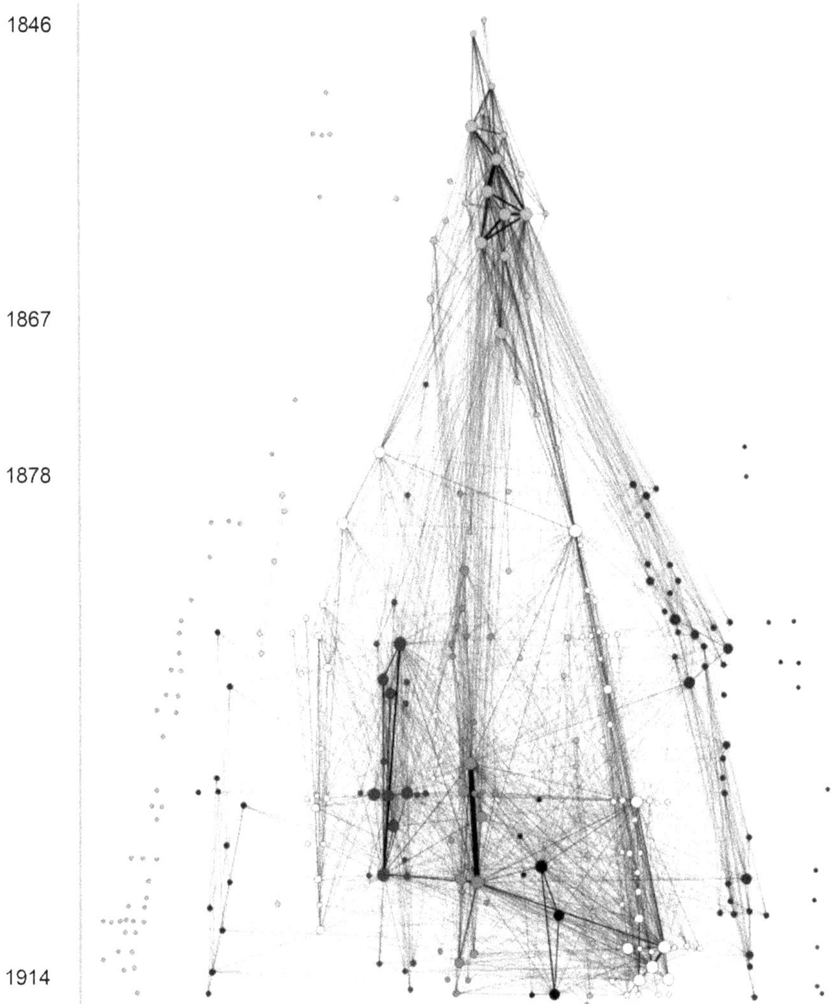

Figure 2.6.2 Socio-cultural reform congress linked by shared visits of Belgian and Dutch reformers, 1846–1914. Nodes: n = 293 congresses, size = number of visitors, colour = modularity class; Edges: n = 7542 shared visitors. Modularity classes: 11.[56]

plotted left) and isolated congresses who were visited by only a few reformers who did not visited other congresses (20 congresses, grey nodes, plotted right). All congresses are ordered chronologically with the oldest on top. The size of the nodes represents the number of Belgian and Dutch delegates each congress had (degree centrality). Although the isolates and pendants do influence the density of the network (0,063), we can clearly see a fairly connected graph, indicating a rather strong presence of Low Country reformers in the network (both synchronically and diachronically), as well as strong shared patterns of congress visits. It is, however, important to note that the congresses that were visited the most, which strongly influence the network, were held in the Low Countries (over 30% of the congresses included in the selection) and were expected to attract larger numbers.

Clusters of co-presence at (or co-membership of) congresses (or congress series) give us a huge insight into the different "causes"[57] that actors (which can be either persons or organisations) were likely to share. As a means to identify these clusters, particular algorithms can be used to structure the network into several subgroups of densely interconnected nodes. An accepted model to calculate this modularity (the strength of division of a network) is the Louvain method for community detection.[58] Applying this algorithm to a specific dataset can help researchers to visually explore their networks and to come up with hypotheses that can be further researched and tested. In our case, modularity calculation means grouping those congresses together that were largely visited by the same Belgian and Dutch reformers. The colour of the nodes indicates the modularity class they belong to. Congresses who share a high amount of Dutch and Belgian participants will have the same colour and will be strongly connected to each other. The modularity structures the network into ten clusters, two before 1878 and eight after. The Paris World's Fair of 1878, which was a major catalyst for the internationalisation of the social question. More than 30 social reform congresses took place that year. For the congresses organised after 1878, the clusters group congresses together that were mainly dominated by liberals and others that were almost exclusively visited by Catholics. For the earlier congresses, the increase in the number of modularity classes follows the expanding network, which is an indicator of the process of specialisation and professionalisation of the fields of social and cultural reform. The narrow lines between the later congresses grouped in different modules indicate that, over time, groups of Belgians and Dutch who visited the same congresses (mostly held on the same or related themes) chose to ignore others, in contrast to the earlier congresses that were generally more strongly linked together (weighted network).

A good example of a cluster revealed by our analysis is a modularity class containing a large number of congresses organised between 1878 and the outbreak of the First World War. This cluster combines a group of congresses on freemasonry and education, with a peak around 1910–12, when five congresses were visited by many Belgians and Dutch and also strong patterns of shared congress visits can be seen. For example, the *Congrès international de l'education populaire* (1910, Brussels) had 49 visitors originating from the Low Countries in common with the *Congrès international de pédologie* (1911, Brussels). Several visitors, especially the prominent figures, can be associated with freemasonry, the *Ligue de l'enseignement belge* or the *Ligue belge des droits des femmes*. Within this "nébuleuse réformatrice", the locally rooted transnational women's movement and feminism occupied a central place, both institutionally, ideologically and in the framing of other issues.[59]

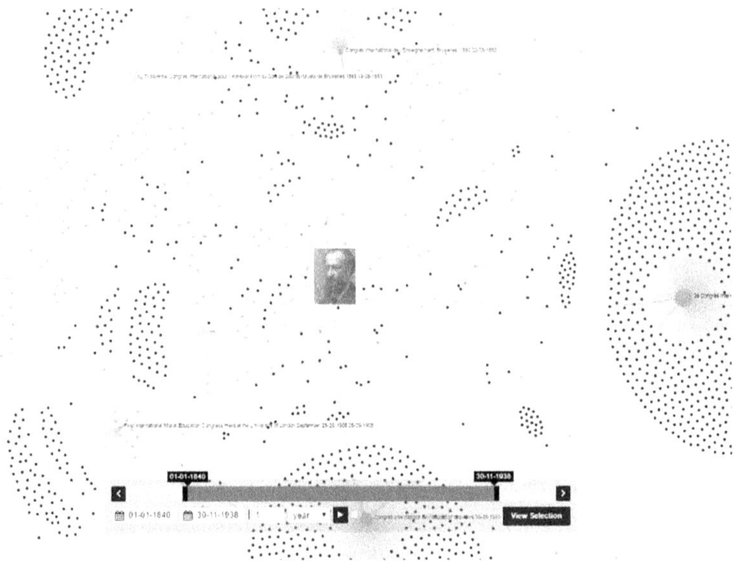

Figure 2.6.3 The personal network (co-membership and congress co-participation) of the Brussels-based educationalist Alexis Sluys (1849–1936). Dark grey nodes = congresses, black nodes = visitors, light grey nodes = organisations, size dark grey nodes = number of visitors. (Nodegoat visualization).

The personal network of educationalist Alexis Sluys (Figure 2.6.3) is a good illustration of a rooted cosmopolitan. The international affiliations in the network constituted a cognitive horizon rooted in local and national organisational milieus. He was not only an anticlerical freemason but also a member of the *Ligue de l'enseignement*, the Liberal Party and many other organisations. The frameworks he used to define different social issues and reform topics clearly echoed one another: "integral" education, the struggle against alcoholism, feminism, mixed freemasonry, secular ethics and "pedology" (the study of the physical and mental development of children).[60]

For the period between 1867 and 1878 the graph (Figure 2.6.2) indicates a stagnation of the ever-expanding network and the "strive for internationalism".[61] This period is also referred to as the transition between the first and second "peak of internationalism",[62] an abrupt but also temporary stop of international mobility mainly due to the Franco-Prussian War (1870–71). For the Belgian case, it has been suggested that the transition between the two periods also meant a transition between generations of congress attendees. Our analysis also confirms this for Dutch reformers. Only a small percentage of the delegates originating from the Low Countries visited both congresses before and after the period of 1878–80. This transition between generations of congress attendees is also reflected in a changing pattern of mobility. Before 1870, reformers originating from the Low Countries only visited congresses organised in the Low Countries, France, the

German States, the United Kingdom, Switzerland and Italy (see Figure 2.6.4). Between 1870 and 1878, a smaller number of conferences were visited, but after 1878 we can see a strong expansion of the mobility radius, as some participants even crossed the Atlantic. However, the strongest mobility is centred around Belgium due to the high number of congresses organised in Belgium between 1890 and WWI (Brussels, Liège and Antwerp).

If we want to understand the hidden mechanisms and internal coherence of groups in the network, we need to include social attributes in our analysis (such as religion, gender, profession, local affiliations). They enable us to explore the internal coherence of clusters in the network, detect divisions and factions and see meaningful trends in the presence or absence of specific social profiles. In other words, social attributes are an important step in interpreting the network and changing opportunity structures. We have to bear in mind that the sources (congress proceedings) our data set relies on do not provide us with a vast amount of biographical information about thousands of individual congress delegates who came from different places and had different socio-professional backgrounds.

Attributes that can be found in the sources include gender, nationality and often also profession and local affiliations (that can also serve as a relational attribute). A comparison between the presence of Dutch and Belgians on international congresses for instance shows that the rise and decline of the mobility of Belgians and Dutch follow a similar pattern, and also the shared patterns of

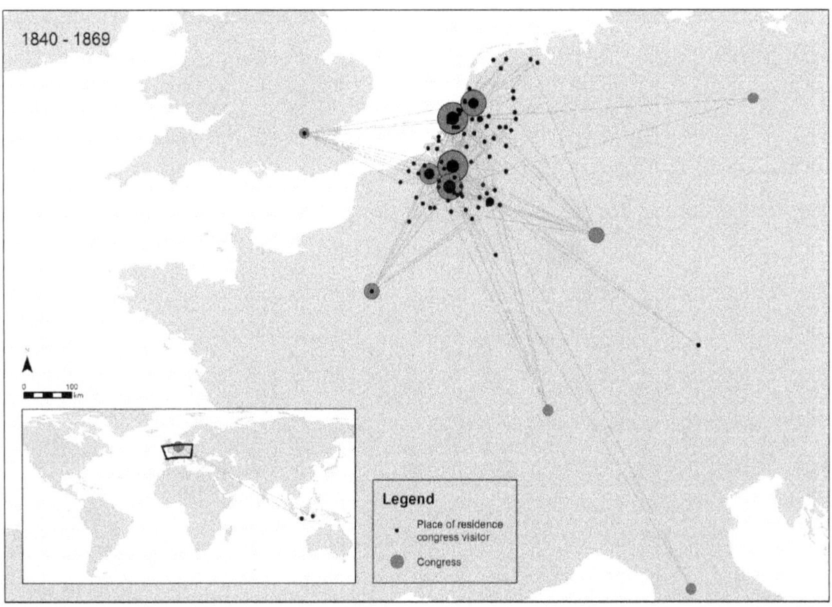

Figure 2.6.4 Congress mobility, 1840–1914.

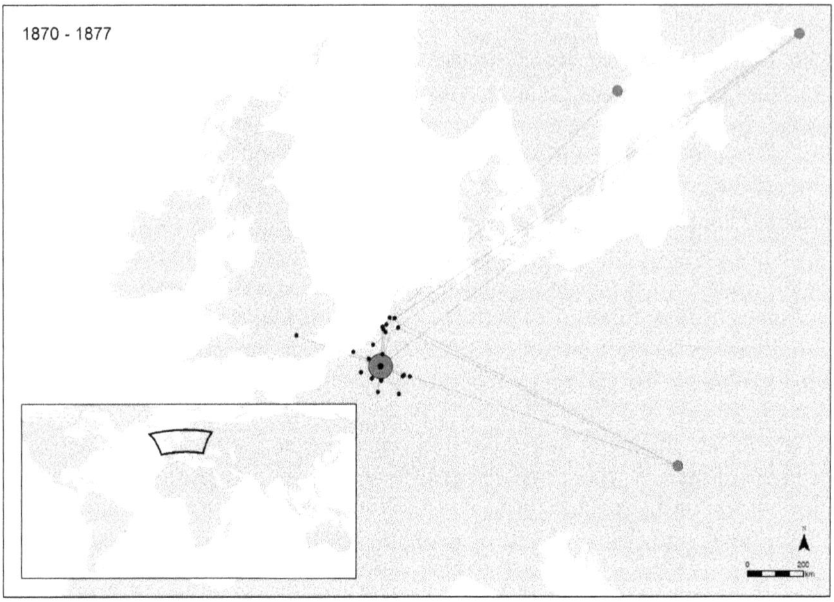

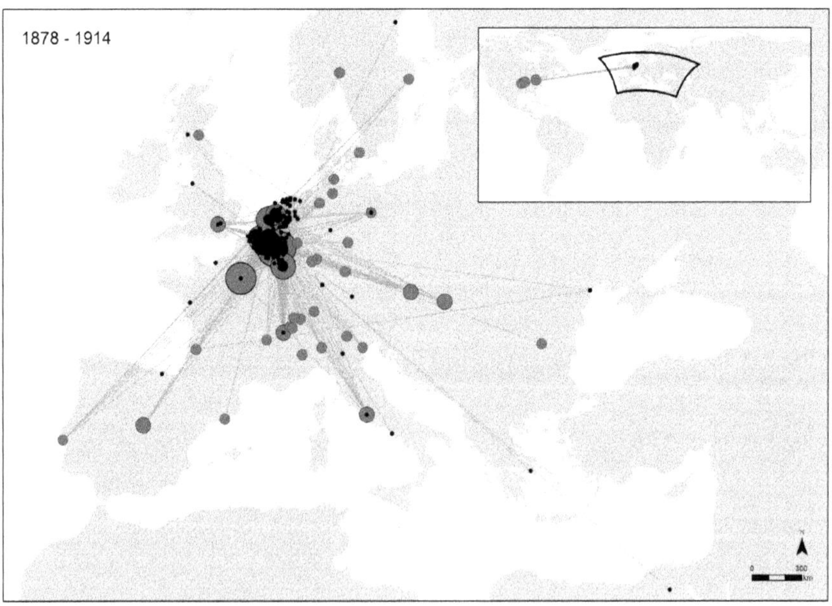

Figure 2.6.4 (*Continued*)

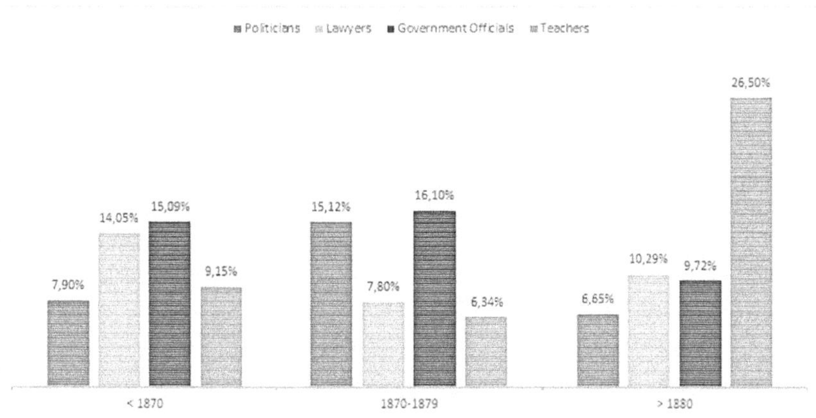

Figure 2.6.5 Known professions of Dutch and Belgian conference attendees, 1853–1914.

congress visits largely overlap. However, the increase of the number of Belgians after 1880 is quite significant and suggests a changing interest in these international gatherings. Furthermore, we can also look at the professions or occupations of the reformers.

In Figure 2.6.5 we compared four frequently occurring professions of Belgian and Dutch persons visiting international congresses, namely politicians, lawyers, government officials and teaching staff across the three periods (before 1870, between 1870 and 1879 and after 1880). The first three attributes only differ a little over the three periods; the most significant change is the strong increase in teaching staff after 1880. The fact that the professors, teachers, educators, instructors and people involved in re-education initiatives almost triple in the last period can be explained by the increasing importance of education at the international congresses, which were part of a wider dynamic of educational internationalism and growing cross-border dynamics between teaching professionals and educational organisations. We can thus trace and compare the professionalisation of welfare in the Low Countries. This illustrates very well the emergence of the "social engineer" and expert formation.

As highlighted before, describing and systematic mapping of the network in terms of centrality or connection presupposes the assumption that these patterns are meaningful. Interrelated objectives, movement dynamics and mobilisation structures can indeed be related to the framing of meanings and issues in different settings. For instance, the framing of temperance or pacifism as related to or sub-causes within the movement for women's rights. Explorative research into the composition of social networks, such as multiple memberships and shared authorship, does not suffice for answering complex research questions about changing discourse or the use of frames. Moreover, ideas and "meaning" (or semantics) are not only present in social networks but can also create such structures and sub-structures.[63]

For the actual analysis of the framing of social and cultural issues and for answering complex research questions about changing discourse, other sources and methods have to be used, such as the actual conference proceedings, publications and ego documents. In forthcoming research, we will analyse the correspondence of social reformers. Recent research has shown the potential of citation and network analysis of correspondence (collection of letters) for mapping and studying the structure of the intellectual field and the evolving conversations.[64] This approach ties in with research rooted in a "scientometric" analysis.[65] The co-occurrence of (title) words, keywords and co-authorship can, for instance, be connected with the author's social attributes and visualized[66] in order to get a grip on discursive changes.[67]

Conclusions

The evolution of networks over time, by tracing clusters in certain sequences or snapshots in time or assigning dates to ties and nodes to include a time dimension, is a good way to include and study network dynamics. Changes over time, in a constantly shifting web of relationships, indicate changes in status from a relational perspective but also changes in personal interests or organisational change. This approach ultimately reveals the relevant social circles in which the creation and circulation of ideas can be interpreted and understood. In our case study, social network analysis was used as a set of data-reduction techniques to summarise and visualize network data instead of formal modelling (as opposed to, for instance, Gould[68]). It has been argued before that this use of network tools does not provide an explanans, but an "interim explanandum", something that has to be explained instead of an explanation in itself. However, it can be very useful in a holistic methodological approach of social structure, social, cultural and intellectual dynamics.

It certainly is a good moment to be a digital historian. However, a lot of progress is still needed and can be expected. The most important change we are currently facing in the arts and humanities is a shift from an individual paradigm for humanities research to a collective one. This change is the result of an organic change in the humanities but is also due to external forces such as the rapid development of the internet, social media, user-friendly cloud solutions, etc.[69] However, this shift is far from complete. We are only slowly moving from a cooperative model towards a collaborative model. Many scholars are not yet taking full advantage of recent developments and the increasing numbers of collaborative visualization platforms facilitating both exploratory and more sophisticated analytical searches. Further infrastructural developments should coincide with the acceptance of a collaborative research paradigm of co-creation and participatory engagement. Early adopters of formal network methods in historical research highlighted the fact that SNA's data requirements are "formidable".[70] Notwithstanding the abundance of digitally available data, this has not changed fundamentally yet. Recent developments in computational linguistics, social network analysis and linked data technology are currently only partially included in the previously mentioned platforms used by cultural and intellectual historians. Increasingly complex interlinked datasets are not yet matched with the right tools to query, annotate and explore them. The heuristic problems in finding "good data" that Bonnie Ericson mentioned are still imminent.[71] Many researchers tend to forget that missing data can result easily result in a high degree of distortion. The only way forward is (international) collaboration and the re-use of data that can be used for social network analysis.

We fully acknowledge the importance of data exploration, which may direct the researcher in a certain direction and may result in unexpected findings. Yet, on a more profound analytical level, social network analysis can only offer significant results when applied following clearly defined research questions and based on a thematically and spatially well-demarcated data collection. Network analysis methods in particular are relevant here as means to identify meaningful links and subgroups in datasets, to reveal shortest paths between entities such as persons and to point researchers in the direction of relevant entities by means of centrality measures and clustering algorithms.[72] Finally, graph visualizations offer powerful means to explore highly complex relational structures, which have hitherto been inaccessible for study. A network analysis should not only re-create network structures but also allow researchers to verify or trigger new research hypotheses.

Notes

1 Dan Edelstein, "Intellectual History and Digital Humanities," *Modern Intellectual History* 13, no. 1 (2016): pp. 237–46.

2 Claire Lemercier, "Formale Methoden der Netzwerkanalyse in den Geschichtswissenschaften: Warum und wie?," *Österreichische Zeitschrift für Geschichtswissenschaft* 23, no. 1 (2012): pp. 16–41.

3 See: www.tic.ugent.be. An earlier version of the description of our overall methodology, workflow and VRE can be found in D'haeninck, Randeraad and Verbruggen 2015. We would like to thank Nico Randeraad and Sally Chambers for their valuable comments.

4 Lemercier, 'Formale Methoden der Netzwerkanalyse in den Geschichtswissenschaften: Warum und Wie?'.

5 Richard White, "What Is Spatial History?," Stanford University Spatial History Project (2010), https://web.stanford.edu/group/spatialhistory/cgi-bin/site/pub.php?id=29 (accessed 6 February 2020).

6 Ann Mische, "Relational Sociology, Culture, and Agency," in Scott; Carrington, *The Sage Handbook of Social Network Analysis*, pp. 80–98.

7 Mary L. Pratt, "Arts of the Contact Zone," *Profession* (1991): pp. 33–40.

8 François Chaubet, "Histoire des intellectuels, histoire intellectuelle: bilan provisoire et perspectives," *Vingtième siècle. Revue d'histoire* 1, no. 101 (2009): pp. 179–90.

9 Stephen Greenblatt, "Cultural Mobility: An Introduction," in *Cultural Mobility: A Manifesto*, ed. Stephen Greenblatt, pp. 1–23 (Cambridge: Cambridge University Press, 2010).

10 Mario Diani, "Social Movements, Contentious Actions, and Social Networks: 'From Metaphor to Substance'?," in *Social Movements and Networks: Relational Approaches to Collective Action,* ed. Mario Diani and Doug McAdam, pp. 1–23 (New York: Oxford University Press, 2003).

11 Michael Werner and Bénédicte Zimmermann, *De la comparaison à l'histoire croisée*, Genre Humain 42 (Paris: Seuil, 2004).
12 Pierre-Yves Saunier, *Transnational History*, Theory and History (New York: Palgrave Macmillan, 2013).
13 Patricia Clavin, "Time, Manner, Place: Writing Modern European History in Global, Transnational and International Contexts," *European History Quarterly* 40, no. 4 (2010): pp. 624–40; Pierre-Yves Saunier, "Les régimes circulatoires du domaine social 1800–1940: projets et ingénierie de la convergence et de la différence," *Genèses* 71, no. 2 (2008): pp. 4–25.
14 Margaret E. Keck and Kathryn Sikkink, *Activists Beyond Borders: Advocacy Networks in International Politics* (Ithaca: Cornell University Press, 1998).
15 Pierre-Yves Saunier, "La secrétaire générale, l'ambassadeur et le docteur. Un conte en trois épisodes pour les historiens du ' monde des causes ' à l'époque contemporaine, 1800–2000," *Monde(s), Histoire, Espaces, Relations* 1, no. 1 (2012): pp. 29–47.
16 Scott Frickel and Neil Gross, "A General Theory of Scientific/Intellectual Movements," *American Sociological Review* 70, no. 2 (2005): pp. 204–32.
17 Sylvain Cuyala, "Mapping the Sources of Diffusion and the Active Movements of Scientists by Using a Corpus of Interviews," *Terra brasilis (nova série). Revista da rede brasileira de história da geografia e geografia histórica*, no. 5 (2015).
18 Robert D. Benford and David A. Snow, "Framing Processes and Social Movements: An Overview and Assessement," *Annual Review of Sociology* 26 (2000): pp. 611–39; Jen I. Allan and Jennifer Hadden, "Exploring the Framing Power of NGOs in Global Climate Politics," *Environmental Politics* 26, no. 4 (2017): pp. 600–20; Jennifer Hadden, *Networks in Contention: The Divisive Politics of Climate Change* (New York: Cambridge University Press, 2015).
19 Charles Camic and Neil Gross, "The New Sociology of Ideas," in *The Blackwell Companion to Sociology*, ed. Judith R. Blau, pp. 236–49 (Oxford: Blackwell Publishing Ltd, 2004).
20 Nick Crossley, *Making Sense of Social Movements* (Buckingham: Open University Press, 2002).
21 David B. Tindall, "Networks as Constraints and Opportunities," in Della Porta; Diani, *The Oxford Handbook of Social Movements*, pp. 231–45; Nick Crossley, "Networks, Interaction, and Conflict: A Relational Sociology of Social Movements and Protest," in *Social Theory and Social Movements: Mutual Inspirations*, ed. Jochen Roose and Hella Dietz, pp. 155–73 (Wiesbaden: Springer Fachmedien, 2016); Florence Passy, "Social Networks Matter: But How?," in *Social Movements and Networks: Relational Approaches to Collective Action,* ed. Mario Diani and Doug McAdam, pp. 21–48 (New York: Oxford University Press, 2003).
22 Mario Diani and Ann Mische, "Network Approaches and Social Movements," in Della Porta; Diani, *The Oxford Handbook of Social Movements*, pp. 306–25; Hadden, *Networks in Contention: The Divisive Politics of Climate Change*; Mario Diani, "Social Movements and Collective Action," in Scott; Carrington, *The Sage Handbook of Social Network Analysis*, pp. 223–35; Ann Mische, *Partisan Publics: Communication and Contention across Brazilian Youth Activist Networks* (Princeton: Princeton University Press, 2008).
23 Naomi Rosenthal et al., "Social Movements and Network Analysis: A Case Study of Nineteenth-Century Women's Reform in New York State," *American Journal of Sociology* 90, no. 5 (1985): pp. 1022–54.
24 Pierre Bourdieu, *Sociology in Question* (London: Sage Publications, 1993).
25 Katherine Giuffre, "Sandpiles of Opportunity: Success in the Art World," *Social Forces* 77, no. 3 (1999): pp. 815–32.
26 Claire Lemercier, "Taking Time Seriously: How Do We Deal with Change in Historical Networks?," in *Knoten und Kanten III: Soziale Netzwerkanalyse in Geschichts-*

und Politikforschung, ed. Marten Düring, Markus Gamper and Linda Reschke, pp. 183–211 (Bielefeld: Transcript Verlag, 2015).

27 Ibid.

28 Mische, *Partisan Publics: Communication and Contention across Brazilian Youth Activist Networks*.

29 Mario Diani and Ann Mische, "Network Approaches and Social Movements," in Della Porta; Diani, *The Oxford Handbook of Social Movements*, pp. 306–25.

30 Donatella Della Porta, "Protest in Social Movements," in *Protest Cultures: A Companion*, ed. Kathrin Fahlenbrach, Martin Klimke and Joachim Scharloth, pp. 13–25 (New York: Berghahn Books, 2016); Donatella Della Porta, "Eventful Protest, Global Conflicts: Social Mechanisms in the Reproduction of Protest," in *Contention in Context: Political Opportunities and the Emergence of Protest*, ed. Jeff Goodwin and James M. Jasper, pp. 256–76 (Stanford: Stanford University Press, 2012).

31 Paul Douglas McLean, *Culture in Networks* (Cambridge: Polity Press, 2017).

32 Paul Douglas McLean, *The Art of the Network: Strategic Interaction and Patronage in Renaissance Florence* (Durham: Duke University Press, 2007); McLean, *Culture in Networks*; Ann Mische, "Partisan Performance: The Relational Construction of Brazilian Youth Activist Publics," in *Social Movement Dynamics: New Perspectives on Theory and Research from Latin America*, ed. Marissa von Bülow and Federico M. Rossi, pp. 43–72 (Farnham: Ashgate Publishing, 2015); Ann Mische, "Relational Sociology, Culture, and Agency," in Scott; Carrington, *The Sage Handbook of Social Network Analysis*, pp. 80–98.

33 Mario Diani and Ann Mische, "Network Approaches and Social Movements," in Della Porta; Diani, *The Oxford Handbook of Social Movements*, pp. 306–25.

34 Marissa von Bülow, *Building Transnational Networks: Civil Society Networks and the Politics of Trade in the Americas* (Cambridge: Cambridge University Press, 2010).

35 Hadden, *Networks in Contention: The Divisive Politics of Climate Change*.

36 Roger V. Gould, "Multiple Networks and Mobilisation in the Paris Commune, 1871," *American Sociological Review* 56, no. 6 (1991): pp. 716–29; Hadden, *Networks in Contention: The Divisive Politics of Climate Change*; Mario Diani, *The Cement of Civil Society: Studying Networks in Localities* (New York: Cambridge University Press, 2009).

37 Michael C. Alexander and James A. Danowski, "Analysis of an Ancient Network: Personal Communication and the Study of Social Structure in a Past Society," *Social Networks* 12, no. 4 (1990): pp. 313–35. Their results show that senators and knights were structurally equivalent.

38 Ulrik Brandes, Natalie Indlekofer and Martin Mader, "Visualization Methods for Longitudinal Social Networks and Stochastic Actor-Oriented Modeling," *Social Networks* 34, no. 3 (2012): pp. 291–308; Ulrik Brandes et al., "Explorations into the Visualization of Policy Networks," *Journal of Theoretical Politics* 11, no. 1 (1999): pp. 75–106.

39 Linton C. Freeman, "Visualizing Social Networks," *Journal of Social Structure* 1, no. 1 (2000): pp. 1–15.

40 Daniel Laqua and Christophe Verbruggen, "Beyond the Metropolis: French and Belgian Symbolists between the Region and the Republic of Letters," *Comparative Critical Studies* 10, no. 2 (2013): pp. 241–58.

41 Jan Fuhse and Mützel Sophie, "Tackling Connections, Structure, and Meaning in Networks: Quantitative and Qualitative Methods in Sociological Network Research," *Quality & Quantity* 45, no. 5 (2011): pp. 1067–89; Mische, *Partisan Publics: Communication and Contention across Brazilian Youth Activist Networks*; Ann Mische, "Cross-Talk in Movements: Reconceiving the Culture-Network Link," in *Social Movements and Networks: Relational Approaches to Collective Action,* ed. Mario Diani and Doug McAdam, pp. 258–80 (New York: Oxford University Press, 2003).

42 Marian-Gabriel Hancean and Matjaž Perc, "Homophily in Coauthorship Networks of East European Sociologists," *Scientific Reports-Nature* 6, no. 36152 (2016).
43 Christopher McCarty et al., "A Comparison of Social Network Mapping and Personal Network Visualization," *Field methods* 19, no. 2 (2007): pp. 145–62.
44 David B. Tindall and Joanna L. Robinson, "Collective Action to Save the Ancient Temperate Rainforest: Social Networks and Environmental Activism in Clayoquot Sound," *Ecology and Society* 22, no. 1 (2017), doi:10.5751/ES-09042-220140; David B. Tindall, "Social Movement Participation Over Time: An Ego-Network Approach to Micro-Mobilization," *Sociological Focus* 37, no. 2 (2004): pp. 163–84.
45 Charles van den Heuvel et al., "Circles of Confidence in Correspondence: Modeling Confidentiality and Secrecy in Knowledge Exchange Networks of Letters and Drawings in the Early Modern Period," *Nuncius* 31, no. 1 (2016): pp. 78–106; Charles van Den Heuvel, "Mapping Knowledge Exchange in Early Modern Europe: Intellectual and Technological Geographies and Network Representations," *International Journal of Humanities and Arts Computing* 9, no. 1 (2015): pp. 95–114; Martin Stuber and Lothar Krempel, "The Scholarly Networks of Albrecht von Haller and the Economic Society: A Multi-Level Network Analysis," *REDES revista hispana para el analisis de redes sociales* 24, no. 1 (2011): pp. 1–25; Laqua and Verbruggen, "Beyond the Metropolis: French and Belgian Symbolists Between the Region and the Republic of Letters," pp. 241–58; Caroline Winterer, "Where Is America in the Republic of Letters?," *Modern Intellectual History* 9, no. 3 (2012): pp. 597–623; Yves Gingras, "Mapping the Structure of the Intellectual Field Using Citation and Co-Citation Analysis of Correspondences," *History of European Ideas* 36, no. 3 (2010): pp. 330–9.
46 Jürgen Osterhammel, *The Transformation of the World: A Global History of the Nineteenth Century* (Princeton: Princeton University Press, 2014).
47 Stéphane Baciocchi, Thomas David and Topalov Christian, eds., *Philanthropes en 1900 (Londres, New York, Paris, Genève)* (Paris: Creaphis editions, 2017); Chris Leonards and Nico Randeraad, "Transnational Experts in Social Reform, 1840–1880," *International Review of Social History* 55, no. 2 (2010): pp. 215–39; Chris Leonards and Nico Randeraad, "Building a Transnational Network of Social Reform in the Nineteenth Century," in Rodogno; Struck; Vogel, *Shaping the Transnational Spere: Experts, Networks and Issues from the 1840s to the 1930s*, pp. 111–31; Christian Topalov, *Laboratoires du nouveau siècle: la nébuleuse réformatrice et ses réseaux en France, 1880–1914* (Paris: Ecole des hautes études en sciences sociales, 1999); Saunier, "La secrétaire générale, l'ambassadeur et le docteur. Un conte en trois épisodes pour les historiens du ' monde des causes ' à l'époque contemporaine, 1800–2000," pp. 29–47.
48 Topalov, *Laboratoires du nouveau siècle: la nébuleuse réformatrice et ses réseaux en France, 1880–1914.*
49 Andrew Jamison et al., *The Making of the New Environmental Consciousness: A Comparative Study of Environmental Movements in Sweden, Denmark and the Netherlands* (Edinburgh: Edinburgh University Press, 1990).
50 Nico Randeraad, "The International Statistical Congress (1853–1876): Knowledge Transfers and Their Limits," *European History Quarterly* 41, no. 1 (2011): pp. 50–65.
51 Eckhardt Fuchs, "Educational Sciences, Morality and Politics: International Educational Congresses in the Early Twentieth Century," *Paedagogica historica* 40, no. 5–6 (2004): pp. 757–84; Damiano Matasci, "International Congresses of Education and the Circulation of Pedagogical Knowledge in Western Europe, 1876–1910," in Rodogno; Struck; Vogel, *Shaping the Transnational Spere: Experts, Networks and Issues from the 1840s to the 1930s*, pp. 218–38.
52 Sidney G. Tarrow, *The New Transnational Activism* (Cambridge: Cambridge University Press, 2005).

53 www.tic.ugent.be (accessed 5 December 2018).
54 https://nodegoat.net (accessed 5 December 2018).
55 Nico Randeraad and Chris Leonards, "Transnational Experts in Social Reform, 1840–1880," *International Review of Social History* 55, no. 2 (2010): pp. 215–39; Chris Leonards and Nico Randeraad, "Building a Transnational Network of Social Reform in the Nineteenth Century," in Rodogno; Struck; Vogel, *Shaping the Transnational Spere: Experts, Networks and Issues from the 1840s to the 1930s*, pp. 111–31; Chris Leonards and Nico Randeraad, "Circulations charitables: les congrès internationaux de réforme sociale," in *Philanthropes en 1900 (Londres, New York, Paris, Genève)*, ed. Stéphane Baciocchi, Thomas David and Topalov Christian (Paris: Creaphis editions, 2017).
56 Made in https://gephi.org (accessed 5 December 2018).
57 Rosenthal et al., "Social Movements and Network Analysis: A Case Study of Nineteenth-Century Women's Reform in New York State," pp. 1022–54.
58 Vincent D. Blondel et al., "Fast Unfolding of Communities in Large Networks," *Journal of Statistical Mechanisms: Theory and Experiment* 10 (2008).
59 Amandine Thiry, Thomas D'haeninck and Christophe Verbruggen, "(Re-)educational Internationalism in the Low Countries, 1850–1914," in *The Civilising Offensive: New Perspectives on Social and Educational Reform in 19th-Century Belgium*, ed. Christoph de Spiegeleer, upcoming (Berlin: De Gruyter, 2018).
60 George Laurent, "Alexis Sluys: un pédagogue engagé au service de l'enseignement officiel en Belgique," *Cahiers bruxellois – Brusselse cahiers* 47, no. 1 (2015): pp. 74–106.
61 Claude Tapia and Jacques Taieb, "Conférences et congrès internationaux de 1815 À 1913," *Relations Internationales* 5 (1976): pp. 11–35; Anne Rasmussen, "Les congrès internationaux liés aux expositions universelles de Paris (1867–1900)," *Mil neuf cent* 7, no. 1 (1989): pp. 23–44; Brigitte Schroeder-Gudehus, *Les congrès scientifiques internationaux* (Paris: SEHRIC, 1990).
62 Jasmien van Daele and Christian Müller, "Peaks of Internationalism in Social Engineering: A Transnational History of International Social Reform Associations and Belgian Agency, 1860–1925," *Revue belge de philologie et d'histoire* 90, no. 4 (2012): pp. 1297–319.
63 Ann Mische, "Cross-Talk in Movements: Reconceiving the Culture-Network Link," in *Social Movements and Networks: Relational Approaches to Collective Action*, ed. Mario Diani and Doug McAdam, pp. 258–80 (New York: Oxford University Press, 2003).
64 Gingras, "Mapping the Structure of the Intellectual Field Using Citation and Co-Citation Analysis of Correspondences," pp. 330–9; Winterer, "Where Is America in the Republic of Letters?," pp. 597–623.
65 Frédéric Vandermoere and Raf Vanderstraeten, "Disciplinary Networks and Bounding: Scientific Communication Between Science and Technology Studies and the History of Science," *Minerva* 50, no. 4 (2012): pp. 451–70.
66 An interesting tool designed for this purpose is Sci2 (https://sci2.cns.iu.edu/ (accessed 5 December 2018).
67 Katy Börner and David E. Polley, *Visual Insights: A Practical Guide to Making Sense of Data* (Cambridge: MIT Press, 2014).
68 Roger V. Gould, "Uses of Network Tools in Comparative Historical Research," in *Comparative Historical Analysis in the Social Sciences*, ed. James Mahoney and Dietrich Rueschemeyer, pp. 241–69 (Cambridge: Cambridge University Press, 2003).
69 Jennifer Edmond, "Collaboration and Infrastructure," in *A New Companion to Digital Humanities*, ed. Susan Schreibman, Ray Siemens and John Unsworth, pp. 54–65 (Chichester: John Wiley & Sons, Ltd, 2016).
70 Charles Wetherell, "Historical Social Network Analysis," *International Review of Social History* 43, no. 6 (1998): pp. 125–44.

71 Bonnie H. Erickson, "Social Networks and History: A Review Essay," *Historical Methods: A Journal of Quantitative and Interdisciplinary History* 30, no. 3 (1997): pp. 149–57.

72 Claire Lemercier, "Formal Network Methods in History: Why and How?," in *Social Networks, Political Institutions, and Rural Societies*, ed. Georg Fertig, Rural History in Europe 11, pp. 281–304 (Turnhout: Brepols Publishers, 2015).

Bibliography

Alexander, Michael C., and James A. Danowski. "Analysis of an Ancient Network: Personal Communication and the Study of Social Structure in a Past Society." *Social Networks* 12, no. 4 (1990): pp. 313–35.

Allan, Jen Iris, and Jennifer Hadden. "Exploring the Framing Power of NGOs in Global Climate Politics." *Environmental Politics* 26, no. 4 (2017): pp. 600–20.

Benford, Robert D., and David A. Snow. "Framing Processes and Social Movements: An Overview and Assessement." *Annual Review of Sociology* 26 (2000): pp. 611–39.

Blondel, Vincent D., Jean-Loup Guillaume, Renaud Lambiotte, and Etienne Lefebvre. "Fast Unfolding of Communities in Large Networks." *Journal of Statistical Mechanisms: Theory and Experiment* 10 (2008).

Börner, Katy, and David E. Polley. *Visual Insights: A Practical Guide to Making Sense of Data*. Cambridge: MIT Press, 2014.

Bourdieu, Pierre. *Sociology in Question*. London: Sage Publications, 1993.

Brandes, Ulrik, Natalie Indlekofer, and Martin Mader. "Visualization Methods for Longitudinal Social Networks and Stochastic Actor-Oriented Modeling." *Social Networks* 34, no. 3 (2012): pp. 291–308.

Brandes, Ulrik, Patrick Kenis, Jörg Raab, Volker Schneider, and Dorothea Wagner. "Explorations into the Visualization of Policy Networks." *Journal of Theoretical Politics* 11, no. 1 (1999): pp. 75–106.

Bülow, Marissa von. *Building Transnational Networks: Civil Society Networks and the Politics of Trade in the Americas*. Cambridge: Cambridge University Press, 2010.

Camic, Charles, and Neil Gross. "The New Sociology of Ideas." In *The Blackwell Companion to Sociology*. Edited by Judith R. Blau, pp. 236–49. Oxford: Blackwell Publishing Ltd, 2004.

Chaubet, François. "Histoire des intellectuels, histoire intellectuelle: bilan provisoire et perspectives." *Vingtième siècle. Revue d'histoire* 1, no. 101 (2009): pp. 179–90.

Clavin, Patricia. "Time, Manner, Place: Writing Modern European History in Global, Transnational and International Contexts." *European History Quarterly* 40, no. 4 (2010): pp. 624–40.

Crossley, Nick. *Making Sense of Social Movements*. Buckingham: Open University Press, 2002.

———. "Networks, Interaction, and Conflict: A Relational Sociology of Social Movements and Protest." In *Social Theory and Social Movements: Mutual Inspirations*. Edited by Jochen Roose and Hella Dietz, pp. 155–73. Wiesbaden: Springer Fachmedien, 2016.

Cuyala, Sylvain. "Mapping the Sources of Diffusion and the Active Movements of Scientists by Using a Corpus of Interviews." *Terra brasilis (nova série). Revista da rede brasileira de história da geografia e geografia histórica*, no. 5 (2015).

Della Porta, Donatella. "Eventful Protest, Global Conflicts: Social Mechanisms in the Reproduction of Protest." In *Contention in Context: Political Opportunities and the*

Emergence of Protest. Edited by Jeff Goodwin and James M. Jasper, pp. 256–76. Stanford: Stanford University Press, 2012.
———. "Protest in Social Movements." In *Protest Cultures: A Companion*. Edited by Kathrin Fahlenbrach, Martin Klimke and Joachim Scharloth, pp. 13–25. New York: Berghahn Books, 2016.
Diani, Mario. "Social Movements, Contentious Actions, and Social Networks: 'From Metaphor to Substance'?" In *Social Movements and Networks: Relational Approaches to Collective Action*. Edited by Mario Diani, and Doug McAdam Diani; McAdam, pp. 1–23, New York: Oxford University Press, 2003.
———. *The Cement of Civil Society: Studying Networks in Localities*. New York: Cambridge University Press, 2009.
———. "Social Movements and Collective Action." In Scott; Carrington, *The Sage Handbook of Social Network Analysis*, pp. 223–35.
Diani, Mario, and Ann Mische. "Network Approaches and Asocial Movements." In Della Porta; Diani, *The Oxford Handbook of Social Movements*, pp. 306–25.
Edelstein, Dan. "Intellectual History and Digital Humanities." *Modern Intellectual History* 13, no. 1 (2016): pp. 237–46.
Edmond, Jennifer. "Collaboration and Infrastructure." In *A New Companion to Digital Humanities*. Edited by Susan Schreibman, Ray Siemens and John Unsworth, pp. 54–65. Chichester: John Wiley & Sons, Ltd, 2016.
Erickson, Bonnie H. "Social Networks and History: A Review Essay." *Historical Methods: A Journal of Quantitative and Interdisciplinary History* 30, no. 3 (1997): pp. 149–57.
Freeman, Linton C. "Visualizing Social Networks." *Journal of Social Structure* 1, no. 1 (2000): pp. 1–15.
Frickel, Scott, and Neil Gross. "A General Theory of Scientific/Intellectual Movements." *American Sociological Review* 70, no. 2 (2005): pp. 204–32.
Fuchs, Eckhardt. "Educational Sciences, Morality and Politics: International Educational Congresses in the Early Twentieth Century." *Paedagogica historica* 40, nos. 5–6 (2004): pp. 757–84.
Fuhse, Jan, and Mützel Sophie. "Tackling Connections, Structure, and Meaning in Networks: Quantitative and Qualitative Methods in Sociological Network Research." *Quality & Quantity* 45, no. 5 (2011): pp. 1067–89.
Gingras, Yves. "Mapping the Structure of the Intellectual Field Using Citation and Co-Citation Analysis of Correspondences." *History of European Ideas* 36, no. 3 (2010): pp. 330–9.
Giuffre, Katherine. "Sandpiles of Opportunity: Success in the Art World." *Social Forces* 77, no. 3 (1999): pp. 815–32.
Gould, Roger V. "Multiple Networks and Mobilisation in the Paris Commune, 1871." *American Sociological Review* 56, no. 6 (1991): pp. 716–29.
———. "Uses of Network Tools in Comparative Historical Research." In *Comparative Historical Analysis in the Social Sciences*. Edited by James Mahoney and Dietrich Rueschemeyer, pp. 241–69. Cambridge: Cambridge University Press, 2003.
Greenblatt, Stephen. "Cultural Mobility: An Introduction." In *Cultural Mobility: A Manifesto*. Edited by Stephen Greenblatt, pp. 1–23. Cambridge: Cambridge University Press, 2010.
Hadden, Jennifer. *Networks in Contention: The Divisive Politics of Climate Change*. New York: Cambridge University Press, 2015.

Hancean, Marian-Gabriel, and Matjaž Perc. "Homophily in Coauthorship Networks of East European Sociologists." *Scientific Reports – Nature* 6, no. 36152 (2016).

Jamison, Andrew, Ron Eyerman, Jacqueline Cramer, and Jeppe Laessøe. *The Making of the New Environmental Consciousness: A Comparative Study of Environmental Movements in Sweden, Denmark and the Netherlands*. Edinburgh: Edinburgh University Press, 1990.

Keck, Margaret E., and Kathryn Sikkink. *Activists beyond Borders: Advocacy Networks in International Politics*. Ithaca: Cornell University Press, 1998.

Laqua, Daniel, and Christophe Verbruggen. "Beyond the Metropolis: French and Belgian Symbolists between the Region and the Republic of Letters." *Comparative Critical Studies* 10, no. 2 (2013): pp. 241–58.

Laurent, George. "Alexis Sluys: un pédagogue engagé au service de l'enseignement officiel en Belgique." *Cahiers bruxellois – Brusselse cahiers* 47, no. 1 (2015): pp. 74–106.

Lemercier, Claire. "Formale Methoden der Netzwerkanalyse in den Geschichtswissenschaften: Warum und wie?" *Österreichische Zeitschrift für Geschichtswissenschaft* 23, no. 1 (2012): pp. 16–41.

———. "Formal Network Methods in History: Why and How?" In *Social Networks, Political Institutions, and Rural Societies*. Edited by Georg Fertig, pp. 281–304. Rural History in Europe 11. Turnhout: Brepols Publishers, 2015.

———. "Taking Time Seriously: How Do We Deal with Change in Historical Networks?" In *Knoten und Kanten III: Soziale Netzwerkanalyse in Geschichts- und Politikforschung*. Edited by Marten Düring, Markus Gamper and Linda Reschke, pp. 183–211. Bielefeld: Transcript Verlag, 2015.

Leonards, Chris, and Nico Randeraad. "Transnational Experts in Social Reform, 1840–1880." *International Review of Social History* 55, no. 2 (2010): pp. 215–39.

———. "Building a Transnational Network of Social Reform in the Nineteenth Century." In Rodogno; Struck; Vogel, *Shaping the Transnational Spere: Experts, Networks and Issues from the 1840s to the 1930s*, pp. 111–31.

———. "Circulations charitables: les congrès internationaux de réforme sociale." In *Philanthropes en 1900 (Londres, New York, Paris, Genève)*. Edited by Stéphane Baciocchi, Thomas David and Topalov Christian. Paris: Creaphis editions, 2017.

Matasci, Damiano. "International Congresses of Education and the Circulation of Pedagogical Knowledge in Western Europe, 1876–1910." In Rodogno; Struck; Vogel, *Shaping the Transnational Spere: Experts, Networks and Issues from the 1840s to the 1930s*, pp. 218–38.

McCarty, Christopher, José Luis Molina, Claudia Aguilar, and Laura Rota. "A Comparison of Social Network Mapping and Personal Network Visualization." *Field Methods* 19, no. 2 (2007): pp. 145–62.

McLean, Paul Douglas. *The Art of the Network: Strategic Interaction and Patronage in Renaissance Florence*. Durham: Duke University Press, 2007.

———. *Culture in Networks*. Cambridge: Polity Press, 2017.

Mische, Ann. "Cross-Talk in Movements: Reconceiving the Culture-Network Link." In *Social Movements and Networks: Relational Approaches to Collective Action*. Edited by Mario Diani, and Doug McAdam, pp. 258–80, New York: Oxford University Press, 2003.

———. *Partisan Publics: Communication and Contention across Brazilian Youth Activist Networks*. Princeton: Princeton University Press, 2008.

———. "Relational Sociology, Culture, and Agency." In Scott; Carrington, *The Sage Handbook of Social Network Analysis*, pp. 80–98.

———. "Partisan Performance: The Relational Construction of Brazilian Youth Activist Publics." In *Social Movement Dynamics: New Perspectives on Theory and Research from Latin America*. Edited by Marissa von Bülow, and Federico M. Rossi, pp. 43–72. Farnham: Ashgate Publishing, 2015.

Osterhammel, Jürgen. *The Transformation of the World: A Global History of the Nineteenth Century*. Princeton: Princeton University Press, 2014.

Passy, Florence. "Social Networks Matter: But How?" In *Social Movements and Networks: Relational Approaches to Collective Action*. Edited by Mario Diani, and Doug McAdam, pp. 21–48, New York: Oxford University Press, 2003,.

Pratt, Mary Louise. "Arts of the Contact Zone." *Profession* (1991): pp. 33–40.

Randeraad, Nico. "The International Statistical Congress (1853–1876): Knowledge Transfers and Their Limits." *European History Quarterly* 41, no. 1 (2011): pp. 50–65.

Randeraad, Nico, and Chris Leonards. "Transnational Experts in Social Reform, 1840–1880." *International Review of Social History* 55, no. 2 (2010): pp. 215–39.

Rasmussen, Anne. "Les congrès internationaux liés aux expositions universelles de Paris (1867–1900)." *Mil neuf cent* 7, no. 1 (1989): pp. 23–44.

Rosenthal, Naomi, Meryl Fingrutd, Michele Ethier, Roberta Karant, and David McDonald. "Social Movements and Network Analysis: A Case Study of Nineteenth-Century Women's Reform in New York State." *American Journal of Sociology* 90, no. 5 (1985): pp. 1022–54.

Saunier, Pierre-Yves. "Les régimes circulatoires du domaine social 1800–1940: projets et ingénierie de la convergence et de la différence." *Genèses* 71, no. 2 (2008): pp. 4–25.

———. "La secrétaire générale, l'ambassadeur et le docteur. Un conte en trois épisodes pour les historiens du ' monde des causes ' à l'époque contemporaine, 1800–2000." *Monde(s), Histoire, Espaces, Relations* 1, no. 1 (2012): pp. 29–47.

———. *Transnational History*. Theory and History. New York: Palgrave Macmillan, 2013.

Schroeder-Gudehus, Brigitte. *Les congrès scientifiques internationaux*. Paris: SEHRIC, 1990.

Stuber, Martin, and Lothar Krempel. "The Scholarly Networks of Albrecht von Haller and the Economic Society: A Multi-Level Network Analysis." *REDES revista hispana para el analisis de redes sociales* 24, no. 1 (2011): pp. 1–25.

Tapia, Claude, and Jacques Taieb. "Conférences et congrès internationaux de 1815 À 1913." *Relations Internationales* 5 (1976): pp. 11–35.

Tarrow, Sidney G. *The New Transnational Activism*. Cambridge: Cambridge University Press, 2005.

Thiry, Amandine, Thomas D'haeninck, and Christophe Verbruggen. "(Re-)educational Internationalism in the Low Countries, 1850–1914." In *The Civilising Offensive: New Perspectives on Social and Educational Reform in 19th-Century Belgium*. Edited by Christoph de Spiegeleer. upcoming. Berlin: De Gruyter, 2018.

Tindall, David B. "Social Movement Participation Over Time: An Ego-Network Approach to Micro-Mobilization." *Sociological Focus* 37, no. 2 (2004): pp. 163–84.

———. "Networks as Constraints and Opportunities." In Della Porta; Diani, *The Oxford Handbook of Social Movements*, pp. 231–45.

Tindall, David B., and Joanna L. Robinson. "Collective Action to Save the Ancient Temperate Rainforest: Social Networks and Environmental Activism in Clayoquot Sound." *Ecology and Society* 22, no. 1 (2017). doi:10.5751/ES-09042-220140.

Topalov, Christian. *Laboratoires du nouveau siècle: la nébuleuse réformatrice et ses réseaux en France, 1880–1914*. Paris: Ecole des hautes études en sciences sociales, 1999.

van Daele, Jasmien, and Christian Müller. "Peaks of Internationalism in Social Engineering: A Transnational History of International Social Reform Associations and Belgian Agency, 1860–1925." *Revue belge de philologie et d'histoire* 90, no. 4 (2012): pp. 1297–319.
van den Heuvel, Charles. "Mapping Knowledge Exchange in Early Modern Europe: Intellectual and Technological Geographies and Network Representations." *International Journal of Humanities and Arts Computing* 9, no. 1 (2015): pp. 95–114.
van den Heuvel, Charles, Scott B. Weingart, Nils Spelt, and Henk Nellen. "Circles of Confidence in Correspondence: Modeling Confidentiality and Secrecy in Knowledge Exchange Networks of Letters and Drawings in the Early Modern Period." *Nuncius* 31, no. 1 (2016): pp. 78–106.
Vandermoere, Frédéric, and Raf Vanderstraeten. "Disciplinary Networks and Bounding: Scientific Communication Between Science and Technology Studies and the History of Science." *Minerva* 50, no. 4 (2012): pp. 451–70.
Werner, Michael, and Bénédicte Zimmermann. *De la comparaison à l'histoire croisée.* Genre Humain 42. Paris: Seuil, 2004.
Wetherell, Charles. "Historical Social Network Analysis." *International Review of Social History* 43, no. 6 (1998): pp. 125–44.
White, Richard. "What Is Spatial History?" Stanford University Spatial History Project (2010), https://web.stanford.edu/group/spatialhistory/cgi-bin/site/pub.php?id=29 (accessed 6 February 2020).
Winterer, Caroline. "Where is America in the Republic of Letters?" *Modern Intellectual History* 9, no. 3 (2012): pp. 597–623.

3

Computational extraction of network data from large corpora

3.1 Utilizing historical network analysis on meta-data to model East German foreign intelligence cycle in the Baltic Sea Region 1975–89

Kimmo Elo

Introduction

The very essence of all intelligence services is related to their networking capacity. A closer look at the so-called intelligence cycle, the cyclical process consisting of information-gathering (single-source collection), exploitation, (all-source) analysis and dissemination,[1] quickly reveals how fundamentally important networking capabilities and networks are for all intelligence services. Each intelligence service attempts to build a functioning, dense, network providing the service with raw data. In intelligence analysis, analysts combine information from the collected raw data with previous knowledge and information from other sources in thematic reports. Finally, these reports are disseminated to a small or large network of recipients at various levels of governmental administration.

The Cold War Nordic constellation offers an interesting object for the study of intelligence networks. Despite the differences in security-political settings of the four Nordic countries, the Nordic region was built upon shared cultural, socio-political and historical elements. But they also shared a wide range of issues in European politics, forcing them into close cooperation and coordination. Therefore, research on East German intelligence in the Nordic countries should shift its focus away from pure single-country studies toward a "Nordic perspective". A comparative perspective could provide an understanding of the role(s) the Nordic countries played in the Cold War European relations and intelligence history as a community on the one hand and as single countries on the other.

Taking into account the importance of intelligence during the Cold War and the centrality of networks for intelligence services, the limited interest in network analysis among Cold War studies is somewhat surprising. One reason might be that Cold War studies are still dominated by historians who are neither primarily interested in intelligence studies[2] nor familiar with network analysis as a research method.[3] Conversely, recent research in terrorism or crime studies has discussed empirical, methodological as well as theoretical questions and problems in relation to uncovering hidden structures or analyzing network dynamics,[4] which are also relevant for intelligence studies. But, outside the humanities, the opposite holds true as network researchers have shown only modest interest in historical sources. There may be a rational explanation for this lack of interest. If empirical data is used just for algorithm testing, the use of materials requiring in-depth knowledge of source criticism and time-consuming preparation simply makes no sense.

The structure of this chapter is as follows. Section two discusses the research method and material used. Section three consists of analysis and visualizations reconstructing in a step-by-step manner the intelligence cycle of East German foreign intelligence for producing intelligence reports on Nordic affairs. The chapter will conclude with a critical assessment of the analysis research[5] and Digital Humanities.[6]

Method and material

Data analysis and visualization methodology

According to Michael Herman, intelligence is an information system serving central government and being "torn between its twin skills of collecting information and evaluating it",[7] carried out and realized in the intelligence cycle. Although the theories of the intelligence process might be helpful for providing general understanding of how intelligence services work, their systemic design is analytically less powerful when structure, dynamics and internal dependencies of actors within an intelligence system are of interest. Scholars holistically interested in the intelligence cycle are more likely to be interested in connection patterns, interactions, information flows and relationships during the information gathering, analysis and dissemination processes. Hence, scholars look for answers to questions like "which topics depend on which sources?", "how are sources connected to each other?" or "how does the intelligence flow through the institutions?" All these questions revolve around relationships and relational aspects, since they touch upon structural relationships between different units. Additionally, the answers to these questions can easily be translated into the language of (social) network analysis and, thus, be presented as a set of nodes (actors, institutions, objects etc.) linked together with edges, forming a network topology.[8]

In this chapter, historical network analysis is applied on intelligence data in order to explore and analyze the Nordic intelligence cycle of East German foreign intelligence. Historical network research is a methodological approach to apply social network analysis in history sciences and is based on an assumption of the importance of relationships among units (people, organizations, concepts etc.). It has been applied to map out, measure and visualize relationships between different units of a network. During recent years, network visualizations have gained a central position in network research, for two main reasons. First, new, easy-to-use visualization software is available, making data visualization much easier. And second, network visualizations are an effective form of presenting complex relationships in an intuitive and fairly easy-to-understand form. Additionally, different visualization layouts offer new possibilities for highlighting network-related attributes or visualizing a node's relevance with regard to its close neighborhood or the complete network.[9]

The network visualizations presented in this chapter are produced by Visone, an open-source software for network analysis and visualizations. Visone offers an extensive set of analytical tools and innovative network visualization integrated in an intuitive, easy-to-use user interface.[10]

Intelligence report data of East German foreign intelligence

The East German foreign intelligence (HV A) was formally a part of the East German State Security Service (*Ministerium für Staatssicherheit*, abbr. MfS, Stasi). Like all intelligence services, the HV A was responsible for running the complete intelligence cycle from information gathering to report dissemination, according to objectives and guidelines decided by the party and state leadership.[11] The main task of the HV A was to gather, evaluate and process technical, scientific, military and political information. It was also responsible for compiling reports on political, economic and strategic issues, disseminated not only to responsible state and party organizations inside the GDR but also to allied services inside the Soviet Empire, most importantly to the Soviet KGB. Put in theoretical terms, the HV A was a huge information system responsible for collecting, storing, analyzing and disseminating information.[12] The backbone of this information system formed an extensive network of "unofficial collaborators", the IMs (*Inoffizielle Mitarbeiter*), organized in human networks operating in all Western European countries. In 1989, approximately 189,000 IMs operated for the Stasi and about 15,000 of them for the HV A.[13]

In 1990, the archive of the HV A was almost completely destroyed. The remaining materials are maintained by The Agency of the Federal Commissioner for the Stasi Records (*Der Bundesbeauftragte für die Unterlagen des Staatssicherheitsdienstes der ehemaligen Deutschen Demokratischen Republik*, abbr. BStU). In total, the Agency maintains approximately 111 kilometers of archived files of which the share of the HV A is just 47 meters.[14] However, in 1998 experts of the BStU succeeded in decrypting an operational database system of the HV A called SIRA (*System der Informationsrecherche der HV A*). This system was instigated in the mid-1970s and used to maintain the intelligence cycle, administer undercover operations and, most importantly, to store a meta-data of information and reports.[15] Because SIRA was developed for the maintenance of daily information flows to and from the HV A, it opens a window into the HV A's daily operational work and provides scholars with operational data related to the intelligence cycle from the perspective of the HV A. The SIRA entries cover the years from 1969 to 1989, but the completeness of the stored information varies a great deal. In addition, SIRA records do not contain original documents and thus do not substitute the destroyed archival files.

The material used in this chapter consists of a selected corpus of dissemination (type "SA"[16]) records on Nordic affairs from 1975 to 1989. Although the Nordic countries were no central operational area for the HV A, the Nordic countries undoubtedly enjoyed certain importance for East German foreign intelligence, and the HV A also conducted many intelligence operations targeting the Nordic countries.[17] The year 1975 is widely regarded as a turning point in the GDR's history but also in European politics. The main reason for this assessment is the Conference on Security and Cooperation in Europe (CSCE) held in Helsinki in August 1975. During the second half of the 1970s, the party leadership in the GDR was increasingly concerned about the destabilizing impact of the CSCE on its power and consequently instructed the HV A to conduct continuous evaluations of the situation in Europe.[18] These developments were boosted by Mikhail Gorbachev's rise to power in the Soviet Union in 1985.[19]

The report corpus, used in the analysis, consists of 69 reports. Almost one third of them, 28 reports, were produced from 1975–77. Another peak in reporting, a total of 18 reports, were completed from 1984–86 (see Figure 3.1.1). The data preparation was carried out in several steps. First, all paper documents obtained from the BStU were scanned to image files. In the second step, these image documents were processed with optical character recognition (OCR) software, *Tesseract*. In the third step, a small Python[20] program was written to process the text files in order to recognize different fields, extract field contents and store the extracted data in a separate database.[21] In the last step, extracted data was carefully controlled for errors. For the objectives of this chapter, the following meta-data was extracted from the documents and stored in the database: 1) original

Figure 3.1.1 HV A reports on Finnish affairs 1975–89. Bars represent the annual number of reports; lines represent the total number of reports since 1975.

Source: Author's calculations based on material from BStU

date of the report (field [URSPRUNGSDATUM]), 2) content related keywords ([SACHVERHALT]), 3) country references ([LÄNDER-HINWEIS]), (4) references to objects (institutions, parties, universities etc.) and persons ([OBJEKT-HINWEIS], [PERSON-HINWEIS]), 5) gathered intelligence on which the report was based ([URSPRUNG]) and 6) receiving institutions for dissemination ([EMPFÄNGER]). In regard to point (5), the meta-data consists of cross-references to all SIRA input records exploited by HV A analysts for the report. Sources that delivered the "raw data" for the reports can, thus, be found in these SIRA records, making a re-construction of the intelligence-gathering network possible.

The intelligence cycle 1975–89

Information-gathering network

Compared to the general thematic structure of East German foreign intelligence gathering with a strong focus on scientific and technical intelligence,[22] Nordic intelligence had a rather different thematic structure dominated by political and military intelligence (see Table 3.1.1). Within the Nordic countries, Swedish intelligence seems to have been more focused on economic issues, whereas Norwegian intelligence revolved around politics and the military. Norway's status as the key NATO country in the European north explains the dominance of security political intelligence. We should, however, keep in mind that the principal actor in Nordic intelligence was not the HV A but the KGB. The East German foreign intelligence had a complementary role only, i.e., the HV A, like all other socialist intelligence services, were expected to support the KGB's activities.

The network responsible for gathering intelligence on Nordic affairs grew continuously between 1960 and 1989 (see Figure 3.1.2).[23] The most active phase was in the mid-1970s, reflecting the historical fact that from 1975 onwards the follow-up process of CSCE and West European integration were the two main concerns for the GDR party leadership, causing it to instruct its foreign intelligence service to gather relevant data and produce reports on

Table 3.1.1 Information flows from Finland, Denmark, Sweden and Norway (1969–89).

	Economy	*Politics and military*	*Intelligence services*	*Other topics*	*Total[a)]*
Finland	98 (7%)	905 (64%)	89 (6%)	313 (22%)	1,405 (100%)
Denmark	209 (9%)	1,370 (60%)	78 (3%)	633 (28%)	2,290 (100%)
Sweden	538 (18%)	1,478 (51%)	184 (6%)	719 (25%)	2,919 (100%)
Norway	130 (7%)	1,527 (81%)	70 (4%)	159 (8%)	1,886 (100%)
Total	975 (11%)	5,280 (62%)	421 (5%)	1,824 (22%)	8,500 (100%)

a) In order to equalize the rounding effect between calculated column percents, percents in the total column are set to 100.

Source: Elo and Müller-Enbergs, 2010, 38.

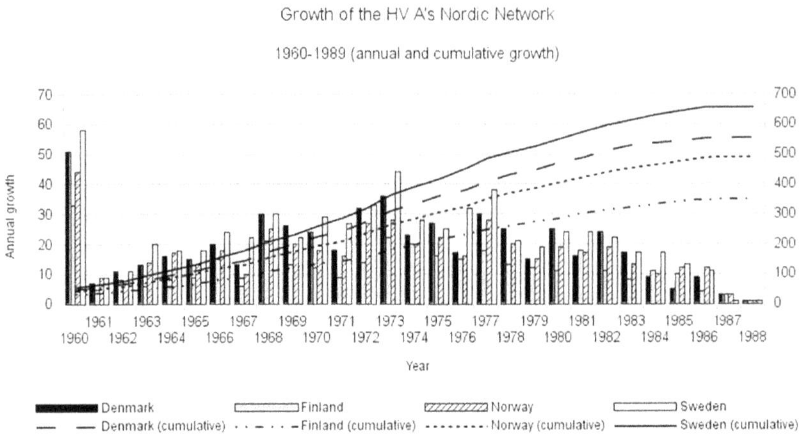

Figure 3.1.2 Annual and Cumulative Growth of the HV A's Nordic Network 1960–89.

these processes on a regular basis.[24] In total, 658 sources provided the HV A with intelligence related to Swedish affairs. With regard to Danish, Norwegian and Finnish affairs, the total amount of sources adds up to 558, 493 and 352, respectively. However, 78 percent of the sources delivered fewer than five intelligences, providing the impression that the bulk of intelligence on Nordic affairs had been a side-product of operations that had their main focus elsewhere.[25]

The network visualization in Figure 3.1.3 enables the formation of a number of observations that can improve understanding of the internal structure and dynamics of East German intelligence gathered in the European north. First, the graph highlights information flows to different regions. On the one hand, there are only a few strong flows within each cluster, i.e., intelligence gathering has been quite fragmented and heterogeneous, and each region has only had a couple of "productive" sources. On the other, there is also a handful of strong cross-regional flows, i.e., some sources have provided the HV A with intelligence on more than one Nordic country. It is worth noting that especially sources belonging to the "Finland" cluster seem to have delivered intelligence on at least one other country. A similar pattern can also be identified among some sources in the "Norway" cluster, having actively delivered on Swedish affairs as well. Another interesting issue is the "Nordic" sources, i.e., sources that have delivered intelligence on all four countries, as a Nordic region. It should, however, be kept in mind that a connection between a source and a region/country does *not* say anything about the nationality or geographical location of the source. Actually, both the existence of strong cross-regional ties and the "Nordic" sources might indicate that most of the intelligence on Nordic affairs was actually gathered *outside* the region, i.e., as a by-product of activities that had their focus outside the geographical area of the Nordic countries.

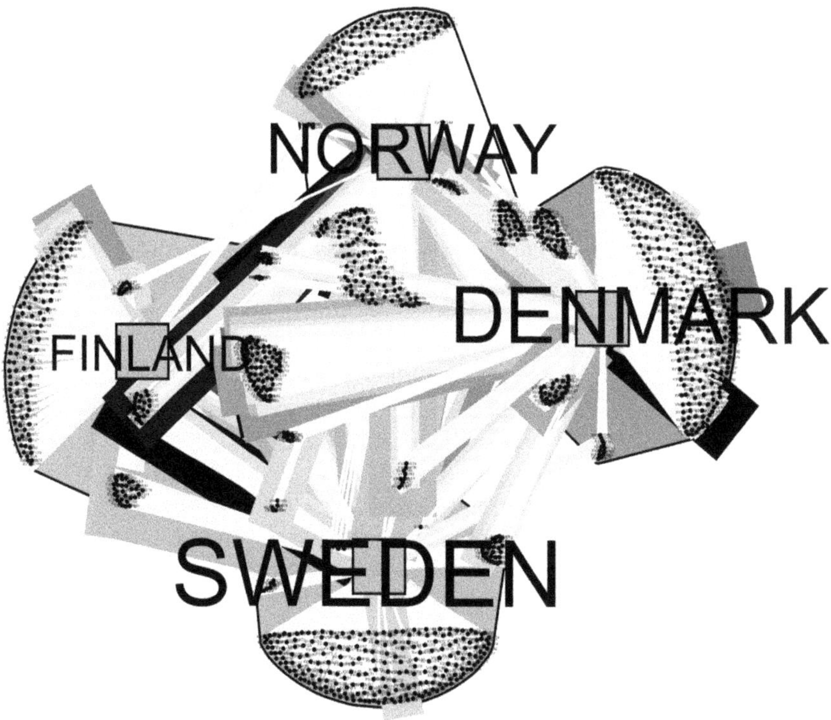

Figure 3.1.3 The Nordic source network of the HV A 1960–89. Nodes represent HV A sources and target regions. Ties represent country-related information flows from the HUMINT network (layout: stress minimizer with Louvain-grouped[26] groups. Node (label) size: proportional to in-degree (i.e., the number of inward directed ties from a given graph node). Link width: # of country-related intelligence delivered by a source. Link color: grayscale interpolation based on link weight).

The content structure of intelligence reports

From the perspective of text network analysis, concepts in a text network can be categorized into four different types depending on their degree and betweenness centrality. In network theory, centrality, in general, indicates a node's position in the network and can be calculated either relative to a node's direct neighbors or the whole network. A node's degree is the simplest centrality measure and equals to the number of connections the node has to other nodes. Betweenness, as the term itself indicates, defines centrality by analyzing where a node is placed within the network. Consequently, a node's betweenness centrality score is computed by taking into consideration the rest of the network and by looking at how many times a node sits on the shortest path linking two other nodes together, thus helping to identify nodes having "a high probability of occurring on a randomly chosen shortest path between two randomly chosen vertices".[27] Compared to eigenvector betweenness, which is quite reliant on the ties of the node's connections, betweenness centrality depends on the node's capability to act as a connection between two or more nodes otherwise disconnected. Considering meaning circulation across the entire network, the latter capability is assumed to be more relevant, and therefore we use betweenness centrality to measure a concept's status in the keyword network. As both point out, analyzing both degree and betweenness centrality is a proper method to identify concepts that play a central role in circulating meaning across the entire network. Concepts with high betweenness and degree centrality play a meaning circulation role, whereas concepts with low centrality measures are peripheral concepts. Between these two extremes are located concepts with high betweenness but low degree centrality and concepts with low betweenness but high degree centrality. The former play an important role as bridging concepts between local communities; the latter, in turn, are local hubs within a cluster.[28]

For the purpose of this chapter *central keywords* are defined as keywords linking several reports together, thus describing a content property typical for reports belonging to the same thematic cluster. These most central keywords should have the highest betweenness centrality and degree centrality values. Additionally, we are interested in changes and dynamics over time resulting in or from historical processes. In this chapter the main interest lies in the final phase of the Cold War (1985–89). In order to capture possible thematic changes in this final stage we have divided our data in snapshots consisting of reports before and after the key year 1985.

For the period between 1975 and 1984 (see Table 3.1.2); the three most central keywords are *domestic affairs* (INNENPOLITIK), *foreign affairs* (AUSZENPOLITIK), *CSCE process* (KSZE-PROZESZ) and *communist party* (KP). This content structure well correlates with the historical fact that, from the second half of the 1970s onward, the CSCE – but especially its political consequences for the Communist camp – began to dominate the GDR's political agenda. The party leadership in the GDR was increasingly worried about destabilizing impacts of the CSCE on its power and, consequently, instructed the HV A to continuous evaluation of the situation in Europe.[29]

A keyword worth being mentioned is *peace movement* (FRIEDENSBEWEGUNG). It has the fourth highest betweenness centrality value but a very low degree centrality value. Accordingly, this keyword is expected to bridge keywords between different local communities. A closer look at keywords co-occurring with FRIEDENSBEWEGUNG confirms this assumption. There are only three reports focusing on the peace movement. Two of these reports focus on dissidents and opposition movements and mention a relatively high number or dissidents and peace activists. The third report revolves around questions related to nuclear weapons and military and security policy. As the peace movement was ranked as opposition movement in the GDR, the Stasi sought to underpin its actions and, thus, followed its activities very closely.[30]

Table 3.1.2 TOP-10 keywords in Nordic reports: 1975–84 vs. 1985–89.

1975–84 (total # of reports: 194)		
Most connected (degree)	*Most influential (betweenness)*	*Most frequent (# of reports)*
INNENPOLITIK (148)	INNENPOLITIK (89,737.89)	MILITAERWESEN (95)
AUSZENPOLITIK (109)	AUSZENPOLITIK (54,884.84)	AUSZENPOLITIK (89)
EINSCHAETZUNG (77)	KP (52,114.13)	BEZIEHUNG (87)
KSZE-PROZESZ (74)	FRIEDENSBEWEGUNG (34,780.65)	STREITKRAEFTE (75)
KP (72)	STREITKRAEFTE (27,889.61)	ZUSAMMENARBEIT (73)
STREITKRAEFTE (67)	RUESTUNG (27,298.16)	EINSCHAETZUNG (63)
RUESTUNG (65)	KSZE-PROZESZ (23,940.75)	WIRTSCHAFT (62)
KSZE (64)	EINSCHAETZUNG (23,704.02)	FEINDTAETIGKEIT (54)
REGIERUNG (63)	FORSCHUNG (22,711.31)	OST-WEST-BEZIEHUNG (52)
RICHTLINIE (50)	REGIERUNG (19,465.19)	SL (52)
1985–89 (total # of reports: 74)		
Most connected (degree)	*Most influential (betweenness)*	*Most frequent (# of reports)*
AUSZENPOLITIK (46)	AUSZENPOLITIK (10,650.58)	AUSZENPOLITIK (22)
OST-WEST-BEZIEHUNG (20)	ROHSTOFF (3,568.62)	OST-WEST-BEZIEHUNG (21)
HALTUNG (19)	KIL (3,559.98)	WIRTSCHAFT (14)
REGIERUNG (18)	HALTUNG (3425,93)	NATOLAND (13)
STREITKRAEFTE (18)	REGIERUNG (2,993.30)	MILITAERWESEN (12)
NATOLAND (17)	OST-WEST-BEZIEHUNG (2,840.76)	HANDEL (11)
ROHSTOFF (15)	STREITKRAEFTE (2,143.50)	SICHERHEITSPOLITIK (11)
ENERGIEWIRTSCHAFT (13)	NATOLAND (2,125.94)	PARTEI (10)
KIL (13)	HANDEL (2,072.60)	ROHSTOFF (10)
MILITAERPOLITIK (13)	PARTEI (1,558.98)	STAATSAPPARAT (10)

Source: Elo and MÃŒller-Enbergs, 2010, 38.

For the period 1985 89 the most central keywords include *foreign policy, position/attitude* (HALTUNG) and *east-west relations* (OST-WEST-BEZIEHUNG). The CSCE process seems to lose its central importance, and the focus seems to shift toward European economic integration and to the development of the East-West relations. Further, the changes also imply that economic issues and questions related to energy production gained in importance. These changes also correlate well with historical developments: in the late 1980s bottlenecks in the GDR's energy sector worsened dramatically – as the Soviet Union cut its oil exports – but also due to increase in international oil price. Together with general lack of raw materials, the question of finding alternative concepts and technologies for energy production and consumption – energy-saving included – gained a high priority also in the HV A's activities. The party leadership expected the HV A to gather know-how and scientific-technical knowledge to a growing extent, an expectation the HV A sought to fulfill by intensifying its scientific-technical intelligence.[31]

Generally speaking the bulk of the reports analyzed for this chapter seem to deal either with security policy or military issues. This is not a big surprise, since both Norway and Denmark were members in NATO. At the same time, however, the results seem to give support for the conclusion that the Baltic Sea Region was primarily a security and/or military political question. This, in turn, is well in line with Michael Herman's conclusion that "the Cold War was substantially about military threats and military competition, and most Cold War intelligence in the West – perhaps also in the East – was military intelligence about military power".[32]

Put in the historical context, the changes visualized in the thematic structure of the reports can be read as a map of challenges the GDR was facing. In the second half of the 1970s, the CSCE clearly dominated the political agenda of the GDR. From the early 1980s onwards, economic issues and questions related to international politics became more significant. The developments in the second half of the 1980s clearly illustrate how East German foreign intelligence became increasingly focused on finding solutions to the GDR's domestic problems, mostly in the technological and economic domain. Since these problems are known to have played an important role in the demise of communism,[33] the HV A seems to have been aware of the danger yet not capable of producing any usable solutions.[34]

Dissemination network

The intelligence cycle is rounded up with report dissemination delivering "the required intelligence to the appropriate user in the proper form at the right time" (U.S. Marine Corps, 2003, 1–1). The dissemination network of the reports analyzed for this chapter supports the idea of targeted dissemination (Figure 3.1.4). The core of this dissemination network is marked with a red circle, located on the upper-left corner of the graph and contains twelve reports, as well as over 70 receivers. The main target group for dissemination is the political leadership of the GDR, including high party and politburo members. This core network is connected by two reports to another interesting sub-network located on the lower-left corner of the graph and having Ministry for Foreign Affairs of the GDR (MFAA) as its hub. Altogether, 16 documents are disseminated within this sub-network, all of them to the Ministry for Foreign Affairs of the GDR. Another dense dissemination sub-network can be

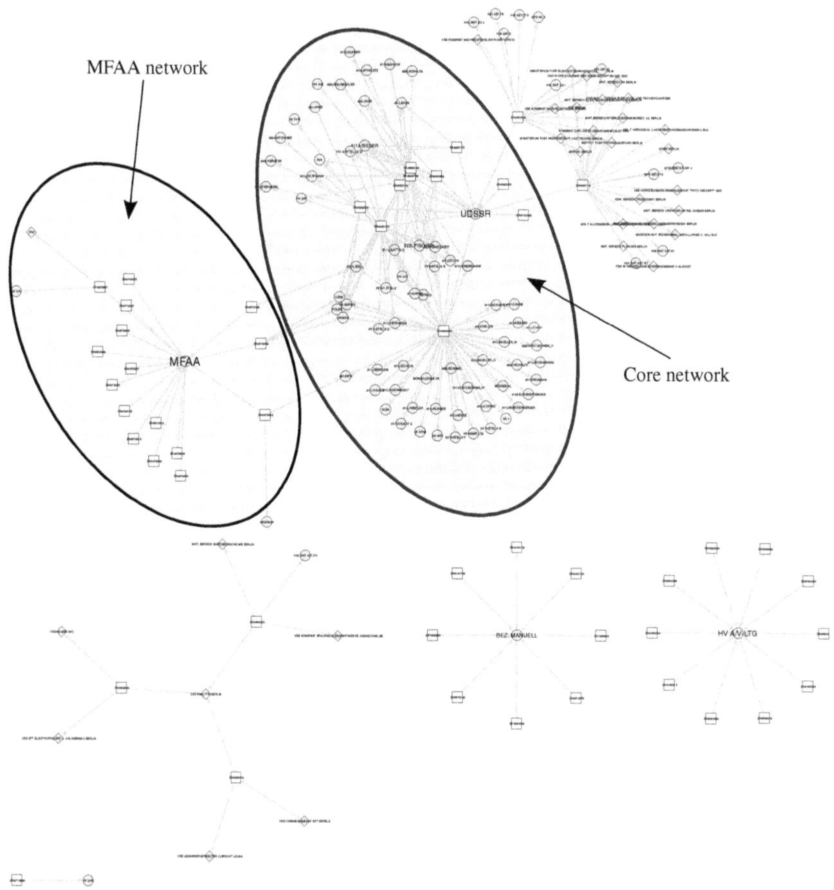

Figure 3.1.4 Report dissemination network (layout: spring embedder. Node label size: proportional to in-degree. Node color: white=report, gray=dissemination target).

found on the lower-mid area of the graph and contains mainly technical and scientific institutions and organs of the GDR. A closer look at the keywords used in these reports include technical and scientific concepts – sensor technology, radio technology or mobile communication, i.e., results from the HV A's technical and scientific espionage. The dissemination sub-network on the upper-right corner of the graph revolves around reports with focus on energy technology and electricity.

In targeted dissemination receivers should get only reports relevant for them. Against this background dissemination sub-networks can indicate thematic differences in reports. The two sub-networks in Figure 3.1.5 give support for this idea. The reports disseminated to the party and state leadership are clearly focused on issues related to European security and co-operation as well as

West-European (economic) integration (see sub-figure (a) in Figure 3.1.5). The core content of the reports disseminated to the Ministry for Foreign Affairs of the GDR was, instead, related to the official visit of the Secretary General Erich Honecker to Finland in 1984. These reports also dealt with the West-European integration and the CSCE, but these topics were not so central as in reports delivered to the party and state leadership (see sub-figure (b) in Figure 3.1.5).

In summary, the analysis of the dissemination network supports the hypothesis that a correlation exists between the content of a report and its dissemination network, seeking to ensure that intelligence reports are delivered to appropriate users and the dissemination system is not overloaded by unneeded or irrelevant information. Additionally, the results of the analysis indicate that the HV A seem to be able to produce reports that tackle actual questions and issues and, thereby, respond to information requests coming from the state leadership. At the same time the limited amount of documents focusing on the European north indicate the relatively marginal role of this region for the HV A.

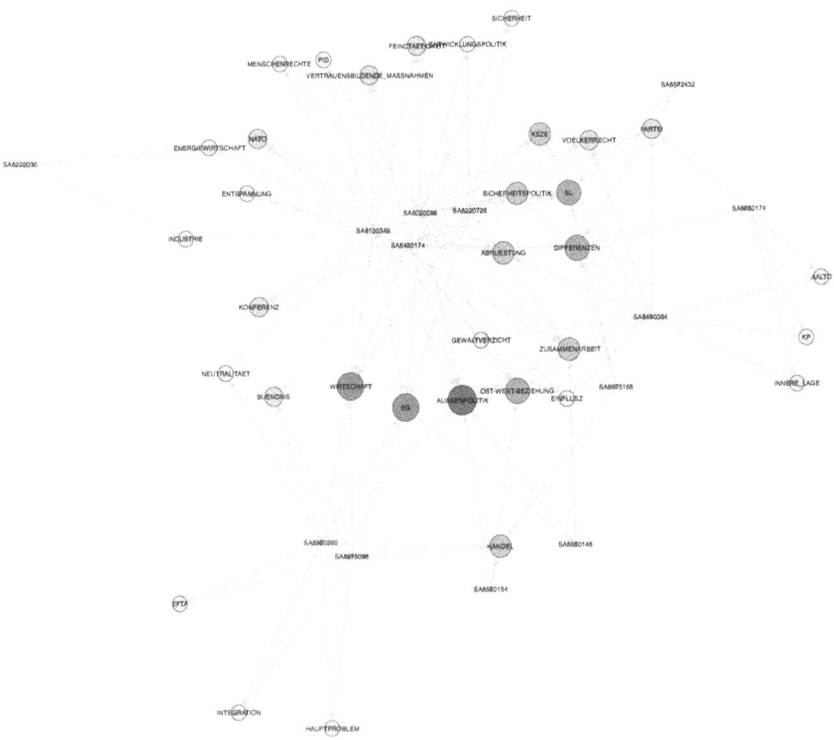

Figure 3.1.5 Keyword-to-report network of (a) the core and (b) MFAA report dissemination networks. Keywords mentioned only in one report excluded. (Layout: spring embedder. Node (label) size: proportional to in-degree. Node color: grayscale interpolation based on in-degree).

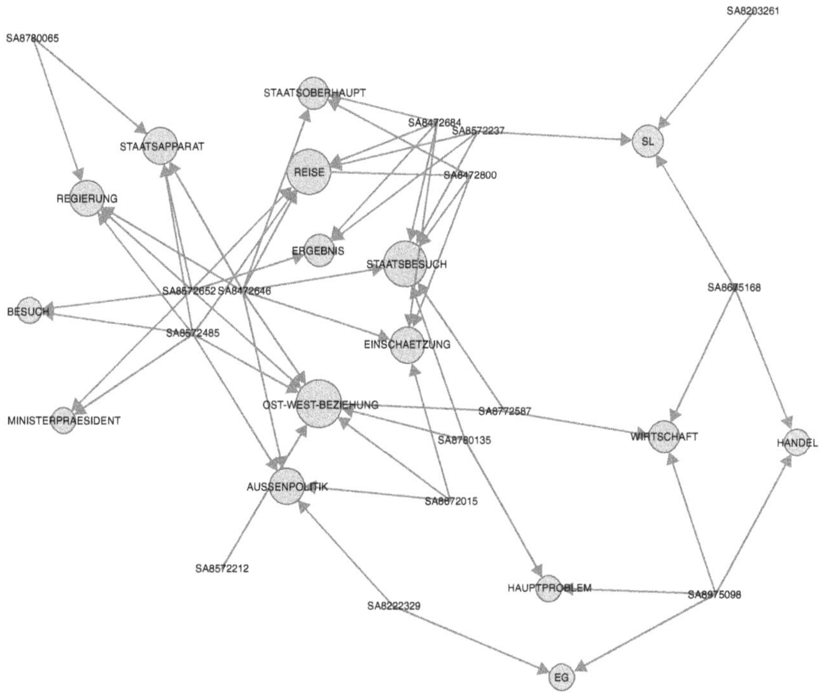

Figure 3.1.5 (*Continued*)

Conclusion

This chapter sought to exemplify how historical network analysis could be utilized on meta-data to explore, re-construct and analyze the East German foreign intelligence cycle on Nordic affairs. The empirical results give at least modest support for the assumption that the HV A succeeded in gathering relevant information focusing on changes in the Baltic Sea region. Although the focus of the HV A's reporting on Nordic affairs and the Baltic Sea region remained on military issues and security policy, problems regarding economic performance and scientific-technical questions and problems gained importance after 1985. However, no clear evidence could be found to support the hypothesis that the HV A was able to foresee the tectonic political changes in the Baltic Sea region (or wider in Europe) finally resulting in the collapse of the bi-polar world order in 1989/90. The increasing importance of East-West relations in the reports, however, seems to indicate a growing awareness about negative consequences of the economic and political competition in Europe.

Regarding methodological aspects, this chapter sought to evidence the usefulness of historical network analysis when it comes to revealing, analyzing and visualizing hidden networks typical for the study of intelligence history. In this sense the results are positive and encouraging. Rather a lot, however, depends on the data available, forcing me to round up this chapter with a word of caution. As pointed out, the material base is, especially in regard to the last years of the 1980s, rather fragmented and biased. The missing data cannot be easily compensated, and therefore there is no easy way to reconstruct what the Stasi really knew and where its blind spots were. Hence, a careful reflection against previous research and a solid contextual knowledge is required to avoid falling in structural holes in the reconstructed networks.

Notes

1 Michael Hermann, *Intelligence Services in the Information Age* (London: Frank Cass, 2001), p. 79; James B. Bruce and Roger Z. George, "Intelligence Analysis: The Emergence of a Discipline," in *Analyzing Intelligence: Origins, Obstacles, and Innovations*, ed. James B. Bruce and Roger Z. George, pp. 1–15 (Washington, DC: Georgetown University Press, 2008), p. 2; Patrick F. Walsh, *Intelligence and Intelligence Analysis* (New York, Abingdon: Routledge, 2011).

2 Raymond L. Garthoff, "Foreign Intelligence and the Historiography of the Cold War," *Journal of Cold War Studies* 6, no. 2 (2004): p. 21.

3 Claire Lemercier, "Formal Network Methods in History: Why and How?," in *Social Networks, Political Institutions, and Rural Societies*, ed. Georg Fertig, pp. 281–310 (Turnhout: Brepols Publishers, 2015), http://halshs.archives-ouvertes.fr/halshs-00521527/fr/ (accessed 13 December 2018). Although network analysis has been used in a wide range of historical case studies (an up-to-date bibliography of historical network research can be found at: www.zotero.org/groups/209983/historical_network_research? (accessed 13 December 2018)), to our knowledge it has not been applied to historical intelligence studies.

4 E.g. Vladis E. Krebs, "Uncloaking Terrorist Networks," *First Monday* 7, no. 4 (2002), http://firstmonday.org/ojs/index.php/fm/article/view/941/863 (accessed 13 December 2018); Jörg Raab and H.B. Milward, "Dark Networks as Problems," *Journal of Public Administration Research and Theory* 13, no. 4 (2003): pp. 413–39; Jennifer Xu and Hsinchun Chen, "Criminal Network Analysis and Visualization," *Communications of the ACM* 48, no. 6 (2005): pp. 100–7; Walter Enders and Xuejuan Su, "Rational Terrorists and Optimal Network Structure," *Journal of Conflict Resolution* 51, no. 1 (2007): pp. 33–57; Daniel Schwartz and Tony Rouselle, "Using Social Network Analysis to Target Criminal Networks," *Trends in Organized Crime* 12, no. 2 (2009): pp. 188–207; Christopher E. Hutchins and Marge Benham-Hutchin, "Hiding in Plain Sight: Criminal Network Analysis," *Computational and Mathematical Organization Theory* 16, no. 1 (2010): pp. 89–111; Aili Malm and Gisela Bichler, "Networks of Collaborating Criminals: Assessing the Structural Vulnerability of Drug Markets," *Journal of Research in Crime and Delinquency* 48, no. 2 (2011): pp. 271–97; Carlo Morselli, "Assessing Vulnerable and Strategic Positions in a Criminal Network," *Journal of Contemporary Criminal Justice* 26, no. 4 (2010): pp. 382–92.

5 Historical network research is a research paradigm seeking to apply network research and analysis in the historical disciplines. The main interest of historical network research lies in analyzing relational patterns like social or institutional relations or interactions in historical context. For a more detailed discussion about historical network research, see e.g. http://historicalnetworkresearch.org/ (accessed 13 December 2018).

6 Digital Humanities is an emerging research area seeking to foster the application of methods, tools and algorithms originating from computational sciences in the

Humanities. Since the whole discipline is subject to continuous, often controversial debates, there is no room for an exhaustive debate about and around Digital Humanities in this chapter. Hence, the following working definition should be sufficient for the purposes of this chapter: "The particular contribution of the digital humanities, however, lies in its exploration of the difference that the digital can make to the kinds of work that we do as well as to the ways that we communicate with one another. These new modes of scholarship and communication will best flourish if they, like the digital humanities, are allowed to remain plural" Fitzpatrick, Kathleen, "The Humanities, Done Digitally," http://dhdebates.gc.cuny.edu/debates/text/30 (accessed 13 December 2018). A reader interested in debates in the Digital Humanities could find this digital publication http://dhdebates.gc.cuny.edu/debates/ (accessed 13 December 2018) as a good starting point for his/her journey into this very interesting field.

7 Hermann, *Intelligence Sevices in the Information Age*, pp. 3–4.

8 See also Malcolm K. Sparrow, "The Application of Network Analysis to Criminal Intelligence: An Assessment of the Prospects," *Social Networks* 13, no. 3 (1991): p. 252; Christopher Durugbo et al., "Modelling Collaboration Using Complex Networks," *Information Sciences* 181, no. 15 (2011): p. 3144; M. Zacarias et al., "A 'Context-Aware' and Agent-Centric Perspective for the Alignment between Individuals and Organizations," *Information Systems* 35, no. 4 (2010): p. 451.

9 See further e.g. J.P. Scott, *Social Network Analysis*, 3rd ed. (London: Sage Publishing, 2013); Barbara Schultz-Jones, "Examining Information Behavior through Social Networks: An Interdisciplinary Review," *Journal of Documentation* 66, no. 4 (2009): pp. 592–631; Marten Düring and Martin Stark, "Historical Network Analysis," in *Encyclopedia of Social Networks*, ed. George A. Barnett (London: Sage Publishing, 2011).

10 See further the homepage of Visone: http://visone.info/html/about.html (accessed 13 December 2018).

11 See Helmut Müller-Enbergs, *Inoffizielle Mitarbeiter des Ministeriums für Staatssicherheit, Teil 2: Anleitungen für die Arbeit mit Agenten, Kundschaftern und Spionen in der Bundesrepublik Deutschland*, 2nd ed. (Berlin: Ch. Links Verlag, 1998), pp. 40–1; Helmut Müller-Enbergs, *Die inoffiziellen Mitarbeiter (MfS-Handbuch)* (Berlin: BStU, 2008), p. 5.

12 On information systems, see e.g. David E. Avison and Michael D. Myers, "Information Systems and Anthropology: And Anthropological Perspective on IT and Organizational Culture," *Information Technology & People* 8, no. 3 (1995): pp. 43–56; Steven Alter, "Defining Information Systems as Work Systems: Implications for the IS Field," *European Journal of Information Systems* 17, no. 5 (2008): pp. 448–69; Timo Hyvönen, Janne Järvinen and Jukka Pellinen, "A Virtual Integration: The Management Control System in a Multinational Enterprise," *Management Accounting Research* 19, no. 1 (2008): pp. 45–61.

13 Helmut Müller-Enbergs, *Hauptverwaltung A (HV A): Aufgaben – Strukturen – Quellen*, MfS-Handbuch (Berlin: BStU, 2011), p. 21.

14 Ibid., pp. 11–12.

15 SIRA was built as a relational database and consists of four main tables (sub-databases) numbered from 11 to 14. Each sub-database is dedicated to a specific domain of intelligence: "Scientific and technical espionage" (sub-database #11), "Problems and operations related to domestic and foreign policies, economy and military politics outside the GDR" (sub-database #12), "Political relations in the operation area" (sub-database #13) and "Counter-intelligence" (sub-database #14). Additionally, administrative information of operations – e.g. supervisor changes, opening of new files – are stored in the sub-database #21. (Stephan Konopatzky, "Möglichkeiten und Grenzen der SIRA-Datenbank," in *Das Gesicht dem Westen zu. DDR-Spionage gegen die Bundesrepublik Deutschland*, ed. Georg Herbstritt and

Helmut Müller-Enbergs, pp. 112–32 (Bremen: Edition Temmen, 2003); Helmut Müller-Enbergs, *"Rosenholz" Eine Quellkritik*, with the assistance of Sabine Fiebig et al., BF informiert 28 (Berlin: Die Bundesbeauftragte für die Unterlagen des Staatssicherheitsdienstes der ehemaligen Deutschen Demokratischen Republik, 2007), p. 13ff.) Researchers cannot directly access the database, but SIRA queries are carried out by BStU according to search criteria defined by the researcher. Since the database is administered in a SQL-based system, complex multi-criteria queries are possible. The results are available in printed form only, and BStU charges a small per-page fee (currently ten cents/page).

16 Each SIRA record has an unique ID starting a two-character string ("SE", "SA" or "SB") describing the information type, followed with two digits identifying the recording year and five digits from the database counter. Records marked with "SE" (*SIRA Eingang*) are input records, i.e., meta-data of intelligence gathered by the HV A. Records marked with "SA" (*SIRA Ausgang*) contain the meta-data of disseminated material (reports, evaluations etc.) the HV A has disseminated to external partners. Finally, records marked with "SB" (*SIRA Bestellung*) are records storing meta-data for intelligence requests from outside. As an example, a SIRA record with the ID "SA7503201" is a dissemination record (type: SA7503201) stored in 1975 (SA7503201).

17 For a good summary, see Thomas Wegener Friis, Kristie Macrakis and Helmut Müller-Enbergs, eds., *East German Foreign Intelligence: Myth, Reality and Controversy* (London, New York: Routledge, 2010).

18 E.g. Klaus Schroeder, *Der SED-Staat. Geschichte und Strukturen der DDR* (München: Bayrische Landeszentrale für politische Bildungsarbeit, 1998), p. 233ff; Jens Gieseke, "East German Espionage in the Era of Detente," *Journal of Strategic Studies* 31, no. 3 (2008): pp. 395–424.

19 Archie Brown, *The Gorbachev Factor* (Oxford: Oxford University Press, 1996); John Lewis Gaddis, *The Cold War: A New History* (New York: The Penguin Press, 2005), p. 229ff.

20 Python enjoys relatively high popularity among Digital Humanists. Python, as a relatively smooth learning curve, makes processing easy to learn, even for researchers with no previous experience with programming languages. Additionally, Python offers powerful and robust tools for text processing. See also: www.python.org/ (accessed 13 December 2018).

21 For text file processing with Python, see e.g. http://pythonnotes.curiousefficiency.org/en/latest/python3/text_file_processing.html (accessed 13 December 2018).

22 Müller-Enbergs, *Hauptverwaltung A (HV A)*, p. 20.

23 Annual growth calculated based on the year of the recruitment.

24 Ibid., p. 134.

25 See also Kimmo Elo and Helmut Müller-Enbergs, "Suomen merkitys DDR: n ulkomaantiedustelun kohteena," *Kosmopolis* 40, no. 4 (2010): p. 44.

26 See Vincent D. Blondel et al., "Fast Unfolding of Communities in Large Networks," *Journal of Statistical Mechanics: Theory and Experiment* (2008), https://arxiv.org/pdf/0803.0476v2.pdf (accessed 13 December 2018). Visione has built-in support for Louvain grouping.

27 Yi-Yu Hsu and Hung-Yu Kao, "Coin: A Network Analysis for Document Triage," *Database* (2013): pp. 1–11. See also Christina Prell, *Social Network Analysis: History, Theory and Methodology* (London: Sage Publications, 2012), pp. 103–4.

28 Junseop Shim, Chisung Park and Mark Wilding, "Identifying Policy Frames through Semantic Network Analysis: An Examination of Nuclear Energy Policy across Six Countries," *Policy Sciences* 48, no. 1 (2015): p. 59f.

29 E.g. Schroeder, *Der SED-Staat. Geschichte und Strukturen der DDR*, p. 233ff; Gieseke, "East German Espionage in the Era of Detente," pp. 395–424.

30 Ehrhart Neubert, *Geschichte der Opposition in der DDR 1949–1989* (Berlin: Ch. Links Verlag, 1998); Kimmo Elo, "The Content Structure of Intelligence Reports," *Connections* 35, no. 1 (2015), doi:10.17266/35.1.2: pp. 20–8.
31 Jens Gieseke, *Mielke-Konzern: Die Geschichte der Stasi 1945–1990* (Stuttgart, München: dva, 2001), pp. 210–14; Elo, "The Content Structure of Intelligence Reports," pp. 20–8.
32 Michael Herman, cited in: John N.L. Morrison, "Intelligence in the Cold War," *Cold War History* 14, no. 4 (2014), doi:10.1080/14682745.2014.950248: p. 576.
33 E.g. Celeste A. Wallander, "Western Policy and the Demise of the Soviet Union," *Journal of Cold War Studies* 5, no. 4 (2003): pp. 137–77; Gaddis, *The Cold War: A New History*; Sascha Rafalzik, *Wirtschaftsspionage der DDR* (Münster: LIT, 2010) Gaddis, *The Cold War: A New History*; Rafalzik, *Wirtschaftsspionage der DDR*.
34 See also Helmut Müller-Enbergs, "Political Intelligence: Foci and Sources, 1969–1989," in *East German Foreign Intelligence: Myth, Reality and Controversy*, ed. Thomas W. Friis, Kristie Macrakis and Helmut Müller-Enbergs, p. 111 (London, New York: Routledge, 2010).

Bibliography

Alter, Steven. "Defining Information Systems as Work Systems: Implications for the IS Field." *European Journal of Information Systems* 17, no. 5 (2008): pp. 448–69.

Avison, David E., and Michael D. Myers. "Information Systems and Anthropology: An Anthropological Perspective on IT and Organizational Culture." *Information Technology & People* 8, no. 3 (1995): pp. 43–56.

Barnett, George A., ed. *Encyclopedia of Social Networks*. London: Sage Publishing, 2011.

Blondel, Vincent D., Jean-Loup Guillaume, Renaud Lambiotte, and Etienne Lefebvre. "Fast Unfolding of Communities in Large Networks." *Journal of Statistical Mechanics: Theory and Experiment* (2008). https://arxiv.org/pdf/0803.0476v2.pdf (accessed 13 December 2018).

Brown, Archie. *The Gorbachev Factor*. Oxford: Oxford University Press, 1996.

Bruce, James B., and Roger Z. George, eds. *Analyzing Intelligence: Origins, Obstacles, and Innovations*. Washington, DC: Georgetown University Press, 2008.

———. "Intelligence Analysis: The Emergence of a Discipline." In *Analyzing Intelligence: Origins, Obstacles, and Innovations*. Edited by James B. Bruce and Roger Z. George, pp. 1–15. Washington, DC: Georgetown University Press, 2008.

Düring, Marten, and Martin Stark. "Historical Network Analysis." In *Encyclopedia of Social Networks*. Edited by George A. Barnett. London: Sage Publishing, 2011.

Durugbo, Christopher, Windo Hutabarat, Ashutosh Tiwari, and Jeffrey R. Alcock. "Modelling Collaboration Using Complex Networks." *Information Sciences* 181, no. 15 (2011): pp. 3143–61.

Elo, Kimmo. "The Content Structure of Intelligence Reports." *Connections* 35, no. 1 (2015): pp. 20–8. doi:10.17266/35.1.2.

Elo, Kimmo, and Helmut Müller-Enbergs. "Suomen merkitys DDR: n ulkomaantiedustelun kohteena." *Kosmopolis* 40, no. 4 (2010): pp. 31–47.

Enders, Walter, and Xuejuan Su. "Rational Terrorists and Optimal Network Structure." *Journal of Conflict Resolution* 51, no. 1 (2007): pp. 33–57.

Friis, Thomas Wegener, Kristie Macrakis, and Helmut Müller-Enbergs, eds. *East German Foreign Intelligence: Myth, Reality and Controversy*. London, New York: Routledge, 2010.

Gaddis, John Lewis. *The Cold War: A New History*. New York: The Penguin Press, 2005.

Garthoff, Raymond L. "Foreign Intelligence and the Historiography of the Cold War." *Journal of Cold War Studies* 6, no. 2 (2004): pp. 21–56.

Gieseke, Jens. *Mielke-Konzern: Die Geschichte der Stasi 1945–1990*. Stuttgart, München: dva, 2001.

———. "East German Espionage in the Era of Detente." *Journal of Strategic Studies* 31, no. 3 (2008): pp. 395–424.

Herbstritt, Georg, and Helmut Müller-Enbergs, eds. *Das Gesicht dem Westen zu. DDR-Spionage gegen die Bundesrepublik Deutschland*. Bremen: Edition Temmen, 2003.

Hermann, Michael. *Intelligence Services in the Information Age*. London: Frank Cass, 2001.

Hsu, Yi-Yu, and Hung-Yu Kao. "Coin: A Network Analysis for Document Triage." *Database* (2013): pp. 1–11.

Hutchins, Christopher E., and Marge Benham-Hutchin. "Hiding in Plain Sight: Criminal Network Analysis." *Computational and Mathematical Organization Theory* 16, no. 1 (2010): pp. 89–111.

Hyvönen, Timo, Janne Järvinen, and Jukka Pellinen. "A Virtual Integration: The Management Control System in a Multinational Enterprise." *Management Accounting Research* 19, no. 1 (2008): pp. 45–61.

Konopatzky, Stephan. "Möglichkeiten und Grenzen der SIRA-Datenbank." In *Das Gesicht dem Westen zu. DDR-Spionage gegen die Bundesrepublik Deutschland*. Edited by Georg Herbstritt and Helmut Müller-Enbergs, pp. 112–32. Bremen: Edition Temmen, 2003.

Krebs, Vladis E. "Uncloaking Terrorist Networks." *First Monday* 7, no. 4 (2002). http://firstmonday.org/ojs/index.php/fm/article/view/941/863 (accessed 13 December 2018).

Lemercier, Claire. "Formal Network Methods in History: Why and How?" In *Social Networks, Political Institutions, and Rural Societies*, pp. 281–310. Edited by Georg Fertig. Turnhout: Brepols Publishers, 2015. https://halshs.archives-ouvertes.fr/halshs-00521527v2/document (accessed 13 December 2018).

Malm, Aili, and Gisela Bichler. "Networks of Collaborating Criminals: Assessing the Structural Vulnerability of Drug Markets." *Journal of Research in Crime and Delinquency* 48, no. 2 (2011): pp. 271–97.

Morrison, John N.L. "Intelligence in the Cold War." *Cold War History* 14, no. 4 (2014): pp. 575–91. doi:10.1080/14682745.2014.950248.

Morselli, Carlo. "Assessing Vulnerable and Strategic Positions in a Criminal Network." *Journal of Contemporary Criminal Justice* 26, no. 4 (2010): pp. 382–92.

Müller-Enbergs, Helmut. *Inoffizielle Mitarbeiter des Ministeriums für Staatssicherheit, Teil 2: Anleitungen für die Arbeit mit Agenten, Kundschaftern und Spionen in der Bundesrepublik Deutschland*. 2nd ed. Berlin: Ch. Links Verlag, 1998.

———. *"Rosenholz" Eine Quellkritik*. With the assistance of Sabine Fiebig et al. BF informiert 28. Berlin: Die Bundesbeauftragte für die Unterlagen des Staatssicherheitsdienstes der ehemaligen Deutschen Demokratischen Republik, 2007.

———. *Die inoffiziellen Mitarbeiter (MfS-Handbuch)*. Berlin: BStU, 2008.

———. "Political Intelligence: Foci and Sources, 1969–1989." In *East German Foreign Intelligence: Myth, Reality and Controversy*. Edited by Thomas W. Friis, Kristie Macrakis and Helmut Müller-Enbergs, pp. 91–112. London, New York: Routledge, 2010.

———. *Hauptverwaltung A (HV A): Aufgaben – Strukturen – Quellen*. MfS-Handbuch. Berlin: BStU, 2011.

Musiał, Kazimierz. "Reconstructing Nordic Significance in Europe on the Threshold of the 21st Century." *Scandinavian Journal of History* 34, no. 3 (2009): pp. 286–306.

Neubert, Ehrhart. *Geschichte der Opposition in der DDR 1949–1989*. Berlin: Ch. Links Verlag, 1998.

Prell, Christina. *Social Network Analysis: History, Theory and Methodology*. London: Sage Publications, 2012.

Raab, Jörg, and H. Brinton Milward. "Dark Networks as Problems." *Journal of Public Administration Research and Theory* 13, no. 4 (2003): pp. 413–39.

Rafalzik, Sascha. *Wirtschaftsspionage der DDR*. Münster: LIT, 2010.

Schroeder, Klaus. *Der SED-Staat. Geschichte und Strukturen der DDR*. München: Bayrische Landeszentrale für politische Bildungsarbeit, 1998.

Schultz-Jones, Barbara. "Examining Information Behavior through Social Networks: An Interdisciplinary Review." *Journal of Documentation* 66, no. 4 (2009): pp. 592–631.

Schwartz, Daniel, and Tony Rouselle. "Using Social Network Analysis to Target Criminal Networks." *Trends in Organized Crime* 12, no. 2 (2009): pp. 188–207.

Scott, J.P. *Social Network Analysis*. 3rd ed. London: Sage Publishing, 2013.

Shim, Junseop, Chisung Park, and Mark Wilding. "Identifying Policy Frames through Semantic Network Analysis: An Examination of Nuclear Energy Policy across Six Countries." *Policy Sciences* 48, no. 1 (2015): pp. 51–83.

Sparrow, Malcolm K. "The Application of Network Analysis to Criminal Intelligence: An Assessment of the Prospects." *Social Networks* 13, no. 3 (1991): pp. 251–74.

U.S. Marine Corps. "Marine Air Ground Task Force Intelligence Dissemination: Tech. Rep. MCRP 2–1C." (2003). https://fas.org/irp/doddir/usmc/mcrp2-1c.pdf (accessed 13 December 2018).

Wallander, Celeste A. "Western Policy and the Demise of the Soviet Union." *Journal of Cold War Studies* 5, no. 4 (2003): pp. 137–77.

Walsh, Patrick F. *Intelligence and Intelligence Analysis*. New York, Abingdon: Routledge, 2011.

Xu, Jennifer, and Hsinchun Chen. "Criminal Network Analysis and Visualization." *Communications of the ACM* 48, no. 6 (2005): pp. 100–7.

Zacarias, M., H. Pinto, R. Magalhães, and R. Tribolet. "A 'Context-Aware' and Agent-Centric Perspective for the Alignment between Individuals and Organizations." *Information Systems* 35, no. 4 (2010): pp. 441–66.

3.2 Social and semantic network analysis in the study of religions

Frederik Elwert

In the study of religions, relational approaches have recently been gaining attention. As the idea of (world) religions as static, a-historical entities is increasingly contested, focus moves towards the dynamic relations that define religious traditions in the first place. Applied consequentially, a relational view of religious history is not only suitable for the study of exchange between religious traditions. It also enables scholars to study the very process that creates what is perceived as a religious tradition. This allows for an empirical – in contrast to a taxonomical – approach to the study of religions.

Network analysis is a promising tool for operationalizing these conceptual ideas. Religious interaction can then be modelled as a network of social actors as well as of the religious ideas they exchange. Methodologically, this poses some challenges. a) Many approaches in historical network analysis focus on the study of historical social networks. The history of religions, however, requires a special focus on the semantic aspects inherent in networks. b) An empirical bottom-up approach to the history of religions requires the development of categories inductively from the material and not the classification of the phenomena using theoretically derived taxonomies, e.g. for types of nodes or relations. c) The source material under study in the history of religions often spans different genres, including large corpora of religious literature. This requires an approach that bridges historical and literary studies and that allows work with large collections.

This chapter discusses different approaches for the identification of social and semantic structures in texts. Instead of manual annotation, techniques adopted from computational linguistics can be applied to identify text structures and build networks. This allows for the automated extraction of relational data, yet the results depend heavily on the algorithms used. Any interpretation of the resulting networks has therefore to consider the implications of methodological choices made during their creation.

Introduction[1]

Religion as a concept is particularly hard to grasp. Postcolonial critique of the study of religions as an academic discipline has highlighted the concept's own troubled history. This has led to serious doubts about the usefulness of the concept of religion as an analytical category.[2] This chapter explores how a relational perspective can be a useful tool in the study of religions on a conceptual as well as empirical level. Its value lies in taking the category's entanglement and dependencies into account, without dismissing it completely. In a second step, the operationalisation of this general framework with the help of network analysis will be discussed. One of the challenges when applying network analysis to the history of religions is the interplay of social structure and content: while the application of social network analysis to historical processes enables the study of the dimension of social structure, it falls short in examining the role of religious ideas. Three ways in which religious semantics can be integrated into historical network analysis will be examined.

Beyond the box: relational approaches in the study of religions

In the study of religions, postcolonial critique has had significant impact on the very concepts of religion and religions. The main critique follows two lines: first, the singular "religion" suggests a concept, even a phenomenon, that is relatively stable across time and space. This assumption makes comparison possible between certain beliefs and practices in e.g. Imperial China, pre-Columbian Meso-America and Early Modern Europe. Among others, Talal Asad has challenged this notion, arguing that the contemporary idea of religion is rooted in a very specific historical and geographical context, namely nineteenth-century Europe.[3] The "discovery" of religion in other times and spaces can then be better understood as an invention or projection of European scholars.

Second, the notion of "religions" (plural) suggests a set of distinct but comparable entities. The operation of comparison is the very heart of the discipline called "comparative religion". This idea of distinct religions, also often called "world religions", is also the most recent state in a historical process that created these entities in the first place. In "The Invention of World Religions", Tomoko Masuzawa[4] shows in a detailed analysis how the European notion of multiple "religions" – in contrast to a single phenomenon called "religion" – came about and how the notion of what was supposed to be a religion in this sense changed drastically. In a similar fashion, David Chidester[5] demonstrates, in "Savage Systems", how the process of defining separable entities that form the object of comparison was also deeply intertwined with colonial practices.

Both refer to the alteration of an older classification scheme that distinguishes Christianity, Judaism, Islam and Paganism as distinct religions from a more recent one that recognises which religions are still defined as world religions, adding Buddhism, Hinduism and Confucianism to the set.[6] Chidester stresses that this is not just a process of gaining more knowledge that allows for the differentiation of the vague category of Paganism, but that this corresponds to a practice of classification. Through taxonomical procedures like genealogy and morphology, colonial contact with native peoples led to the creation of taxonomies of distinct religions.[7] This is not limited to the observation that the notion

changes of what the entities were that one would compare. Even more severe is the claim that the very idea of defining separable entities is itself a construct. Masuzawa refers to this idea as the "box model" of world religions: clear cut borders are assumed to divide monolithic entities that can be described by a set of defining characteristics, such as founders, beliefs or rituals.[8] Historically, the World Parliament of Religions at the World Exhibition in Chicago in 1893 is often referred to as the crystallisation point of this idea: representatives of all "World Religions" were thought to speak in the name of their religion as a whole.[9] This model is extremely powerful and is still the way religion is presented in school books, popular lexicons and even Religious Studies curricula. But this way of thinking about religions stresses the differences between religions so that they appear as clear-cut borders, and it neglects, to a large extent, the differences within the various traditions.

Now while some scholars have argued that this background renders "religion" useless as a scholarly concept,[10] we can observe that these concepts, as constructed they may be, have been powerful in history and remain so today. Thus one major challenge in our field is to study the history of religions without falling into the trap of ethnocentrism, i.e., applying a European concept to cultures that did or do not have a notion of religion. One approach for dealing with this is the use of empirical studies of the processes through which religious traditions emerge in the first place and define their borders. Using the example of Christianity and Judaism in late antiquity, Daniel Boyarin[11] has shown how the very category of religion came into existence and how it was then used to define and enforce borders between groups. He suggests borrowing ideas from historical linguistics when talking about religious history: the formation of "religions" can then be described in a similar way to the formation of languages from various local dialects. The definition of separable languages, as well as the definition of the category of "language" in contrast to "dialect", is better understood as a socio-political process than a linguistic one.[12] Analogous to this approach, he challenges the idea of a religious *Stammbaum* (family tree) and argues for a wave model to describe Christian-Jewish history. While the former assumes a process of differentiation from a common root, the latter is able to describe various modes of entanglement and of divergence, as well as convergence.[13]

In a similar line of argumentation, one could use a network model instead of the wave metaphor. jimi adams[14] suggests conceptualising religions as nodes and the conversions between them as edges. But this still requires the researcher to define religions as monolithic entities that can be represented by a single node. Instead, a different approach has been followed in this chapter. In the past, Krech[15] has argued for studying the emergence of local religious fields. He imagines religions as networks of religious goods and religious actors who act as consumers or producers of religious goods (and often both at the same time).[16] Religious transfer is then the exchange of goods between two such networks.[17] But following Boyarin, one step further can be taken: it is barely possible to tell when a religious good travels between two traditions and not within a tradition, because the process of exchange connects the two religious networks. As soon as there are connections between network components, it is no longer possible to talk of two religious networks but only of one. As a consequence, two opposed processes can be observed. On the one hand, religious traditions emerge by drawing boundaries between interconnected systems, as in the case of Judaism and early Christianity. On the other hand, with the emergence of a global religious field, new connections are established between formerly separate fields, blurring the borders between religions.[18] Santería with its fused elements of Christianity and the Yoruba religion is a prominent example of this, but also more subtle forms of entanglement are frequent in the history of religions. The act of classification, of drawing borders, is then part of the policing actions of the religious field in the sense of Boyarin.[19] And, as shown by Chidester, the discipline of comparative religion has had its own share in these classification processes, especially during the colonial era.

A network model of entangled religious traditions allows for the refrain from defining borders between religions. It supports a highly empirical study of religious traditions, following a similar heuristic approach, such as community detection methods in network analysis. Community detection is the task of identifying parts of a network in which the nodes are densely connected with each other, although they only have few relations to nodes outside the community.[20] The classificatory act of detecting communities in a larger network does not reveal definitive borders, and two community detection methods (or even two runs of the same method) will result in slightly different outcomes.[21] Often, these algorithms contain a random element that causes a certain level of unpredictability, but this instability is not simply a flaw in the analysis: the question concerning which groups of nodes are more densely connected with each other than with others is in itself blurry and allows only for an approximation. At the same time, the identification of communities is not a mere process of chance. There is an empirical basis for it in the relation structure under study. If we use this heuristic approach, religions could be understood as highly connected parts of a larger – and possibly global – socio-semantic religious network at a certain point in time. Religious traditions appear, change their shape and vanish through social processes. Their borders are highly contested and can be drawn differently, highlighting various aspects of similarity and difference. But they constitute a social reality that can be subject to empirical study.

From theory to method: network analysis and religious history

These elaborations use the idea of a network rather metaphorically. While they introduce network concepts to the study of religions, these ideas are still largely theoretical, and their methodological operationalisation remains unclear. In the second part of this chapter, some concrete ways in which these ideas can be applied will be introduced.

One of the distinctive characteristics of the approach outlined previously is the inclusion of ideas, culture and content into network thinking. Social network analysis and its application to history has traditionally focused largely on *structural* properties of actor networks, e.g. central positions in the network that govern information spread. As a consequence, social network analysis per se tends to ignore *cultural* aspects of interaction, e.g. the value that actors ascribe to certain positions. In relational sociology, however, the opposition of structure and culture is increasingly overcome, and culture is regarded as a central aspect of network studies.[22] This approach is critical for an application of network analysis in the study of religions. A reductionist structural view of religions simply as social structures will miss some of their most distinctive properties. Therefore, it is crucial to include the content of religious thought, i.e., religious "meaning" or "semantics", into a network model of religion.

Generally speaking, there are three modes of adding "content" or "semantics" to networks: first, semantics can be seen as information flows in networks. In this model, the networks themselves are social networks between people or groups. The structure of these social networks, however, determines the information flow in these networks and thus the spread of new ideas. Changes in religious semantics would then be a result of the spread of information in a social network. This approach is applied e.g. by Anna Collar[23] who uses an information diffusion approach to model the spread of monotheism in antiquity. Second, semantics can be encoded as the type of relations. In a multigraph, social actors act as the nodes, but the relations between these actors are semantically meaningful, and actors can have multiple relations with each other. This enables differentiation between networks of, e.g. friendship or economic relations. Nagel[24] describes a procedure of extracting this kind of network from texts. Third, semantic concepts themselves can act as nodes in a network. Following this approach, multi-modal networks that relate social actors and semantic items can be seen. Carley and Diesner[25] describe a method that applies this concept.

These three approaches have conceptual as well as methodical implications. The first and second approaches are basically applications of social network analysis. In the "information flow model", semantics are kept out of the network itself. Network characteristics are instead used to explain certain patterns in the spread of ideas. The "typed relations model" adds semantic differences to the network model but as characteristics of a social network. Methodically, this allows the analysis of sub-networks, like the friendship network or the economy network. But the nodes of the network are still only social actors.

The "semantic nodes model" has quite a distinct view on semantics: it takes semantic items, e.g. concepts or notions, as network nodes. Thus, the resulting network is not just a social network with semantic information. Instead, semantic concepts themselves are part of the network. Of course, this kind of network represents something different from a social network. It is rather a representation of the social and semantic entities present in a certain set of texts. But as a result semantic information can be analysed using statistical procedures like centrality measures or community detection.

Digging for data: applying text network analysis to religious corpora

Following these general methodological considerations, additional steps have to be taken when applying these approaches to concrete sources and research questions. The starting point for the application of semantically aware network analysis techniques is the definition of a corpus. Any empirical study that applies network analysis methodology will work on a smaller scale than the idea of a global religious field. It is particularly through well-defined case studies that one can track the local developments that produce religious thought and shape the religious field. The scope of study will focus on relations and divisions that are more finely-grained than what is presently understood to be a religious tradition, e.g. "Christianity" or "Buddhism". But pragmatic issues also shape the domains that allow for the application of network analysis. When it comes to the study of semantics, most computational methods cannot easily be applied to multilingual corpora.

The remainder of this chapter discusses concrete analyses that were conducted as part of the project "Social and semantic network analysis as a means to study religious contact" (SeNeReKo). The project was a joint enterprise of the Trier Center for Digital Humanities (TCDH) and the Center for Religious Studies at Ruhr University Bochum (CERES). It was funded by the German Ministry of Education and Research from 2012 to 2015. Examples from two of the corpora studied in the SeNeReKo project are presented later: The Mahābhārata – an ancient Sanskript epic – and a collection of ancient Egyptian texts from the database "Thesaurus Linguae Aegyptiae".[26] Using these case studies, examples of how the religious semantics of historical knowledge systems can be studied will first be discussed. They cover two of the approaches of networks semantics outlined earlier: the first example reconstructs a (fictional) social network from a literary text – the Mahābhārata – and includes network semantics through the "typed relations model". The second one models the relational structure of ancient Egyptian religious thought itself as a network, following the "semantic nodes" approach. What both approaches share is an inductive approach that tries to trace the categorisations present in the material itself, rather than applying external scholarly taxonomies.

The Mahābhārata is an ancient Indian epic that presumably took its current shape in the time between the third century BCE and the third century CE. Of around one million words, it is fairly large for a single literary work. It tells the story of a family feud, but it combines this with elaborations on various aspects of religious thought. Today, it is referred to as one of the central texts of Hindu thought. Its volume and content make it a worthwhile source for computational analysis. To this end, a digital version of the text was used that includes part-of-speech and lemma information. On this basis, a social network of the main protagonists could be extracted using a straightforward procedure: whenever two characters appear in the same verse, an edge is added between them.[27]

A social network alone is interesting and reveals some patterns of interaction between the main characters. It shows the social dimension of the text: the families and factions. However, it only reveals a little about the content of the text, the meaning associated with the relations. To this end, nodes and edges in the network were supplemented with semantic information. The aim was not to use a predefined taxonomy of relations but to make sense of them on the text's own terms. This was achieved by employing a topic modelling algorithm called "nubbi".[28] Topic modelling aims at detecting latent meaning structures in texts that form semantically coherent word groups or topics.[29] Nubbi extends this idea by combining network structure and text content: Using the text that surrounds the mention of a character or relation in the source as context information, it tries to detect recurring types of entities (nodes) and relations. Each node and relation is then described by a topic distribution, i.e., by the varying degree to which the topics are characteristic of the persons or their connections.

The result of the computational analysis consists of two types of information: first, one gets two sets of "topics", one describing nodes and one describing edges. Each topic takes the form of a list of words that are related to the topic. In Figure 3.2.1, each topic is represented by a colour with an associated label in the legend that was assigned to the topic by a researcher based on the word list. The second type of result is the strength of association of each node and edge with the topics. Each node and edge is associated with multiple topics but to a different degree. The actors as the nodes of the network are represented by their names at the bottom of the diagram. The multi-coloured pie above their name represents the share of the different entity topic that is associated with the character. E.g. Arjuna as the main hero has a large share of battle and fighting, while the gods Brahma and Śiva are associated mostly with religion. The large arcs that connect the actors represent the relations. They are coloured according to the prevalent topic of that relation (for readability reasons, the edges between the nodes show only the most significant topic for each edge). Arjuna is related to Viṣṇu-Kṛṣṇa mostly through discussion, while he has a fighting relationship to his opponent Karṇa. Where the word lists for entity and relation topics were similar, the same label was assigned twice, e.g. Family I and Family II. Strictly speaking, however, they are not identical. The results show that among the topics detected by the algorithm, one of the entity and one of the relation topics (Religion I and Religion II) seem to be especially close to what one could describe as the religious dimension. Both topics consist of words like *deva* (god), *dharma* (law), *karman* (karma) but also specialist designations like *ṛṣi* (seer) or *brāhmaṇa* (brahmin). And these topics are usually stronger for gods and their relations to each other than for human actors. This finding is rooted in the topical structure of the text itself: regardless of whether one wants to use the label "religion" for it or rejects it as a specifically European concept, we can say that these terms were related to each other and describe a semantic field that is relatively distinct from other fields like family or battle.

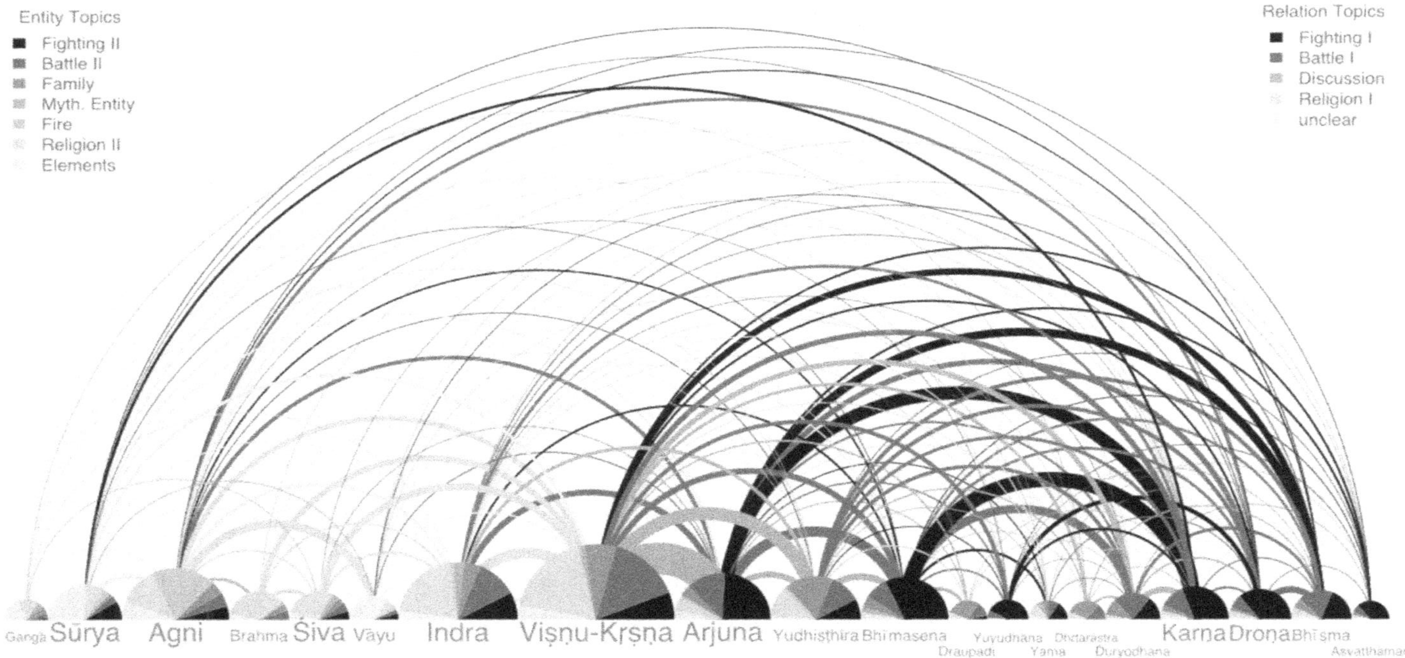

Figure 3.2.1 A social network of the Mahābhārata. The visualization is inspired by Steinweber's "Similar Diversity" project. The code builds on the R implementation by G. Sanchez, "Star Wars Arc Diagram" (2012), http://gastonsanchez.com/work/starwars/ (accessed 27 May 2015).

The second example draws from a different field: Ancient Egypt. Here, our study directly focused on semantic analyses. In the lines of a critical reflection on terms like "religion", Bernd-Christian Otto[30] has brought forward a critique of the distinction between "religion" and "magic" that is common in Egyptological work. These two terms are often assumed to correspond to the ancient Egyptian words *maat* (mꜣꜥ.t) and *heka* (ḥkꜣ.w). While the first denotes something like justice and order, the latter is often interpreted as power, or: "magic". However, the term "magic" is not only loaded with pejorative connotations. Based on evolutionary thought of the late nineteenth and early twentieth centuries, it establishes a distinction between two realms that might not be separable in the case of ancient Egyptian thought.[31]

Network analysis reveals the semantic fields and interrelatedness of historical concepts. Thus, a text network analysis of the terms *maat* and *heka* was used in order to find out more about the semantic structure that differentiates but also connects the two. The corpus of the Thesaurus Linguae Aegyptiae, which is lemmatized and contains morphological annotations,[32] served as the data source.

The methodology is based on the "semantic nodes model": first, all text passages from the corpus were extracted in which the terms *maat* and *heka* appear together. Then, the local context of the two terms was constructed by linking all nouns appearing within a certain distance from each other.[33] The target terms *maat* and *heka* themselves were removed from the network, since they were the selection criteria and as a consequence have a lot of connections that skew the network. However, the shades of grey chosen for the visualization indicate which node has a direct link to which of the target terms (see the following figure). As the final step, closely connected areas in the network were identified using a community detection algorithm.[34]

The network in Figure 3.2.2 shows the community structure of the network. Some words are only connected to *maat* (dark grey), some only to *heka* (white) and some to both (light grey). Taking the clustering into account, one can see that some word clusters are mainly made up of words that are connected to one of the key terms or to both. This allows the differentiation of contexts that outline a distinctive domain for both concepts but also areas that show interaction between the two. The analysis demonstrates that the two common clusters of *maat* and *heka* are related to the cosmic and political order. The large cluster on *the left* is arranged around the concept of "heaven(s)" and other terms related to cosmology (e.g. "circumpolar star" and ritual ("offering"). The shared cluster at the bottom highlights the importance of the pharaoh ("palace") and his role to repel "evil" and the "enemy". This suggests that both concepts interact in a way that helps establish and maintain order. This discovery challenges the idea of two strictly separable domains as suggested by the concepts "religion" and "magic" and points to the centrality of their interaction in ancient Egyptian thought. Definitive conclusions cannot be derived from the network itself. But when taken together with a closer reading of the corresponding passages in the sources, the analysis supports the idea of *heka* serving as a central means to establish *maat* in the sense of order.[35]

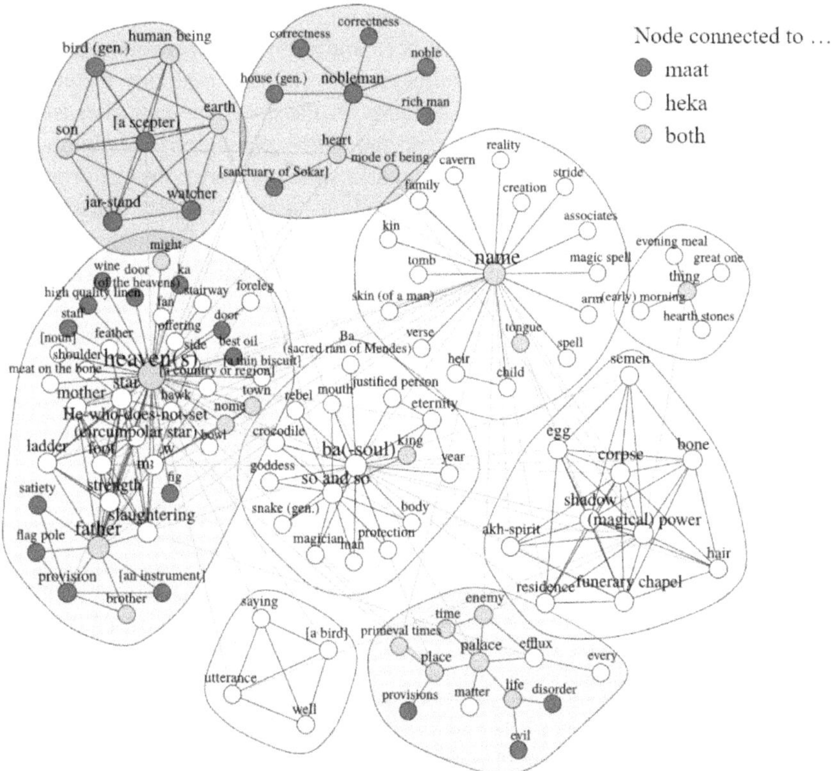

Figure 3.2.2 Semantic network of *maat* and *heka*. The original analysis worked with the Ancient Egyptian lemmas; the English translation is provided for clarity. The visualization was created with the igraph software package using a semi-custom layout algorithm that highlights community structure.

Conclusion

The study of religions and religious history has received fundamental critique in the light of postcolonial research. The application of the category of "religion" as an a-historical scientific concept to other times and places is, at least, problematic, as is the idea of separable "world religions". Network approaches can help meet these challenges. First, they can do so by understanding religious traditions not as monolithic entities but as a network of entangled fields the boundaries of which are blurred. Second, they can do so by trying to reconstruct the historical semantic fields that we today would label as "religion" and by studying their internal structure instead of imposing modern notions on them.

However, in order to do so, network analysis must take meaning into account: religious history, like all cultural history, cannot be reduced to a mere social structure; it is also loaded with distinctive semantics. The application of social network analysis alone therefore tends to reduce religious phenomena to social processes but neglects the content of religious exchange. This might in part explain why network analysis has been adopted rather reluctantly in mainstream religious studies. Models of religious networks that explicitly include the semantic dimension might be more suitable for the discipline and increase the exchange between network research and the study of religions. At this point, three approaches that add meaning to networks were presented: the "information flow model", the "typed relations model" and the "semantic nodes model". The examples gave an idea as to how these approaches can be used to study the internal logic of historical semantic fields. This allows connections to be made between recent theoretical advances with concrete empirical case studies. However, the studies presented in this chapter serve only as a starting point; further application and critical evaluation of the methods and tools involved in this process is still required.

Notes

1 This chapter is based on research in the context of the project "Social and Semantic Network Analysis as a Means to Study Religious Contact" (SeNeReKo). SeNeReKo was a joint research project of the Center for Religious Studies, Ruhr-University Bochum and the Trier Center for Digital Humanities, University Trier. It was funded by the German Federal Ministry of Education and Research between 2012 and 2015 under the project number 01UG1242A. The author of this chapter is responsible for its content. I would like to thank Beate Hofmann, Florian Kerschbaumer, Linda Keyserlingk and Sven Sellmer for their comments on an earlier draft of the chapter.

2 See e.g. Talal Asad, "The Construction of Religion as an Anthropological Category," in *A Reader in the Anthropology of Religion*, ed. Michael Lambek, Blackwell Anthologies in Social and Cultural Anthropology 2, pp. 114–32 (Malden, MA: Blackwell Publishers, 2002).

3 Ibid.

4 Tomoko Masuzawa, *The Invention of World Religions: Or, How European Universalism Was Preserved in the Language of Pluralism* (Chicago, London: University of Chicago Press, 2005, cop. 2015).

5 David Chidester, *Savage Systems: Colonialism and Comparative Religion in Southern Africa*, [Reprint.], Studies in Religion and Culture (Charlottesville: University Press of Virginia, 1996).

6 Ibid.; Tomoko Masuzawa, "Theory without Method: Situating a Discourse Analysis on Religion," in *Religion and Society: An Agenda for the 21st Century*, ed. Gerrie T. Haar and Yoshio Tsuruoka, International Studies in Religion and Society 5, pp. 173–204 (Leiden, Boston: Brill, 2007), p. 183.

7 Chidester, *Savage Systems*, p. 17.

8 The box metaphor was prominent in a talk titled "Troubles with Pluralism: Conceptualizing Religious Diversity Outside the Box" that Masuzawa gave at Ruhr-University Bochum, Germany, in October 2008.

9 Masuzawa, "Theory without Method: Situating a Discourse Analysis on Religion," pp. 196–7, 173–204.

10 See e.g. Russell T. McCutcheon, *Manufacturing Religion: The Discourse on Sui Generis Religion and the Politics of Nostalgia* (New York: Oxford University Press, 1997).

11 Daniel Boyarin, "Hybridity and Heresy: Apartheid Comparative Religion in Late Antiquity," in *Postcolonial Studies and Beyond: Hybridity and Heresy: Apartheid Comparative Religion in Late Antiquity*, ed. Ania Loomba, pp. 339–58 (Durham, NC: Duke University Press, 2005).
12 Ibid., p. 347.
13 Ibid., pp. 344–5.
14 jimi adams, "Network Analysis," in *The Routledge Handbook of Research Methods in the Study of Religion*, ed. Michael Stausberg and Steven Engler, Routledge Handbooks, pp. 323–32 (London, New York: Routledge, 2014), p. 329.
15 Volkhard Krech, "Dynamics in the History of Religions: Preliminary Considerations on Aspects of a Research Programme," in *Dynamics in the History of Religions between Asia and Europe: Encounters, Notions, and Comparative Perspectives*, ed. Volkhard Krech and Marion Steinicke, Dynamics in the History of Religions 1, pp. 15–70 (Leiden: Brill, 2012).
16 Ibid., pp. 41–8.
17 Ibid., p. 46.
18 For a study of the genesis of Jewish ethnic identity through the lens of network theory, see Anna Collar, "Re-Thinking Jewish Ethnicity through Social Network Analysis," in *Network Analysis in Archaeology: New Approaches to Regional Interaction*, ed. Carl Knappett, pp. 223–45 (Oxford: Oxford University Press, 2013), p. 224ff.
19 Boyarin, "Hybridity and Heresy: Apartheid Comparative Religion in Late Antiquity," pp. 339–40, 339–58.
20 Vincent D. Blondel, Renaud L. Jean-Loup Guillaume and Etienne Lefebvre, "Fast Unfolding of Communities in Large Networks," *Journal of Statistical Mechanics: Theory and Experiment 2008*, no. 10 (2008): p. 2.
21 In contrast to the term "community", the notion of a "clique" in network analysis does have a deterministic definition: a clique is a set of nodes where each node is connected to every other in the clique.
22 See Ronald L. Breiger, "Dualities of Culture and Structure: Seeing through Cultural Holes," in *Relationale Soziologie: Zur kulturellen Wende der Netzwerkforschung*, ed. Jan A. Fuhse, 1st ed., Netzwerkforschung 2, pp. 37–47 (Wiesbaden: VS Verlag für Sozialwissenschaften, 2010).
23 Anna Collar, "Network Theory and Religious Innovation," *Mediterranean Historical Review* 22, no. 1 (2007): pp. 149–62.
24 Alexander-Kenneth Nagel, *Analysing Change in International Politics: A Semiotic Method of Structural Connotation* (Bremen: University, Sfb 597 Staatlichkeit im Wandel, 2008).
25 Jana Diesner and Kathleen M. Carley, "Revealing Social Structure from Texts: Meta-Matrix Text Analysis as a Novel Method for Network Text Analysis," in *Causal Mapping for Research in Information Technology*, ed. V.K. Narayanan and Deborah J. Armstrong, pp. 81–108 (Hershey, PA: Idea Group, 2005).
26 Thesaurus Linguae Aegyptiae is a project by the Berlin-Brandenburg Academy of the Sciences and Humanities, http://aaew.bbaw.de/tla/ (accessed 13 December 2018).
27 Network extraction approaches like this are quite common in computational literary studies (see e.g. Franco Moretti, "Network Theory, Plot Analysis," *New Left Review*, no. 68 (2011): pp. 80–102), but as a result of their simplicity there are a couple of issues worth noting. The pure appearance of two character names in the same section does not necessarily mean that there is a personal relation nor does it define the importance or type kind of relation. Indirect references instead of names pose further challenges.
28 Jonathan Chang, Jordan Boyd-Graber and David M. Blei, "Connections between the Lines: Augmenting Social Networks with Text," in *Proceedings of the 15th ACM SIGKDD International Conference on Knowledge Discovery and Data Mining: KDD*, pp. 169–78 (New York: ACM Press, op. 2009).

29 See: Megan R. Brett, "Topic Modeling: A Basic Introduction," *Journal of Digital Humanities* 2, no. 1 (2012).
30 Bernd-Christian Otto, "Zauberhaftes Ägypten – Ägyptischer Zauber? Überlegungen zur Verwendung des Magiebegriffs in der Ägyptologie," in *Ägypten, Kindheit, Tod. Gedenkschrift für Edmund Hermsen*, ed. Florian Jeserich, pp. 39–70 (Vienna, Cologne, and Weimar: Böhlau Verlag, 2012).
31 Ibid., p. 40ff.
32 Beate Hofmann and Frederik Elwert, "Heka und Maat. Netzwerkanalyse als Instrument ägyptologischer Bedeutungsanalyse," in *"Vom Leben umfangen": Ägypten, das Alte Testament und das Gespräch der Religionen Gedenkschrift für Manfred Görg*, ed. Stefan J. Wimmer and Georg Gafus, Ägypten und Altes Testament 80, pp. 235–45 (Münster: Ugarit-Verlag, 2014, cop. 2014).
33 Here, we used a window of five words, following the method of Dmitry Paranyushkin, "Identifying the Pathways for Meaning Circulation Using Text Network Analysis," (2011), http://noduslabs.com/research/pathways-meaning-circulation-text-network-analysis/ (accessed 13 December 2017) that increases the weight of relations between close words. We excluded proper nouns like personal names in order to highlight the conceptual structure independent of personal attribution and social structure.
34 Jean-Loup Guillaume Blondel and Etienne Lefebvre, "Fast Unfolding of Communities in Large Networks".
35 For a more detailed elaboration, see Hofmann and Elwert, "Heka und Maat. Netzwerkanalyse als Instrument ägyptologischer Bedeutungsanalyse," pp. 243, 235–45.

Bibliography

Adams, Jimi. "Network Analysis." In *The Routledge Handbook of Research Methods in the Study of Religion*. Edited by Michael Stausberg and Steven Engler, pp. 323–32. Routledge Handbooks. London, New York: Routledge, 2014.

Asad, Talal. "The Construction of Religion as an Anthropological Category." In *A Reader in the Anthropology of Religion*. Edited by Michael Lambek, pp. 114–32. Blackwell Anthologies in Social and Cultural Anthropology 2. Malden, MA: Blackwell Publishers, 2002.

Blondel, Vincent D., Renaud Lambiotte Jean-Loup Guillaume, and Etienne Lefebvre. "Fast Unfolding of Communities in Large Networks." *Journal of Statistical Mechanics: Theory and Experiment 2008*, no. 10 (2008).

Boyarin, Daniel. "Hybridity and Heresy: Apartheid Comparative Religion in Late Antiquity." In *Postcolonial Studies and Beyond: Hybridity and Heresy: Apartheid Comparative Religion in Late Antiquity*. Edited by Ania Loomba, pp. 339–58. Durham, NC: Duke University Press, 2005.

Breiger, Ronald L. "Dualities of Culture and Structure: Seeing Through Cultural Holes." In *Relationale Soziologie: Zur kulturellen Wende der Netzwerkforschung*. Edited by Jan A. Fuhse. 1st ed., pp. 37–47. Netzwerkforschung 2. Wiesbaden: VS Verlag für Sozialwissenschaften, 2010.

Brett, Megan R. "Topic Modeling: A Basic Introduction." *Journal of Digital Humanities* 2, no. 1 (2012).

Chang, Jonathan, Jordan Boyd-Graber, and David M. Blei. "Connections between the Lines: Augmenting Social Networks with Text." In *Proceedings of the 15th ACM SIGKDD International Conference on Knowledge Discovery and Data Mining: KDD*, pp. 169–78. New York: ACM Press, op. 2009.

Chidester, David. *Savage Systems: Colonialism and Comparative Religion in Southern Africa*. [Reprint.]. Studies in Religion and Culture. Charlottesville: University Press of Virginia, 1996.

Collar, Anna. "Network Theory and Religious Innovation." *Mediterranean Historical Review* 22, no. 1 (2007): pp. 149–62.

———. "Re-Thinking Jewish Ethnicity through Social Network Analysis." In *Network Analysis in Archaeology: New Approaches to Regional Interaction*. Edited by Carl Knappett, pp. 223–45. Oxford: Oxford University Press, 2013.

Diesner, Jana, and Kathleen M. Carley. "Revealing Social Structure from Texts: Meta-Matrix Text Analysis as a Novel Method for Network Text Analysis." In *Causal Mapping for Research in Information Technology*. Edited by V.K Narayanan and Deborah J. Armstrong, pp. 81–108. Hershey, PA: Idea Group, 2005.

Fuhse, Jan A., ed. *Relationale Soziologie: Zur kulturellen Wende der Netzwerkforschung*. 1st ed. Netzwerkforschung 2. Wiesbaden: VS Verlag für Sozialwissenschaften, 2010.

Haar, Gerrie ter, and Yoshio Tsuruoka, eds. *Religion and Society: An Agenda for the 21st Century*. International Studies in Religion and Society 5. Leiden, Boston: Brill, 2007.

Hofmann, Beate, and Frederik Elwert. "Heka und Maat. Netzwerkanalyse als Instrument ägyptologischer Bedeutungsanalyse." In *"Vom Leben umfangen": Ägypten, das Alte Testament und das Gespräch der Religionen Gedenkschrift für Manfred Görg*. Edited by Stefan J. Wimmer and Georg Gafus, pp. 235–45. Ägypten und Altes Testament 80. Münster: Ugarit-Verlag, 2014, cop. 2014.

Jeserich, Florian, ed. *Ägypten ' Kindheit ' Tod*. Vienna, Cologne, and Weimar: Böhlau Verlag, 2014.

Knappett, Carl, ed. *Network Analysis in Archaeology: New Approaches to Regional Interaction*. Oxford: Oxford University Press, 2013.

Krech, Volkhard. "Dynamics in the History of Religions: Preliminary Considerations on Aspects of a Research Programme." In *Dynamics in the History of Religions between Asia and Europe: Encounters, Notions, and Comparative Perspectives*. Edited by Volkhard Krech and Marion Steinicke, pp. 15–70. Dynamics in the History of Religions 1. Leiden: Brill, 2012.

Krech, Volkhard, and Marion Steinicke, eds. *Dynamics in the History of Religions between Asia and Europe: Encounters, Notions, and Comparative Perspectives*. Dynamics in the History of Religions 1. Leiden: Brill, 2012.

Lambek, Michael, ed. *A Reader in the Anthropology of Religion*. Blackwell Anthologies in Social and Cultural Anthropology 2. Malden, MA: Blackwell Publishers, 2002.

Loomba, Ania, ed. *Postcolonial Studies and Beyond: Hybridity and Heresy: Apartheid Comparative Religion in Late Antiquity*. Durham, NC: Duke University Press, 2005.

Masuzawa, Tomoko. "Theory without Method: Situating a Discourse Analysis on Religion." In *Religion and Society: An Agenda for the 21st Century*. Edited by Gerrie t. Haar and Yoshio Tsuruoka, pp. 173–204. International Studies in Religion and Society 5. Leiden, Boston: Brill, 2007.

———. *The Invention of World Religions: Or, How European Universalism Was Preserved in the Language of Pluralism*. Chicago, London: University of Chicago Press, 2005, cop. 2015.

McCutcheon, Russell T. *Manufacturing Religion: The Discourse on Sui Generis Religion and the Politics of Nostalgia*. New York: Oxford University Press, 1997.

Moretti, Franco. "Network Theory, Plot Analysis." *New Left Review*, no. 68 (2011): pp. 80–102.

Nagel, Alexander-Kenneth. *Analysing Change in International Politics: A Semiotic Method of Structural Connotation*. Bremen: Univ., Sfb 597 Staatlichkeit im Wandel, 2008.

Narayanan, V.K., and Deborah J. Armstrong, eds. *Causal Mapping for Research in Information Technology*. Hershey, PA: Idea Group, 2005.

Otto, Bernd-Christian. "Zauberhaftes Ägypten – Ägyptischer Zauber? Überlegungen zur Verwendung des Magiebegriffs in der Ägyptologie." In *Ägypten, Kindheit, Tod. Gedenkschrift für Edmund Hermsen*. Edited by Florian Jeserich, pp. 39–70. Vienna, Cologne, and Weimar: Böhlau Verlag, 2014.

Paranyushkin, Dmitry. "Identifying the Pathways for Meaning Circulation Using Text Network Analysis." (2011). http://noduslabs.com/research/pathways-meaning-circulation-text-network-analysis/ (accessed 13 December 2017).

John F. Elder IV, Françoise Fogelman-Soulié, Peter A. Flach, and Mohammed Javeed Zaki: *Proceedings of the 15th ACM SIGKDD International Conference on Knowledge Discovery and Data Mining, Paris, France, June 28 – July 1, 2009. ACM 2009.*

Stausberg, Michael, and Steven Engler, eds. *The Routledge Handbook of Research Methods in the Study of Religion*. Routledge Handbooks. London, New York: Routledge, 2014.

Wimmer, Stefan Jakob, and Georg Gafus, eds. *"Vom Leben umfangen": Ägypten, das Alte Testament und das Gespräch der Religionen Gedenkschrift für Manfred Görg*. With the assistance of Manfred Görg. Ägypten und Altes Testament 80. Münster: Ugarit-Verlag, 2014, cop. 2014.

4

Infrastructures for data collection and exploration

4.1 Deep networks as associative interfaces to historical research

Charles van den Heuvel, Ingeborg van Vugt, Pim van Bree and Geert Kessels

Introduction

The most widely used software tools for network analysis have the explicit goal of creating patterns that visualize features of the underlying data that are regarded as representative for answering research questions or testing hypotheses. Their underlying algorithms often support quantitative analyses and visualization of large data sets. In general statistical methods are predominantly used to explain to which extent these visualizations are representative for the underlying data. This contribution is based on the assumption that full data integration, in particular in the humanities, is impossible for reasons of incompleteness, complexity, ambiguity and uncertainty in data. Therefore the focus should not be on pattern recognition in combination with statistical methods of network representations alone. We need to include approaches that allow users to explore and to interact with these incomplete and complex data. In short, we do not need just networks as representations but also networks as interactive interfaces. These interfaces must enable users to explore, to interact and to make associations with their own selections of data that can combine data-driven and research question-driven approaches. To this end we first discuss problems experienced within the data-driven project *Circulation of Knowledge/ePistolarium* of labelling automatically generated topics. We focus on the tension between explicit and implicit terms that biased our interpretations of network representations of our research case dealing with the role of confidentiality in the correspondences present in ePistolarium. This discussion is followed by a brief exploration of potential solutions to overcome the previously mentioned tension in representing computer generated terms in the future. These introductory paragraphs on large scale data-driven research are followed by a discussion of the experiences with a small scale experiment *Mapping Notes and Nodes: Exploring potential relationships in biographical data and cultural networks in the creative industry in Amsterdam and Rome in the Early Modern Period*, in which a bottom-up approach was followed to create manually multi-layered historical, intellectual and technological networks with the software application Nodegoat. These discussions of experiences with the development of computer-generated and manually created historical networks are followed by an exploration of their implications for user interfaces of historical network research and by an introduction to the term "deep networks." After a discussion about deep networks in relation to our experiences with the *Circulation of Knowledge/ePistolarium* and the *Mapping Notes and Notes* projects we contextualise the term by similar notions in literature in order to explore its potential in the wider contexts of historical network research and digital humanities. We return to our experiences with mappings of unstructured and structured data by introducing in a more generic way classifications of terms that change in meaning over time (concept-drift) and visualization of uncertainty. We claim that current experiments with the representation of big data from multiple perspectives can be read as computer-assisted deep networks that can be used to create associative interfaces to historical research networks similar to the manually created deep networks of the Nodegoat tool. We conclude that the creation of such deep networks not only can contribute in bridging the gap between unstructured and structured data, between distant and close reading, between qualitative hermeneutic approaches and quantitative statistical methods but holds a promise for future historical network research and digital humanities at large.

Experiences with ePistolarium

The aim of The Circulation of Knowledge project was to map the dissemination and appropriation of themes of interest of Early Modern Dutch scholars and scientists working in the Dutch Republic as part of the Republic of Letters.[1] The term Republic of Letters is often used to describe exchanges of knowledge between scholars in informal, sometimes more formal, communities and societies in the Early Modern Period between the fifteenth and eighteenth centuries. In recent years several projects have begun digitally reconstructing sections of the Republic of Letters to analyse the exchange of knowledge in this fictitious republic more comprehensively. The best known initiatives are the project *Mapping the Republic of Letters* of Stanford University, *Cultures of Knowledge* of Oxford University and the SKILLNET project of Utrecht University. These projects focus on metadata of scholars to reconstruct and visualize intellectual networks and geographies in knowledge exchange in letters in Early Modern Europe and beyond. The Circulation of Knowledge project focused both on metadata and the letters themselves. Within the context of that project the ePistolarium tool was developed to extract concepts and ideas from a dataset of approximately 20,000 letters of Dutch scholars and scientists working in the Netherlands in the Early Modern Period, such as Constantijn and Christiaan Huygens, René Descartes, Hugo Grotius, Isaac Beekman, Jan Swammerdam, Antoni van Leeuwenhoek and Caspar Barlaeus. Mapping and analysing the dissemination and appropriation of themes of interest of scholars in these large number of letters in Latin, French, Dutch and English required the application of Natural Language Processing (NLP) techniques based on interactions between computers and natural languages. The result of implementing a combination of NLP techniques such as topic modelling, keyword analysis, language identification, spelling normalisation and named entity recognition with various search facilities allows the data base of the Circulation of Knowledge project to be queried by typing in keywords, by selecting given search options (facets) in combination (facetted search) and similarity search.[2] The topic modelling method enables one to automatically extract words that in combination can be interpreted as being representative for specific topics in the correspondences, while the other NLP techniques were applied to handle the complexities of multi-linguality within the corpus and even within single letters and of spelling variation to improve the quality of content extraction. Similarity search, based on topic modelling, presents letters in the database similar to the letter found after a query. However, it also possible to upload a text to find a similar letter in the database. Furthermore, topic modelling is used to generate alternative keywords for search. The results of all keyword and facetted search queries can be visualized on geographical maps, in time lines, in correspondence networks and in co-citation networks to map the people that are mentioned in relation to specific topics and debates (Figure 4.1.1).

During and after the project that started in 2008 and finished in 2013, several experiments with the ePistolarium tool were set up. First the open source Latent Dirichlet Allocation (LDA) topic modelling tool Mallet, developed at Stanford, was used. Tests by historians of science during an international workshop held in 2010 revealed a limited success of the Mallet tool. Moreover, the historians criticised the lack of control over the extraction of themes of interest in the database of letters that in their view functioned too much as a "black box."[3] For that reason it was decided to compare the first used topic modelling methods LDA (Mallet) with two other methods Latent Semantic Analysis (LSA) and Random indexing (R.I.).[4]

Each of these approaches is derived from the so-called vector space model. If text fragments of two documents address similar topics, it is highly possible that they share many substantive terms. Conversely, if two terms occur in many documents together, the terms are likely to be related. After pre-processing of the letters (e.g. elimination of opening and closing phrases, formulas, removal of stop words etc.) a vector representation of each document was created. Conceptually, LDA and LSA use a similar vector space model in which the entire document is treated as the context of a word being analysed. Terms that occur together very often in the same document are merged into a single dimension of the so-called feature space. Instead of first constructing a co-occurrence matrix and then using a separate dimension reduction phase, Random Indexing builds an incremental word space model in two steps. First, each context (for instance each word or each document) in the data is assigned a unique and randomly generated representation called an index vector. Second, context vectors are produced by scanning the text, and each time a word occurs in a context (e.g. in a document) that context's index vector is added to the context vector for the word in question. Words are thus represented by context vectors that are effectively the sum of the words' contexts. In the end the latter Random Indexing was implemented for various reasons. In the comparative tests RI

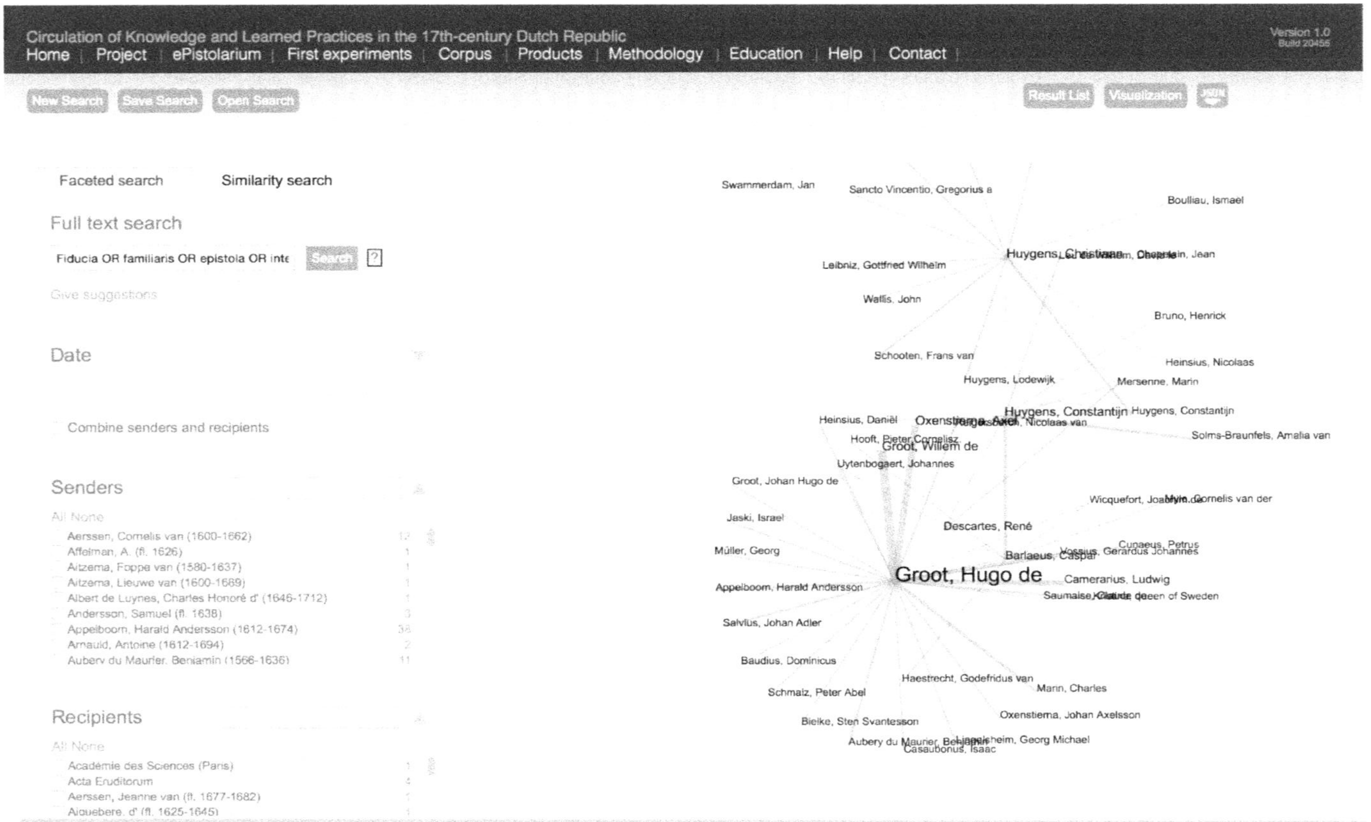

Figure 4.1.1 Correspondence networks of authors using the confidentiality related terms: *fiducia, familiaris, epistola, inter nos, geheim, secreet, secretum, sodalitas, tectus, tacitus, vertrouwelijkheid* [Charles van den Heuvel]

Source: © Huygens ING

performed best in the task of reproducing topic labels assigned by human experts for a randomly selected subset of letters. Different from LDA and LSA, Random Indexing allows proximity of words to be taken into account. From a computational point of view, RI has the benefit that it does not rely on computationally intensive matrix operations as, for example, LSA. Instead, RI builds an incremental word space model that scales very well with increasing corpus size. The latter is crucial given the aim to include far more letters in the ePistolarium to optimize the functionality and use of the tool.

When the tool was launched in June 2013, a couple of new experiments were set up with mixed results. Both the successful and the less successful experiments are all fully documented.[5] Successful was the case developed by Huib Zuidervaart to automatically discover actors who played a crucial role in the discovery of the ring structure and moons around the planet Saturn. The tool returned not only names already discussed in the publications on this topic by the expert Van Helden but also additional relevant historical figures. However, less successful were the results of an experiment by Charles van den Heuvel and Henk Nellen.[6] Their research question was whether the theme of confidentiality, which recurs regularly in the letters of the Dutch scholar Hugo Grotius and which is considered an important characteristic of knowledge exchange in the historiography of the Republic of Letters, could be retrieved automatically from the corpus of correspondences in the Circulation of Knowledge project.[7] Some unknown references to confidentiality in the work of Grotius were found, but very few were found in other correspondences. Nevertheless, manual checks revealed that confidential information did indeed also appear in the letters of other scholars. In short the automatic detection of letters that expressed confidentiality more or less failed. This did not completely come as a surprise. In small datasets the "most similar" letter can still have quite different content. A query for words, (confidentiality-related or not) based on similarity search in a set of 20,000 letters, of which more than a third were written by or to Grotius, almost invariably leads to his correspondence. But apart from the size and composition of the dataset, implicit language use can also explain the low recall. While in the Saturn case words like sun, moons, planets and stars immediately provide explicit associations with astronomy (or perhaps astrology), the many ways to ask recipients of letters handling specific controversial information in confidence are far more implicit.[8] When continuously different word combinations are used to express confidentiality, such as "Interea quas a me tenes litteras tibi habe et Vulcano sacrifica" (Meanwhile, keep the letters received from me for yourself and offer them to the god of fire.) or "Haec inter nos dicta sunto" (Consider this information as exchanged between the two of us.), it becomes less clear which automatically extracted strings of words are representing the same concept or topic.[9] This has implications for the way computer-generated words are tagged. Different levels of abstraction, which often are implicit, bias the quality of the representations of the topics. They affect, for instance, the quality of visualizations of correspondence network and co-citation networks that respectively represent which authors discuss or are named in relation to specific topics. In short, different levels of abstraction (such as in the mentioned astronomy case with explicit terms and the hard to tag implicit references to confidentiality) bias the use of topic modelling and of historical network analysis. Here we sketch experiences with other projects and present some ideas that in the future might result in a better combination of computer-generated unstructured data and manually generated structured data in historical network research.

Historical Network Research and the confidentiality in the Republic of Letters

Not only the semantic analysis of confidentiality-related words in correspondences (such as *fiducia, familiaris, epistola, inter nos, geheim, secreet, secretum, sodalitas, tectus, tacitus, vertrouwelijkheid)* but also the analysis of relationships between people can provide insight for our example of confidentiality and secrecy in knowledge exchange. Several historians have already described the Republic of Letters in terms of networks. Historical interpretations of knowledge exchanges within the Republic of Letters are often based on small sets of letters of one correspondent or a small group of correspondents, and findings are extrapolated to "explain" certain phenomena. In other studies we pointed to that problem in various ways. We questioned the homogeneity of networks and explored differences in the concept of confidentiality in knowledge exchange in intellectual and technological networks.[10] In a later study some hypotheses were formulated for the development of a model to assess the impact of confidentiality in networks of knowledge distribution and exchange.[11] These hypotheses assume 1) a correlation between the nature of personal and professional relationships and the degree of confidentiality for which personal and professional epistolary knowledge networks should be compared; 2) a difference between direct and indirect transfer of confidentiality via intermediaries for which it would be useful to observe evolution in co-citation networks and analyses of introduction letters; 3) changes in expressions of confidentiality and secrecy over time that have an impact on the structure of the networks; 4) an increase in the number of correspondences both in the public sphere (rehabilitation) and private sphere (safeguarding private goods and care for family) as result of exile and finally 5) differences between the dissemination of knowledge in intellectual and technological networks that might have an impact on confidentiality. Testing these hypotheses would require for most of the statements close readings of many letters. However, some statements can already be put to the test by an analysis of the structure of the various networks in which letters were exchanged. To this end we set up a very small pilot in which the concept of reciprocity in the personal and official networks of the Dutch statesman and scholar Grotius, "father of international law," was analysed[12] in a substantial large correspondence of 7,725 letters.[13] His large correspondence of 7,725 letters was divided into epistolary exchanges in personal networks concerning letters to and from well-known scholars and to friends and family and in professional networks related to his administrative functions as a secretary to the Advocate-Fiscal of the States of Holland and an ambassador of Sweden in Paris resulting in official and diplomatic letters. We tested reciprocity across correspondence types as an indicator of confidentiality in Grotius' networks. Using the network analytical tool Gephi, a simple ego-network was created with Hugo Grotius, at the centre (Figure 4.1.2) of which every node represents one of Grotius' top 69 correspondents.[14]

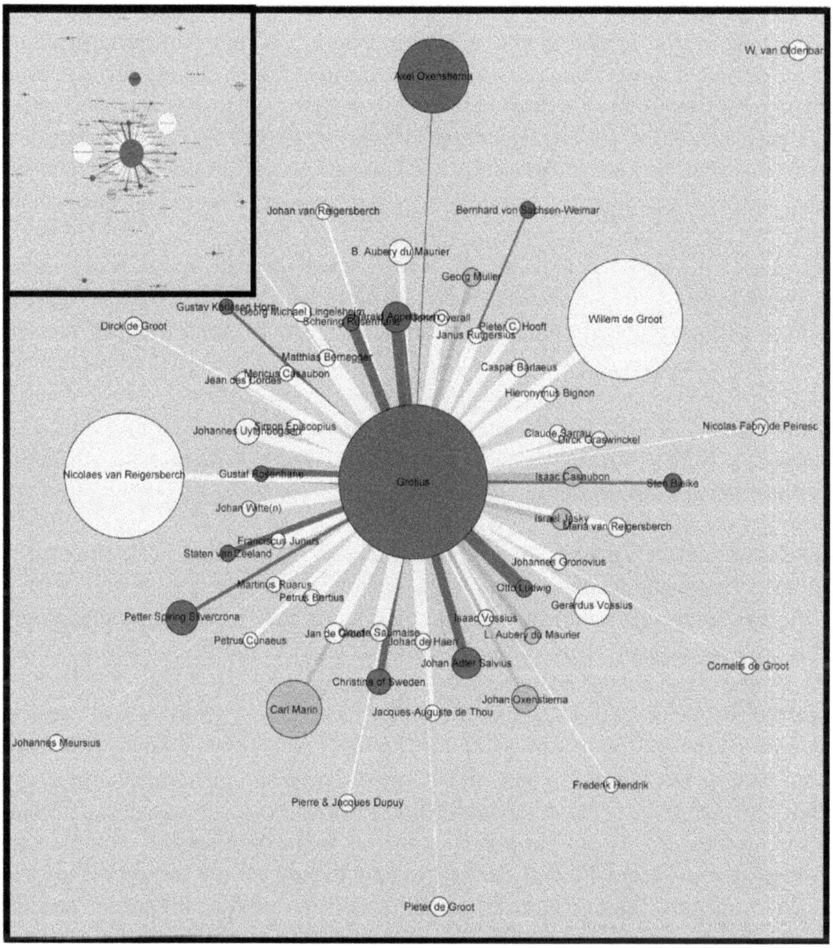

Figure 4.1.2 Ego network Hugo Grotius with top 69 correspondents. Dark grey nodes are professional respondents of Hugo Grotius; white nodes are personal respondents. Light grey ones belong to both networks. A node's distance from Grotius and the thickness of the edge (edge weight) both represent the degree of reciprocity. Detail and insert with complete network.

Source: ©Nils Spelt, Erasmus University Rotterdam

The size of each node corresponds to the amount of letters exchanged with him. A node's distance from Grotius and the weight of the edge (expressed in thickness of the edge) represent the degree of reciprocity (i.e., the number of letters returned by a recipient corresponded). Finally, the colours of the nodes represent professional respondents (dark grey), personal respondents (white) or the ones that belong to both networks (light grey). This is of course an arbitrary choice, but sometimes inevitable, because it was not always possible to decide whether a correspondent was more aligned to the personal network or the professional one. Friends for example could become foes after a specific political event or change of religious affinity, but quantifying that would require a temporal dimension in the representation of both networks. Within the short period of the experiment it was impossible to pursue this idea that would require assessing changes in friendships in relation to mutations in political and religious affinities for each person in Grotius' networks. However, despite all these limitations, even this first very small experiment adds support to the claim that the degree of reciprocity was higher in Grotius' personal network than in his professional one. We found that, on average, for every 100 letters sent or received by Grotius to his personal network, 48 were answered, while in his professional network only 31 in 100 were answered, a difference of about 33%. With that said, a large amount of correspondences in Grotius' personal network showed virtually no reciprocity, while many exchanges with professional contacts were actively reciprocated.[15] On the basis of this small case study only, it was impossible to come to solid conclusions. More attributes, such as (changes in) social status or religious affinity need to be coded to create a more nuanced model of the role of confidentiality and secrecy in the networks of Grotius. Further, more letters of other scholars need to be included as well. For this reason the Reassembling the Republic of Letters initiative to bring representations of distributed European databases of correspondents together in a repository with a Virtual Research Environment around it will be crucial.[16] Although we probably still have to wait for many years before we can extract sufficient data automatically from such combined corpora of letters to analyse and visualize the multilayered networks of the Republic of Letters in a comprehensive way, we can already formulate the requirements for the tools we need to analyse and these networks in the future. To this end it might be useful to explore advantages and disadvantages of existing tools. In the previously discussed experiment not only the lack of data but also facilities of Gephi to represent multiple networks in one image limited the potential outcome of the network analysis. Even with just three parameters, reciprocity in private, in public and in private/public networks we had to scale down to measuring only those correspondents with whom Grotius exchanged ten or more letters to obtain a readable representation. Bringing in more dimensions needed for understanding the complexity of networks of knowledge exchange in Early Modern Europe requires new modes of representation. Here we discuss a second experiment in which multi-layered intellectual and technological networks were created using the software application Nodegoat.[17]

Experiences with mapping notes and nodes

In a private-public project, Mapping Notes and Nodes in Networks, a small interdisciplinary team experimented for nine months for two days a week with Nodegoat to create multi-layered networks of actors and documents that are potentially relevant for the history of the creative industries in Amsterdam and Rome in the Early Modern Period.[18] The project started with the integration of three complementary but heterogeneous (meta)datasets: the full text searchable Biographical Reference Works of the Huygens Institute for the History of Netherlands; Ecartico, a comprehensive database that allows analysing and visualizing metadata concerning painters, art consumers, art collectors, art dealers and others involved in the cultural industry of Amsterdam and the Low Countries in the Early Modern Period and finally Hadrianus, a database of Dutch artists and scholars from the Middle Ages up to the twentieth century that stayed in Rome developed by the Royal Netherlands Institute in Rome.[19] Researchers who heard of the initiative offered their own data.[20] In this project the emphasis was not on quantitative analyses but on a qualitative approach to allow scholars to connect and complement these divergent datasets and to create and visualize networks to see and interact with connections. To this end, it was decided to develop a viewer for Nodegoat that visualizes the co-presence of historical actors (i.e., people of various professions and in different roles (nodes) and of (meta-)data of sources and documents (notes) in Amsterdam and Rome at the same location and at the same moment in partially overlapping, multi-layered networks. The assumption is that the co-presence of artists, artisans, scientists, art-agents, ambassadors, patrons, sponsors and entrepreneurs in a certain location in a given time period allows for recognizing potential networks, while contextual topical information (introduction letters, commissions or artefacts, such as personal gifts) enable users to assess whether there were actually contacts between these persons.[21] When new data is added, the overlap of the multi-layered networks changes, resulting potentially in new answers and other questions. The building of hybrid networks implies perhaps a less systematic research of the history of the creative industry of Rome and Amsterdam, but at the same time both practical and theoretical arguments support this more explorative approach. If complete data-integration is impossible, a tool that on the basis of meta-data enables one to assess the likelihood that one combination of information about historical actors might lead to better results than another could at least support prioritising the digitisation program necessary for research. The second, theoretical argument is that the proposed incremental approach in which researchers overlay multiple networks of data in flux and interact with them from changing perspectives stands closer to hermeneutic methods.[22] Such methods appeal to many humanities scholars who try to give sense to data from multiple perspectives in continuous processes of reinterpretation.[23] For that reason we chose a couple of cases that could demonstrate the multidimensional relationships from data about the cultural industry of Amsterdam and Rome from different perspectives.[24] One of those cases concerned the

Dutch civil engineer Cornelis Meijer (1629–1701). Meijer created on commission of three successive popes constructions against floods near the Via Flaminia in Rome to make the river Tiber navigable and for draining the Pontine Marches between the city and the coast. As such we might perhaps not link Meijer at first sight with the creative industry of Rome. However, he commissioned several Dutch and Flemish artists in Rome to make engravings after his designs, such as Caspar van Wittel (italicised as Gaspare Vanvitelli). He had connections with artistic academies and painters groups in Rome, such as the Accademia di San Luca and the Bentveughels group. Moreover, he was a member himself of an academy of scientists, the Accademia Fisico-matematica. All these networks of people and academies (nodes) and of contextual artefacts such as books, drawings, archivalia etc. (notes) in Rome were mapped (Figure 4.1.3).

Nodegoat not only allows a mapping of relations between actors and artefacts but also to express the nature of those relationships; for instance the tie between two nodes can express whether actors were friends or foes. Given this possibility to indicate the nature of relationships, the competitors of Cornelis Meijer were also mapped in the ties of his network. For instance Carlo Fontana, distantly related to the famous Roman architect Domenico Fontana, had made alternative plans for the defence of the Via Flaminia against the Tiber. According to the advisors of Pope Clement X, Fontana's plan was too expensive in comparison to Meijer's construction that deflected the current of the river. This feature to express the nature of relationships in the ties between the nodes of the networks is important because we often just try to analyse the impact of a specific actor in a surrounding network but not the positive or negative features (such as collaboration and competition in this case) that condition the forces of the network in question.

The mapping of overlaying networks of actors and artefacts yielded interesting results. Meijer did not only work in Rome but received commissions for advice from the court of the Grand duke of Tuscany, Cosimo III de'Medici in Florence. The two networks of the people around Cornelis Meijer were mapped on top of each other. It resulted in two interesting findings. First, the overlay of networks, in combination with archival research, made clear that artefacts played an important role in establishing contacts between actors in the network (Figure 4.1.4).[25]

Ingeborg van Vugt found between the letters of Cornelis and his son, Otto Meijer, to the Grand duke a coloured engraving of a skeleton of a dragon, (similar to the engraving of dragon that adorned the title page of some of his publications but in full flesh), which the engineer claimed to have found during his civil engineering works in the Pontine Marshes.[26] Otto probably wanted to get the attention of Cosimo by reminding him of his father's reputation in the hope to acquire new commissions.[27] Although the testimonies of Cornelis Meijer, who often posed himself in roles he never had, need to be read with the greatest care, the overlay of the Roman and Florentine networks of engineers (including older examples) revealed another unexpected feature of the

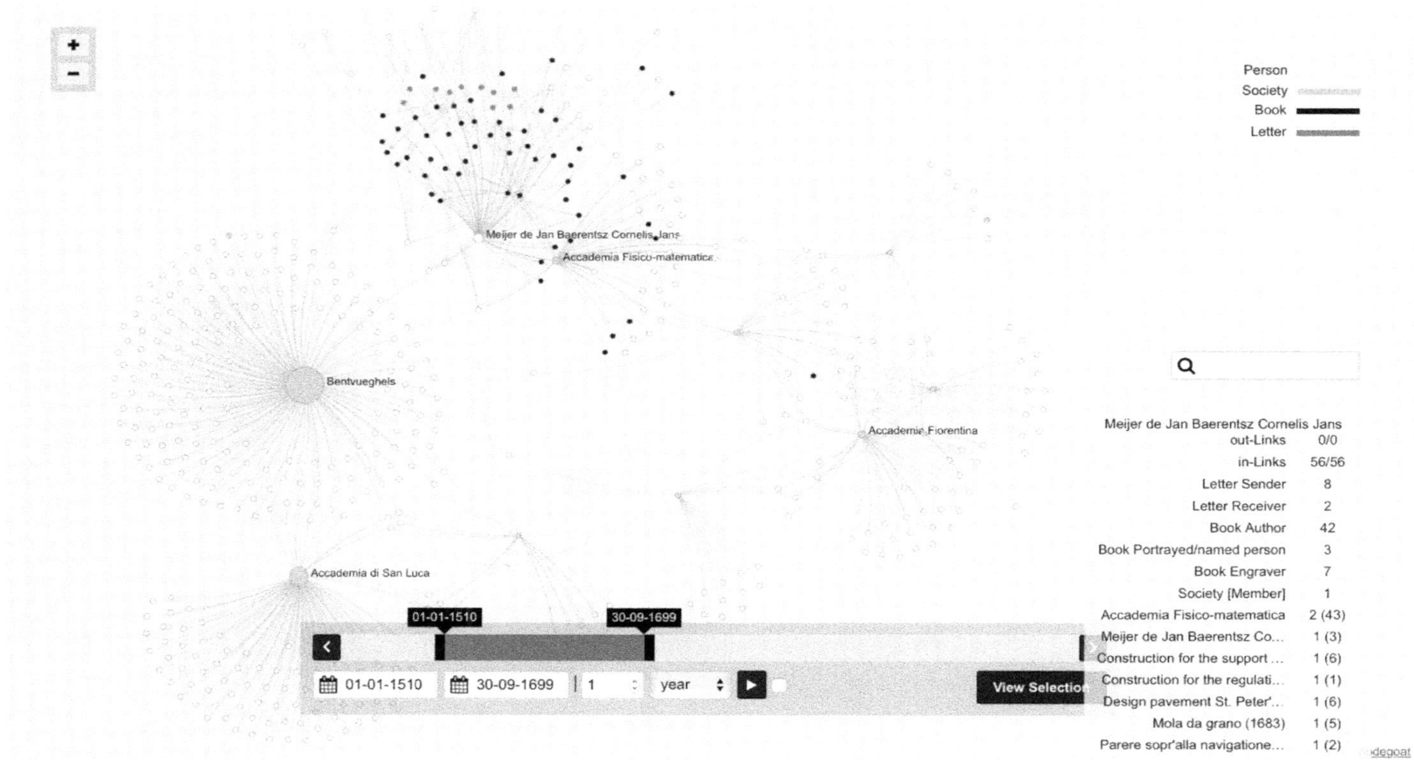

Figure 4.1.3 A visualization of networks around the Accademie or Societies (light grey) and Cornelis Meijer. Meijer, at the centre of this visualization, is surrounded by artefacts (e.g. publications and engravings in black), persons (white) and letters (dark grey). Hovering over the nodes and ties opens an overview with the different connections and specifies the nature of the relationships.

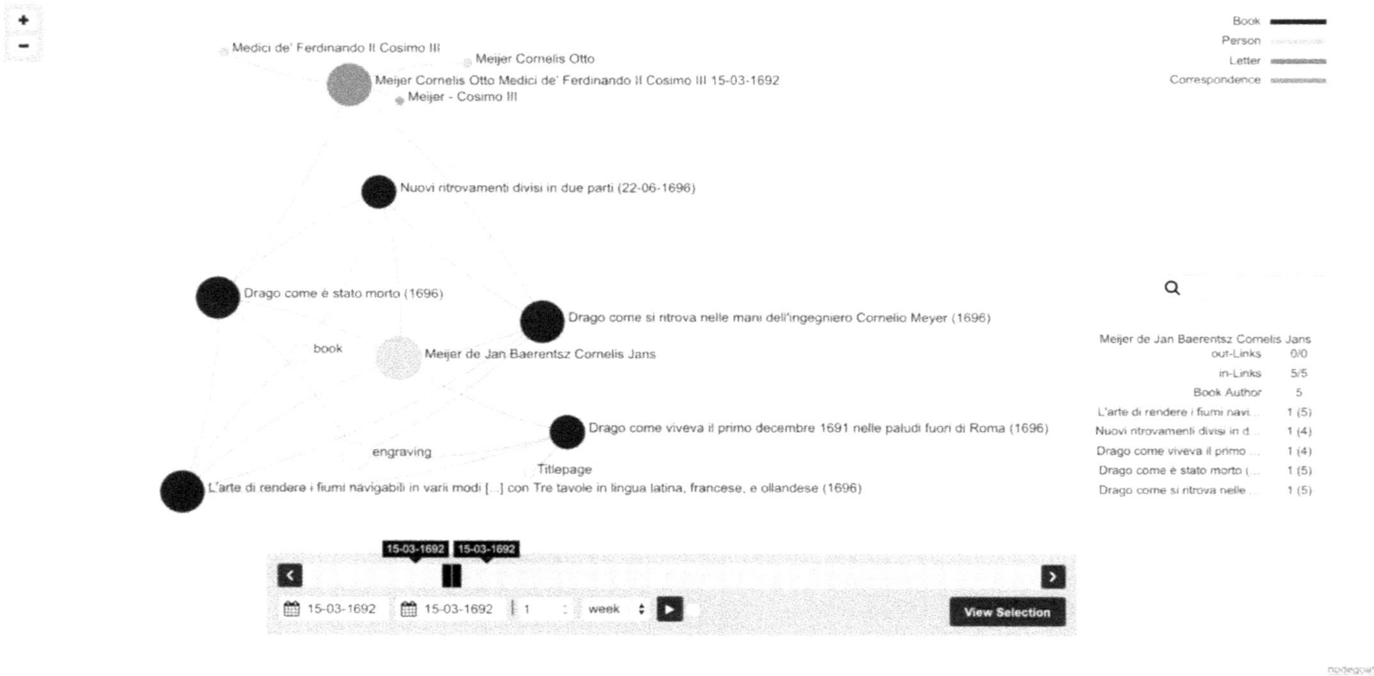

Figure 4.1.4 This visualization shows Cornelis Meijer (grey node in the centre) in relation to the engravings and publications he produced on the dragon he found in the marshes nearby Rome (black). Two of these engravings were enclosed to a letter (dark grey) of his son, Otto Meijer, to Cosimo III. Hovering over the nodes provides a more detailed description of the object.

composition of networks in which technological knowledge was exchanged. Cosimo sent out various engineers to spy on inventions in industries, civil engineering and fortification. From the correspondence of the engineer Pietro Guerrini it becomes clear that Cosimo let his secretary write down in detail which places and people the engineer had to visit to obtain the required, often secret, information about the latest developments in fortification and technical inventions, such as in the linen industry during his tour through Germany, the Low Countries, England and France. Finally, he indicated how Guerrini had to report this information in the form of drawings.[28] This gathering of technical knowledge was certainly not reciprocal, and Guerrini even reports during his espionage tour that he was not always able to provide the requested information "because the Dutch were reluctant to show it."[29] Apart from the fact that this research supported our claim that technological knowledge exchange had a different, more hierarchical nature than the intellectual one, the overlay of more networks also revealed at first sight unexpected relationships. It was not technical people or engineers who introduced Guerrini to key figures in fortification and linen industries in the Low Countries but foremost merchants as Giovacchino Guasconi or book publishers such as Pieter Blaeu (Figure 4.1.5), member of the famous Dutch publisher and cartographers dynasty Blaeu.[30]

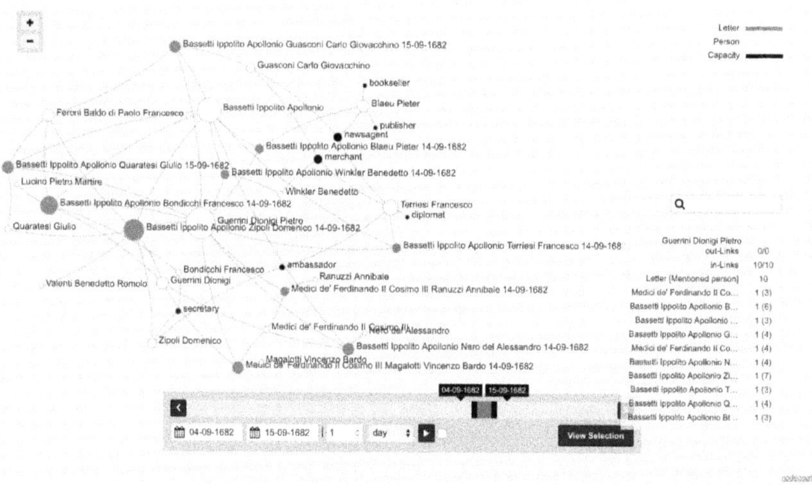

Figure 4.1.5 Introduction network of Pietro Guerrini. At the beginning of September 1682, Apollonio Bassetti and Cosimo III sent several introduction letters (grey) to their contacts (white) in the Dutch Republic, Germany and England to introduce Guerrini. In the Dutch Republic, Guerrini relied mostly on the help of the bookseller Pieter Blaeu, who received Bassetti's letter on the 14th of September 1682, as shown in this visualization. In the introduction letters several other people (white) were mentioned that assisted in the organization of Guerrini's stay in the Dutch Republic. Francesco Feroni, for example, sent money from Florence to maintain Guerrini during his trip.

Multi-layered networks as associative interfaces to historical research

Overlaying different networks of actors and artefacts resulted at first sight in the case of the engineers in Rome and Florence in surprising combinations. In hindsight, we can bring in a plausible argument as to why Pieter Blaeu played such a crucial role in the technological spy missions of Cosimo III de'Medici in the Low Countries. Before sending out his engineers to Northern Europe, Cosimo had already visited the Low Countries twice in the years 1667–68 and 1669.[31] "Low Countries": The use of network analysis to reconstruct and analyze the epistolary relations between the Grand Duchy of Tuscany and the Dutch Republic during the reign of Cosimo III has been extensively studied in Ingeborg van Vugt, *The structure and dynamics of scholarly networks between the Dutch Republic and the Grand Duchy of Tuscany in the 17th century*, Unpublished dissertation, Scuola Normale Superiore di Pisa/University of Amsterdam, 2019. https://hdl.handle.net/11245.1/94502a28-e642-4ecc-81e2-100fda93ecba (accessed 2 March 2020). During these visits Blaeu introduced the grand duke to important magistrates and the collections of important scientists, scholars and publishers. This additional historical information makes clear why it is important that a researcher can use interfaces that allow making associations during the whole process in which different metadata of actors and artefacts are combined. Nodegoat enhances the interoperability of the dataset, making it possible to add, edit and to remove data where needed.

While the first stage publishers were singled out in a network of professions, the further contextualisation of Blaeu, who could be trusted because he had accompanied Cosimo in the Low Countries, seems to suggest that he moved from a professional into a more personal network. For a further understanding of how these personal networks worked at a distance when Cosimo and his secretary were back in Florence again, introduction letters were mapped and added as an extra network on top other ones (Figure 4.1.6).

This example makes clear that the choice of which relationships to combine in networks was rather arbitrary at the beginning and their meaning hard to assess in this stage. Rather than trying to come up immediately with results based on statistical probabilities, it is better that researchers are equipped with a tool that allows them to build up multi-layered networks step by step to explore and to interact with the metadata in a continuous process of associations. Explorations in networks and in research go in this phase hand in hand. During these explorations annotation and contextualisation are crucial to allow for explanations in retrospect. However, there are also limitations to this explorative and interactive approach. It was possible to map the introduction letters manually since their number was limited. However, if we want to map those introduction letters for the whole Republic of Letters as we suggested for one of the hypotheses in our data-model and if we want create networks to test the four other ones, we are in need of different, more quantitative approaches based on probability and statistical methods. The question is how we can close the gap between the qualitative, exploratory and quantitative explanatory methods to optimise

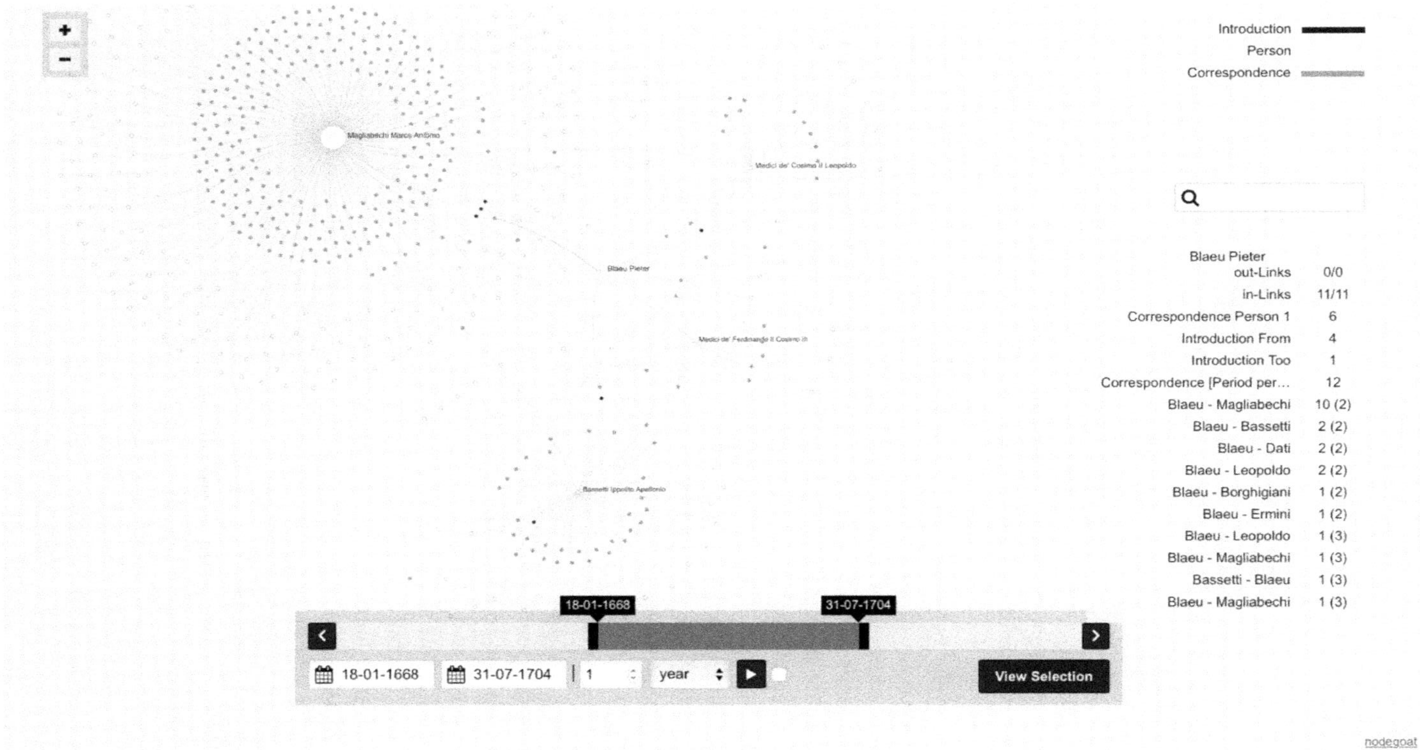

Figure 4.1.6 A visualization of Blaeu's position in the correspondence networks of his main contacts in Florence (Antonio Magliabechi, Apollonio Bassetti, Leopoldo de' Medici and Cosimo III de' Medici). In this image, the correspondents are shown in white, while the correspondence is represented in grey. The filter and scope function of Nodegoat select those correspondents that were introduced by Blaeu (black)

results. We consider the creation of deep networks using both qualitative, manual and quantitative, computer-assisted methods approaches.

Creating deep networks

In another article the term deep networks was introduced to discuss the potential of multi layered networks, similar to David Bodenhamer's deep maps based on multiple layers of maps in Geographical Information Systems (GIS) systems.[32] Bodenhamer stated:

> That the layers of a deep map need not be restricted to a known or documentary record, but could be opened, wiki-like, to anyone with a memory or artefact to contribute. However, structured, these layers would operate as do other layers within a GIS, viewed individually or collective as a whole.[33]

There is a similarity between Bodenhamer's description of the GIS layers that need to be enriched with contextual information or artefacts to create deep maps and our earlier example of network representations of the ePistolarium tool. As Bodenhamer points out,[34] for many humanities scholars, GIS appears reductionist. The maps that are created give an overview of the landscape on the basis of some generic features but do not represent its full richness. They result in one-dimensional, flat maps. The ePistolarium tool allowed us to visualize all sorts of keyword and similarity searches in correspondence and co-citation networks. However, when discussing experiences with the ePistolarium tool it was observed that some topics got easier to the foreground than other, more implicit ones, leading to different levels of abstraction. It implied that we could see patterns but not the depth of those patterns. In short it resulted sometimes in flat networks. In the Mapping Notes and Nodes in Networks project the manual creation of multi-layered networks allowed for observing their depth in every moment of their construction. The scope function of Nodegoat allows users to capture all objects and their subclasses in one visualization, but it is hard to assess how they relate to an overall pattern in networks (Figure 4.1.7).

In particular, this poses problems to analyse transitions in network structures. Without any visible hierarchy in the relations between subclasses and objects, it is difficult to assess which specific event causes a change in the overall network structure. As is said before, letters of introduction were often necessary to be admitted into an epistolary network. Following the evolution of introduction letters alongside a citation network and determining whether there is a shift taking place in the number of intermediates between correspondents (demonstrating shrinking degrees of separation) could reveal the importance of introduction in the establishment of epistolary networks. The following example will clarify that multi-layered networks overlook the importance of cross-sections between different kinds of networks. Andries Fries (1630–75), a Dutch bookseller working for a Venetian publishing firm, maintained for many years a correspondence with the librarian of the Medici family, Antonio Magliabechi.[35]

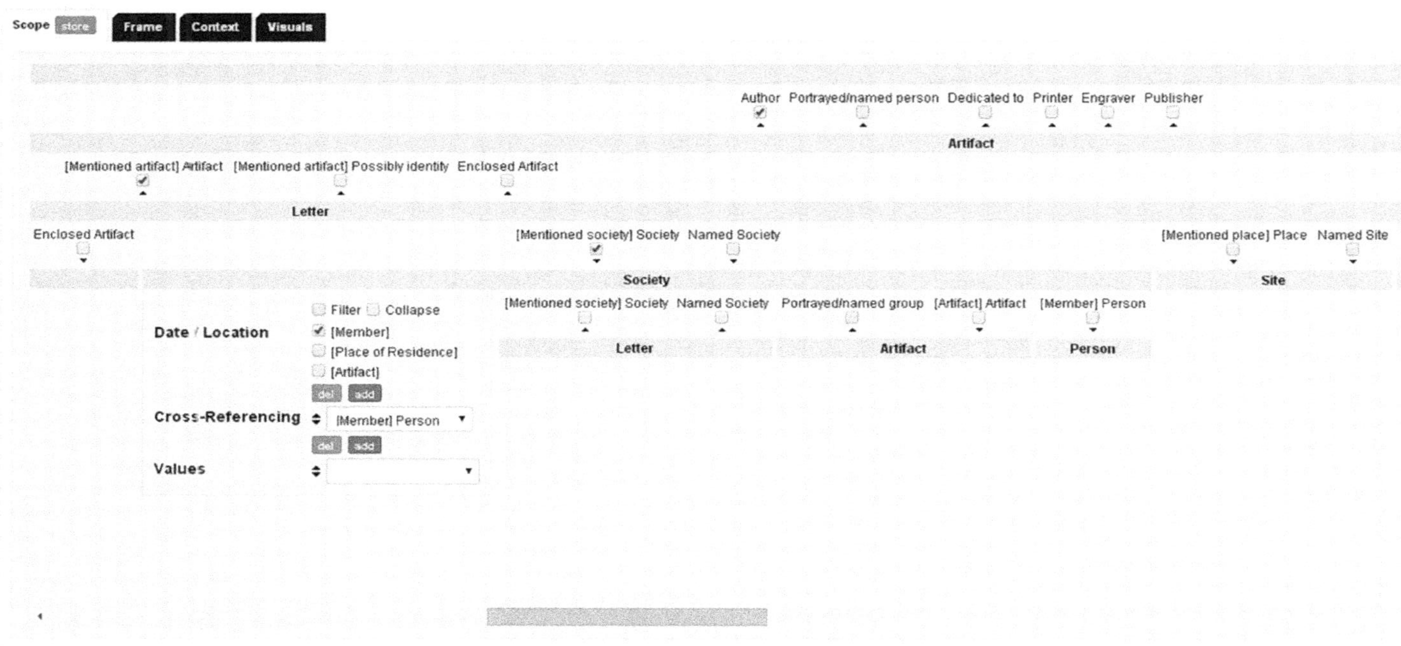

Figure 4.1.7 The scope function of Nodegoat enables the user to select both object and sub-objects, combining them into one visualization

Source: ©Ingeborg van Vugt, Scuola Normale Superiore di Pisa

Consequently, Fries was able to introduce Pieter Blaeu to the Medici court. Blaeu, as we have said before, became one of the most reliable contacts in the spy missions of Cosimo III. This introduction constitutes therefore an essential event to understand further the nature of Blaeu's network. A citation network of this correspondence illustrates indeed that Blaeu was frequently mentioned in the letters of Fries until the moment he was being introduced to the Medici court (Figure 4.1.8). From that moment on, Pieter Blaeu is incorporated in the direct epistolary network of the Florentine court. We are able to follow this transition if we compare these networks along each other (Figure 4.1.9). However, if one captures these networks in one visualization, comparison between these layers becomes problematic. It becomes difficult, even impossible, to assess the role of Pieter Blaeu, present in three different kinds of networks (the citation network, the introduction network and, ultimately, his direct epistolary network), in relation to the overall network visualized. To use the map-landscape metaphor once again: the maps created with Nodegoat provide a very detailed layered representation of parts of the landscape but do not show their relations with the landscape, that is the integration in one network, as a whole.

The reason is that Nodegoat, different from visualization-oriented tools such as Gephi, is built around data entry, management and curation processes. This allows Nodegoat to be used for exploration of multi-modal historical networks in various configurations. Instead of relying on an algorithmic (top-down) approach, Nodegoat pursues a bottom-up approach by applying contextual analyses. By means of path aware filtering and visualization functionalities, selections can be created and highlighted in a broader context within the dataset. Nodegoat contextualises these selections in time and space by means of diachronic geographic and social network visualizations. The diachronic social network visualization is driven by a force layout algorithm for the distribution of the nodes to enhance the readability of networks. No algorithms have been implemented yet to support the statistical analyses of network structures.[36] In the relatively small datasets in Nodegoat, such as in the Mapping Notes and Nodes project, layered networks are a very useful interface for researchers to interact with data in an exploratory way. However, with the inclusion of large datasets, it becomes more important to enhance them with tools to analyse and visualize topological network structures. While Nodegoat allows for the manual creation of deep networks by overlaying networks op top of each other, other ways need to be developed to create computer-assisted, machine-readable deep networks.

Our notion of deep networks was inspired by Bodenhamer deep maps intended to enrich existing maps with observations of users from multiple perspectives. However, in the literature of computer science the term deep networks, more precisely deep belief networks (dbns) is already in use. Deep belief networks are created within the context of machine learning. Multiple layers of latent variables are trained via algorithms that attempt to model high-level abstraction in such a way that hidden units can be revealed (deep or hierarchical learning).[37] These methods are also used in the context of Natural Language Processing and might in the future solve the problem of different levels of abstraction

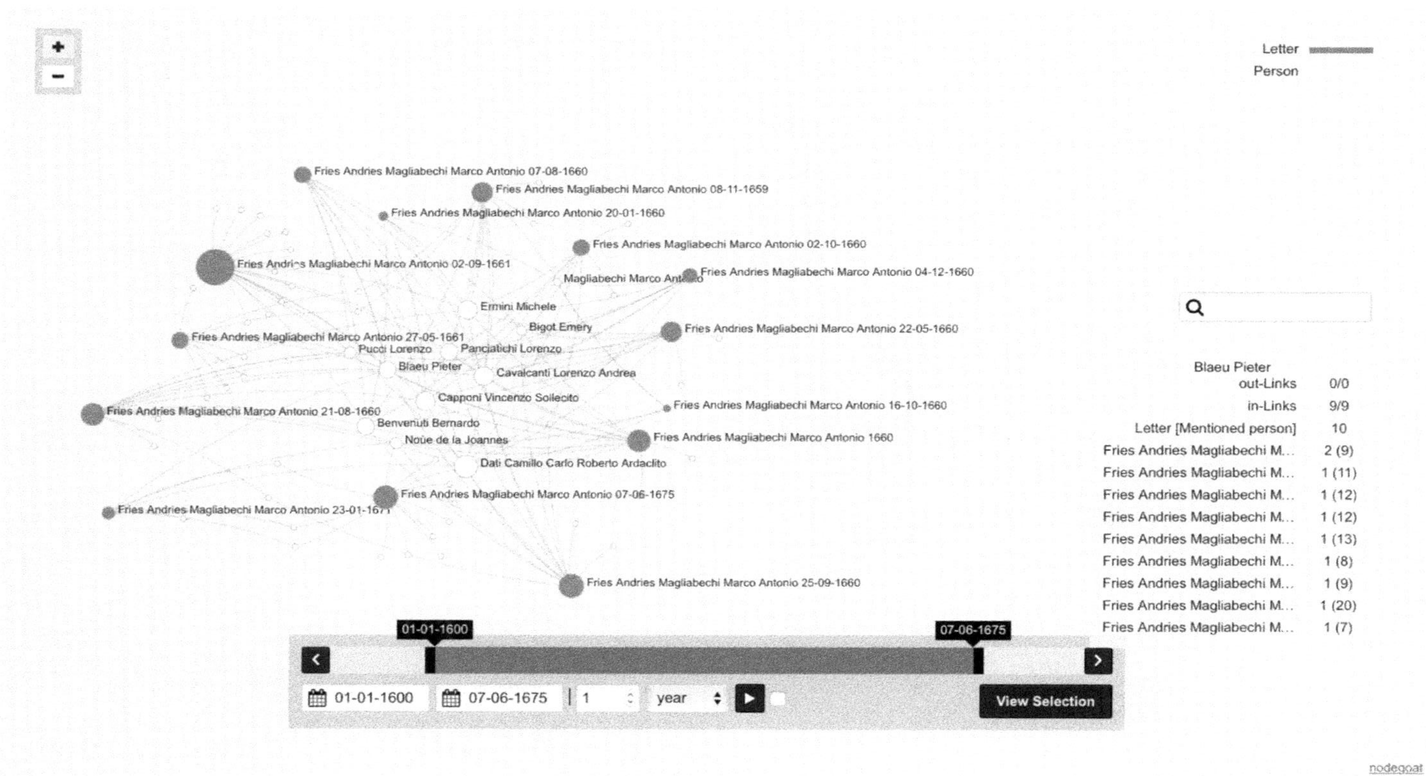

Figure 4.1.8 A (co)-citation network of the correspondence Fries-Magliabech. The white nodes represent cited persons in the letters (grey) between Fries and Antonio Magliabechi, showing the central role of Pieter Blaeu.

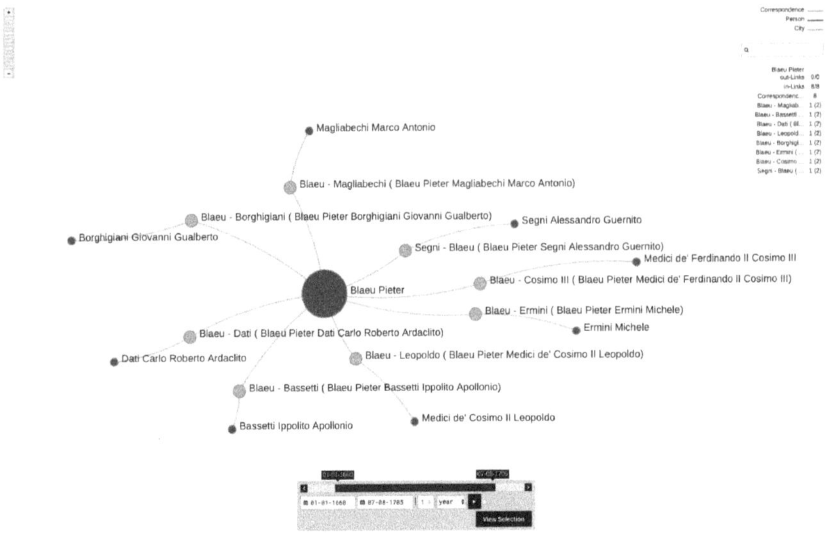

Figure 4.1.9 This graph illustrates the direct epistolary network of Blaeu after he was introduced to the Medici court. Most of the main contacts of Fries (Figure 4.1.8) are incorporated in Blaeu's network (e.g. Magliabechi, Dati, Ermini), and new contacts were added in (Leopoldo, Borghigiani, Bassetti, Segni).

between more and less implicit concepts in topic modelling that we observed in the experiments with the ePistolarium tool. However, there is another problem to solve. Apart from different levels of abstraction there is the problem of changes of topics over time.[38] Whereas in predictive analytics or machine learning "concept drift" affects the statistical properties of the variables the model tries to predict, it also affects our interpretation of patterns in historical data. The historian of English literature, Ted Underwood, addressed this problem eloquently as follows:

> Once you create a set of topics, plotting their frequencies is simple enough. But plotting the aggregate frequency of a group of words is not the same thing as "discovering a trend," unless the individual words in the group correlate *with each other* over time. And it's not self-evident that they will.[39]

In order to understand differences in abstraction that might bias our interpretation of networks we have to mine words in context (context-mining) and represent them in context. A promising experiment set up in collaboration between Princeton University and the Carnegie-Mellon University might in the future be useful both to detect hidden (implicit) concepts and assist in detecting concept drift. David M. Blei and John D. Lafferty developed a 75-topic dynamic topic model to browse the journal *Science* over the period 1880–2002 as supplement to their research, *Modeling the Evolution of Science*.[40] A click on one of the top five words from each topic, taken every ten years, leads to a page that contains the top 100 words from that topic, links to its distribution for the previous and future years and the articles that exhibit that topic with the highest proportion. They used the browser to explore the entire collection of *Science* according to the hidden topical structure that the dynamic topic model had uncovered. However, the layers of increasing topics words not only reveal the (partly hidden) content of articles according to the topics they exhibit, but the browser also enables one to see how a topic has changed over time. The continuation of such research will make it possible in the future to create and to explore patterns in machine-readable, dynamic, deep networks of big data.

Classification and alignment in networks

Earlier, we discussed how content could be hidden in different ways. Especially in large data sets (for instance in ePistolarium) we could see overall patterns in the content but only for some features (representing a part for the whole), whereas in small data sets (such as in Nodegoat) we could see the complete depth but not the relation to the overall network (representing the whole for a part). We also discussed the problem of changes in semantic meaning over time, "concept drift". Here we introduce a third problem, linking data sets of various media structured in different ways. These three problems together lead to the question of how parts of the whole or the whole of the parts best can be brought together taking into account differences between structured and unstructured data and temporal change. To solve these problems for large datasets in the future, historical network research might benefit from the expertise build up in the communities around Linked Open Data. Linked Open Data has emerged as the largest collection of interlinked structured datasets on the Web using formal ontologies, such as hierarchical classifications. Although there is still a long way to go before computer-generated topics with different levels of abstraction can be automatically identified and mapped to ontologies with formal hierarchical classes, automatic domain identification and ontology alignment are promising approaches within Linked Open Data.[41] These approaches might be explored for historical network research in combination with topic modelling and text and document classification to bridge the gap between networks based on more and less structured datasets. Current research on linked data and the semantic web holds the promise of the development of large flexible structures of big data that allow for exploring interactively multiple representations of the same object from various perspectives.[42] As such they can be read as computer-assisted deep networks that can be used to create associative, explorative interfaces for historical research networks similar to the manually created deep networks of the Nodegoat tool. For the problem of concept drift, we already mentioned the topic modelling experiment developed in the context of the *Modeling the Evolution of Science* project. Within Linked Open Data the Memento project developed in the context of web archiving is of great interest for dealing with temporal change in large datasets.[43] For temporal change in classification of metadata in networks based on small datasets the Nodegoat tool provides an interesting approach, named "reversed classification." It assists the user in handling fuzzy datasets over time. It is based on the principle that predefined classifications are counterintuitive when looking at data from an object-oriented perspective. Instead it enables the user to define a multi-facetted filter of data in multiple sets in which they can define any number of parameters that are associated with a classification. This will reverse the classifying process because classes are defined by the exchange between the parameters of the classification and the attributes of the chosen objects. A person like Pieter Blaeu in our case study and the insight he represents within the networks surrounding Cosimo's engineering endeavours can be identified and further explored by creating a reversed classification that maps persons labelled with a capacity other than engineer and related capacities who have introduced two or more people in the complete network. With the introduction of the node of this reversed classification, a new force is applied in the network that gathers and subsequently centres the persons it classifies. Their first order or second order network within a visualization of the complete network provides an entrance from which detailed networks can be contextualised in a greater whole.

Visualizing uncertainty

From the Semantic Web and Linked Open Data initiatives we have learned that bringing together computer-generated and user-generated data results in many uncertainties. We expect the same for a potential future combination of machine-readable dynamic, deep networks of big data with manually created multi-layered networks with small data. Rather than trying to minimise these uncertainties in a reductionist way, we need to acknowledge and incorporate these in our models and make them visible when creating deep networks. To refer to David Bodenhamer in the context of deep maps once again:

> The deep map is meant to be visual and experimental, immersing users in a virtual world in which uncertainty, ambiguity, and contingency are ever-present, influenced by what was known (or believed) about the past and what was hoped for or feared in the future.[44]

Similar to deep maps, deep networks need to be able to express complex relationships full of ambiguities and uncertainties. A first step is to make these complex relationships explicit, which can be supported by visualizations. In the Mapping Notes and Nodes in Networks project we dealt with uncertainties as a result of conflicting metadata of biographical, geographical or temporal data by visualizing the provenance of the data base collection from which these metadata were extracted. Users themselves can assess and express the level of uncertainty using deeper and lighter colour codes (Figure 4.1.10).

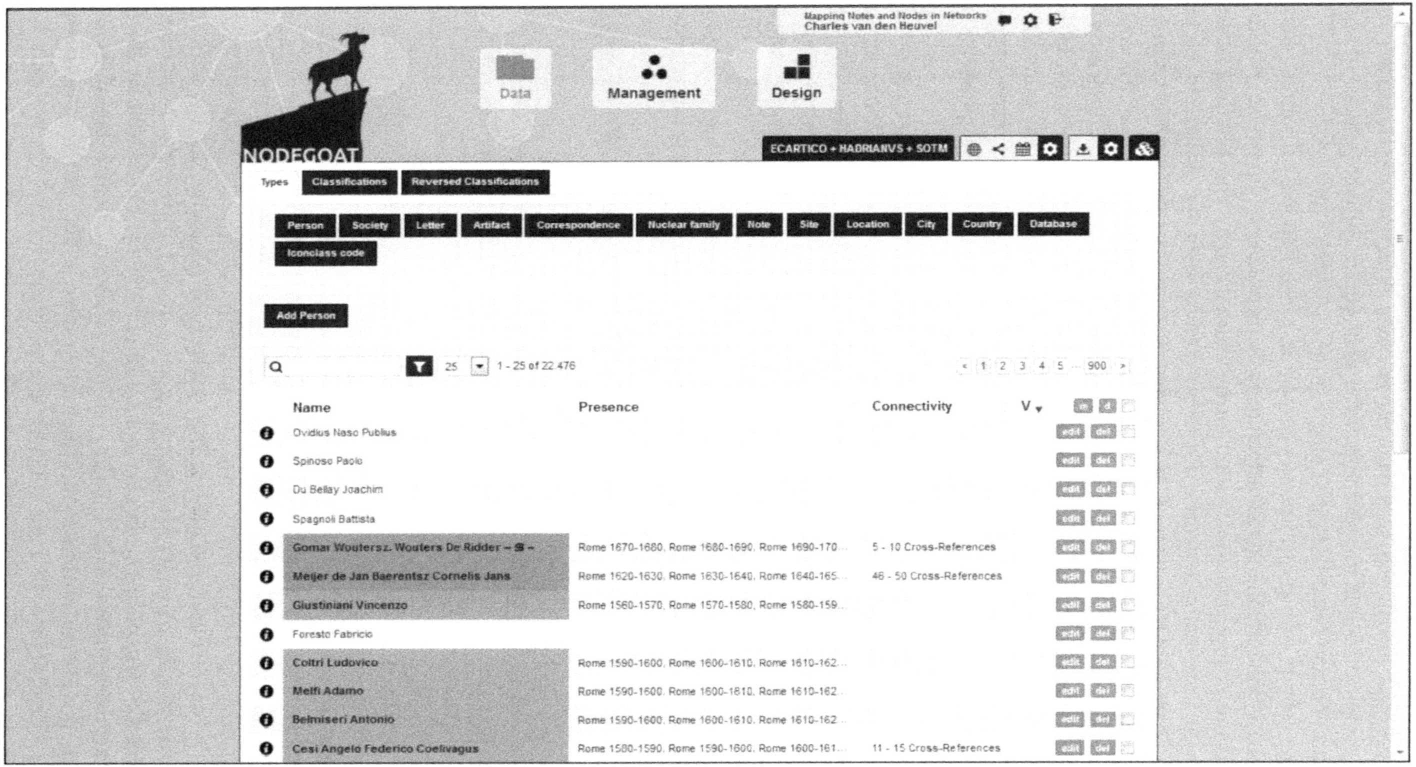

Figure 4.1.10 Nodegoat: visualization degrees of uncertainty based on provenance of metadata

Furthermore, historical network research can benefit from the experience with analysing and visualizing uncertainties in other disciplines.[45] Depending on scale, complexities and uncertainties differ and require different modes of contextualisation and visualization of single or multi-layered networks.

In the context of historical network research of the Republic of Letters, Scott Weingart (forthcoming) suggested large-scale networks to juxtapose multiple networks to enable contextualisation (semi-) automatically and handling uncertainties in large-scale networks. He experimented with the projection of the corpus of the Dutch Circulation of Knowledge project on the Catalogus Epistularum Neerlandicarum[46] (Figure 4.1.11). Nicole Coleman of Stanford University set up various experiments with the Density Design Group of the Politecnico di Milano to visualize uncertainties manually in single small-scale networks. In the project KNOT they created an interface to manipulate and annotate graphically the nodes in a network while retaining the data-driven links that expressed an explicit connection.[47] (Coleman and Heuvel, submitted for publication) (Figure 4.1.12).

The approaches of Scott Weingart and Nicole Coleman are completely different. In his article "The Networked Structure of Scientific Growth" (forthcoming) Weingart does not explicitly use the term uncertainty but in his discussion of risks in the process of data gathering questions Drucker's deconstruction of data and plea for a pure qualitative approach set out in her seminal article "Humanities Approaches to Graphical Display." Instead Weingart (forthcoming p. 10) suggests to employ more rigor in humanities research: "We may lose some of the uniquely human information relevant to humanities research, but what we lose in specificity we gain in rigor".[48]

Weingart further quotes an early expert of agent modelling Joshua Epstein to stress his point:

> You can in fact calibrate to historical cases in there are data, and can test against current to data to the extent that exist. And, importantly, you can incorporate the best domain … expertise in a rigorous way"[49]

The emphasis on rigor and statistical methods is also recognisable in visualizations of uncertainties in research of the Republic of Letters that Scott Weingart presented together with Charles van den Heuvel in the examples of maps showing Bayesian data analysis and probability distributions.[50]

This approach is completely the opposite from the one underlying the visualizations developed by Nicole Coleman in collaboration with the Density Design Group to express uncertainties in the research of Mapping the Republic of Letters project. In her study *Visualizing Uncertainty and Complexity: Humanistic Methods For Mapping The Intellectual Geography of the Early Modern World* Coleman, co-authored with Charles van den Heuvel, Coleman questions the methods of GIS and information visualization experts that aim at representing data as visual objects that are concrete and accurate and that try to assess the boundaries of uncertainty and variability of graphic display to express

Figure 4.1.11 Scot Weingart, visualization of letter correspondents in Catalogus Epistolarum Neerlandicarum (detail) – Circulation of Knowledge project

Source: © Huygens ING

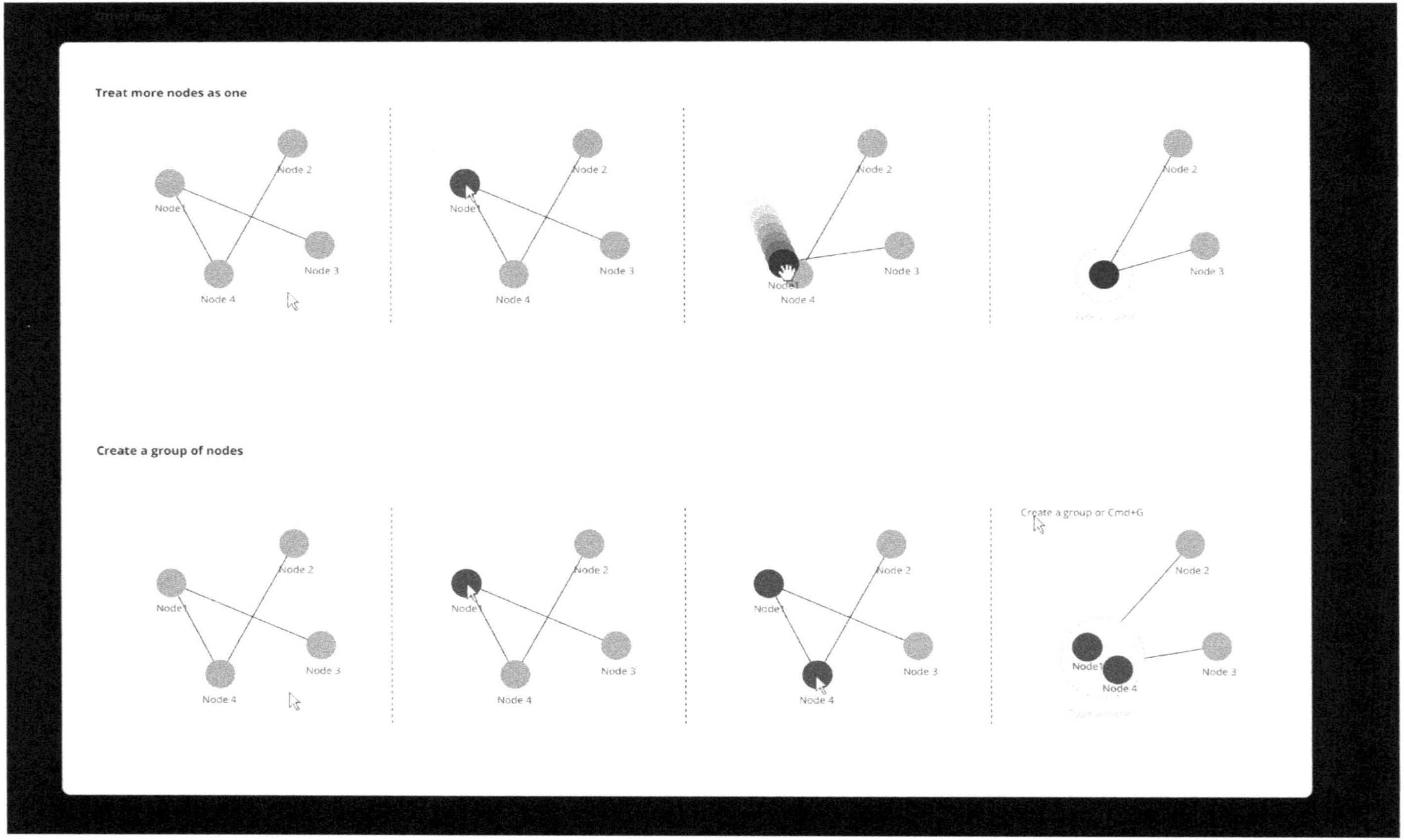

Figure 4.1.12 Illustration by DensityDesign of interaction design for nodes project Knot

Source: Courtesy N. Coleman (©Stanford University)

complexity by measurement. Instead of the development of a graphic language based on deviations of an assumed standard model, she proposes graphic interfaces to contextualise and manipulate visualizations on the basis of the following "Design Principles for Humanist Inquiry: 1) reveal complexity 2) think through data 3) engage with dynamics and 4) let the user structure data."

The two different approaches of Weingart and Coleman to handle data and to analyse and visualize uncertainty in networks of knowledge exchange in the Republic of Letters seem at first sight mutually exclusive, but they are not. The Knot experiment was set up as a node-link graph model to reveal the degree of connectedness between individuals in a correspondence network, and the visual layouts were created according to algorithms based on measurable values. Coleman admits that statistical methods are required to help us aggregate the metadata across the rapidly growing cultural heritage institutions in a consistent and efficient manner but states that other methods are needed to carry out human-scale inquiry of large datasets. And also Weingart proposes a "mixed method" and concludes his article on "The Networked Structure of Scientific Growth" with the announcement of the important next step to combine distant and close reading and mixing traditional historical research with newer quantitative studies to come to a synthesis.[51]

The two different approaches of handling the networks of the Republic of Letters are complementary, rather than contradictory. Analysis and design, exploration and explanation make part of a continuous research process. We believe that the creation of deep networks by overlaying computer-generated patterns and manually created networks for contextualisation and manipulation can play a role in bridging these different approaches and hold a promise for the future of historical network research and a networked humanities.

Notes

1 The full name of the project *was Circulation of Knowledge and Learned Practices in the 17th-century Dutch Republic A web- based Collaboratory around Correspondences*. It was a consortium of institutes of the Royal Netherlands Academy of Arts and Sciences (Huygens Institute and DANS), Dutch universities (Utrecht University and University of Amsterdam) and the National Library of the Netherlands (KB).

2 For an explanation of the various used NLP techniques see http://ckcc.huygens.knaw.nl/?page_id=13 (accessed 5 December 2018).

3 In the the NIAS-Lorentz workshop *Mathematical Life in the Dutch Republic* organised in Leiden in 2010 participated ca. 20 historians of science in the experiment with early version of the ePistolarium tool.

4 Peter Wittek and Walter Ravenek, "Supporting the Exploration of a Corpus of 17th Century Scholarly Correspondences by Topic Modeling," in *Proceedings of Supporting Digital Humanities: Answering the Unaskable*, ed. B. Maegaard (Copenhagen, 2011). ; Ravenek, Walter, Charles van den Heuvel, Guido Gerritsen, "The ePistolarium: origins and techniques." In Jan Odijk, and Arjan van Hessen (eds.) *CLARIN in the Low Countries.* London: Ubiquity Press. pp. 317–323. 2017

5 "ePostolarium: de Causus Saturnus Text," http://ckcc.huygens.knaw.nl/?page_id=1030 (accessed 4 March 2020] Charles van den Heuvel, "Modelling Texts and topics," In: Howard Hotson and Thomas Wallnig (Eds.) Reassembling the Republic of Letters in the Digital Age. Standards, Systems, Scholarship, Göttingen: University Press,

pp. 265–280. 2019 Doi:10.17875/gup2019-1146 and Marlon Cesar Alcantara Marco Braga and Charles van den Heuvel, "Historical Networks in Science Education. A Case Study of an Experiment with Network Analysis by High School Students," Science and Education 29, pp. 101–121, Dordrecht: Springer Netherlands 2020 DOI HYPERLINK "https://doi.org/10.1007/s11191-019-00096-4" https://doi.org/10.1007/

6 Ibid.

7 Henk Nellen, "'In Strict Confidence': Grotius' Correspondence with his Socinian Friends," in *Self-Presentation and Social Identification: The Rhetoric and Pragmatics of Letter Writing in Early Modern Times*, ed. T. van Houdt, G. Tournoy and C. Matheeusen, Supplementa humanistica lovaniensia 18, pp. 227–45 (Leuven: University Press, 2002); Henk Nellen, "The Correspondence of Hugo Grotius," in *Les grands intermédiaires culturels de la République des Lettres: Études de réseaux de correspondances du XVIe au XVIIIe siècles*, ed. C. Berkvens-Stevelinck, Les dix-huitièmes siècles no 91, pp. 127–64 (Paris: Honoré Champion, 2005).

8 Charles van den Heuvel, "Netwerken van vertrouwelijkheid en geheimhouding in brieven en tekeningen," in *In vriendschap en vertrouwen. Cultuurhistorische essays over confidentialiteit*, ed. J. Gabriëls et al. (Hilversum: Verloren, 2014); Charles van den Heuvel et al., "Modeling Confidentiality and Secrecy in Knowledge Exchange Networks of Letters and Drawings in the Early Modern Europe," *Nuncius*, no. 31 (2016): pp. 78–106.

9 P.C. Molhuysen, B.L. Meulenbroek, P.P. Witkam, H.J.M. Nellen and C.M. Ridderikhoff, eds., *Briefwisseling van Hugo Grotius [Rijksgeschiedkundige Publicatiën, Grote Serie]*, Vols. 1–17 (The Hague: Martinus Nijhoff, later Instituut voor Nederlandse Geschiedenis, 1928–2001).

10 Charles van den Heuvel, "Netwerken van vertrouwelijkheid en geheimhouding in brieven en tekeningen"; Charles van den Heuvel, "Mapping Notes and Nodes: Building a Multi-Layered Network for a History of the Cultural Industry," 2015, http://dh2015.org/abstracts/xml/HEUVEL_Charles_van_den_Mapping_Notes_And_Nodes__B/HEUVEL_Charles_van_den_Mapping_Notes_And_Nodes__Buildin.html (accessed 5 December 2018).

11 Charles van den Heuvel et al., "Modeling Confidentiality and Secrecy in Knowledge Exchange Networks of Letters and Drawings in the Early Modern Europe," pp. 78–106.

12 Ibid.

13 The ePistolarium mentions the number of 8,040 but that is based on the inclusion of letters attached the ones of Grotius, not written by or to him and copies.; P.C. Molhuysen, B.L. Meulenbroek, P.P. Witkam, H.J.M. Nellen and C.M. Ridderikhoff, eds., *Briefwisseling van Hugo Grotius [Rijksgeschiedkundige Publicatiën, Grote Serie]*.

14 Mathieu Bastian, Sebastien Heymann and Mathieu Jacomy, "Gephi: An Open Source Software for Exploring and Manipulating Networks." (2009) *International AAAI Conference on Weblogs and Social Media.* https://gephi.org (accessed 5 December 2018).

15 We are indebted to Nils Spelt, student of the Erasmus University Rotterdam, who during an internship in 2014 at the Huygens ING (Netherlands) analyses and visualized reciprocity in the networks of Grotius. The tables with the numbers of exchanges are published in Charles van den Heuvel et al., "Modeling Confidentiality and Secrecy in Knowledge Exchange Networks of Letters and Drawings in the Early Modern Europe," pp. 78–106.

16 COST Action IS 1310 Reassembling the Republic of Letters, www.republicofletters.net/ (accessed 5 December 2018). Howard Hotson and Thomas Wallnig (Eds.) Reassembling the Republic of Letters in the Digital Age. *Standards, Systems, Scholarship,* Göttingen University Press 2019, pp. 237–264. Doi:10.17875/gup2019-1146.

17 P. van Bree, and G. Kessels. "Nodegoat: A web-based data management, network analysis & visualisation environment," (2013), http://nodegoat.net from LAB1100; http://lab1100.com.

18 The full title of the project was: *Mapping Notes and Nodes in Networks, Exploring potential relationships in biographical data and cultural networks in the creative industry in Amsterdam and Rome in the Early Modern Period*. This section is an abridged version of the extended abstract for the Digital Humanities 2015 conference: http://dh2015.org/abstracts/xml/HEUVEL_Charles_van_den_Mapping_Notes_And_Nodes__B/HEUVEL_Charles_van_den_Mapping_Notes_And_Nodes__Buildin.html (accessed 5 December 2018).

19 Biographical Reference Works, www.huygens.knaw.nl/tools-en-data/?lang=en; Ecartico, www.vondel.humanities.uva.nl/ecartico/ and Hadrianus, www.hadrianus.it/about (this website of the KINR in Rome has been stopped, but its data will be incorporated in the website of the RKD in 2019: https://rkd.nl/en/ (all accessed 5 December 2018).

20 Frits Scholten (Rijksmuseum; Vrije University of Amsterdam) and Arjan de Koomen (University of Amsterdam) provided the data about the migration of Dutch sculptors throughout Europe that their students had assembled in the project Sculptors on the Move. Susanna de Beer (Leiden University), with a small group of students, developed in her Mapping Visions of Rome project a typology of descriptions of Rome in humanist poetry.

21 Compare Paul H. Windolf, "Social Capital and Social Inequality: Corporate Networks in Germany and the United States (1896–1938)," *SSRN Electronic Journal* (2014), doi:10.2139/ssrn.2529698: p. 4 (accessed 5 December 2018).

22 Rafael Capurro, "Digital Hermeneutics: An Outline" (2010), www.capurro.de/digital-hermeneutics.html (accessed 5 December 2018).

23 Chiel Akker et al., "Agora Digital Hermeneutics: Online Understanding of Cultural Heritage," *Proceedings of the 3rd International Conference on Web Science (WebSci' 11)* (Koblenz, Germany, 2011), www.cs.vu.nl/~guus/papers/Akker11a.pdf (accessed 5 December 2018).

24 For a full description of the project and cases see the end report, available at: www.huygens.knaw.nl/wp-content/uploads/2015/05/EndReportMNN.pdf (accessed 5 December 2018).

25 For an example on the role of books as dynamic actors within the early modern network by means of multi-layered visualizations see Ingeborg van Vugt, "Using Multi-layered Networks to Disclose Books in the Republic of Letters," *Journal of Historical Network Research* 1 (2017): 25–51. Moreover, Matteo Valleriani et al. used multi-layered networks to model the emergence of early modern epistemic communities: Matteo Valleriani, "The Emergence of Epistemic Communities in the *Sphaera* Corpus: Mechanisms of Knowledge Evolution," Journal of Historical Network Research 3 (2019): pp. 50–91. The importance of textual objects as participants in networks has also been stressed by Bruno Latour, *Reassembling the Social: An Introduction to Actor-Network-Theory* (Oxford: University Press, 2005).

26 An engraving with the same composition but with the dragon in full flesh we find on the title page of the second volume of Cornelia Meijer, *Racolta di vari segreti* published in Rome in 1691 with the text: *Drago come è stato morto*. (Dragon as it was killed). A different composition of the dragon we find on the title page of Cornelio Meijer, L'arte di rendere i fiumi navigabili published in Rome of 1696 with the subscription: *Drago come viveva il primo di Decembre 1691 nelle paludi fuori di Roma* (Dragon as it lived on the 1st of December of 1691 in the marshes outside Rome).The connections between these works are visualized in Figure 4.1.3.

27 Letter from Otto Meijer to Cosimo III, 15 March 1692. Mediceo del Principato 1133, ff. 851–2. Florence: State Archive.

28 Francesco ed. Martelli, *Il viaggio in Europa di Pietro Guerrini (1682–1686): Edizione della corrispondenza e dei disegni di un inviato di Cosimo III dei Medici*, Documenti di storia italiana Serie II 11 (Florence: Leo S. Olschki, 2005), p. 2.

29 Ibid., p. LXXIII.

30 Alfonso Mirto and Henk Th. van Veen, *Pieter Blaeu: lettere ai Fiorentini, Antonio Magliabechi, Leopoldo e Cosimo III de' Medici e altri* (Firenze, Amsterdam, Maarssen: Centro Di, 1993).

31 Godefridus Joannes Hoogewerff, "De twee reizen van Cosimo de' Medici, prins van Toscane door de Nederlanden (1667–1669)," *Journalen en Documenten* (Amsterdam: J. Müller, 1919); Lodewijk Wagenaar, *Een Toscaanse prins bezoekt Nederland. De twee reizen van Cosimo de' Medici, 1667–1669* (Amsterdam: Uitgeverij Bas Lubberhuizen, 2014).

32 Charles van den Heuvel, "Mapping Knowledge Exchange in Early Modern Europe: Intellectual and Technological Geographies and Network Representations," *International Journal of Humanities and Arts Computing* 9, no. 1 (2015), doi:10.3366/ijhac.2015.0140: pp. 95–114.

33 David J. Bodenhamer, "The Potential of Spatial Humanities," in *The Spatial Humanities: GIS and the Future of Humanities Scholarship*, ed. David J. Bodenhamer, Spatial humanities (Bloomington: Indiana University Press, 2010), Ch.2., p. 33.

34 Ibid., p. 24.

35 For more information about Antonio Magliabechi (1633–1714) and his network in the Dutch Republic, see Ingeborg van Vugt, "Storia e Geografia di una rete epistolare." In *Antonio Magliabechi nell'Europa dei Saperi*, edited by P. Boutier, M.P. Paoli and C. Viola (Pisa: Edizioni della Normale, 2017), pp. 259–292.

36 At the moment the developers are implementing algorithms that allow for analysis of topological features of network structures similar to Gephi.

37 Geoffrey Hinton, Simon Osidero and Yee Whye The, "A Fast Learning Algorithm for Deep Belief Nets," *Neural Computation*, 18, no. 7 (July 2006): pp. 1527–54.

38 For this problem see for instance Eyal Sagi, Stefan Kaufmann and Brady Clark, "Semantic Density Analysis Comparing Word Meaning across Time and Phonetic Space," *Proceedings of the EACL 2009 Workshop on GEMS: Geometrical Models of Natural Language Semantics*, Athens, Greece, pp. 104–11 (31 March 2009); Rada Mihalcea and Vivi Nastase, "Word Epoch Disambiguation: Finding How Words Change Over Time," *Proceedings of the Association for Computational Linguistics*, Jeju, Republic of Korea (8–14 July 2012); and Yoon Kim, Yi-I Chiu, Kentaro Hanaki et al., "Temporal Analysis of Language through Neural Language Models," *Proceedings of the ACL 2014 Workshop on Language Technologies and Computational Social Science*, Baltimore, Maryland, USA, pp. 61–5 (2014).

39 Ted Underwood, "The Stone and the Shell," (2011), http://tedunderwood.wordpress.com/2011/09/16/topics-are-also-trends (accessed 5 December 2018).

40 The links http://topics.cs.princeton.edu/Science/ and for the browser http://topics.cs.princeton.edu/Science/browser/ are broken, but references to the experiment are published in David M. Blei and John D. Lafferty, "Dynamic Topic Models," in *Dynamic Topic Models: Proceedings of the 23rd International Conference on Machine Learning*, ed. William Cohen and Andrew Moore, Association for Computing Machinery, New York, NY, United States, pp. 113–20 (2006).

41 Sarasi Lathisena et al., "Automatic Domain Identification for Linked Open Data: Proceedings of the 2013 IEEE/WIC/ACM International Joint Conferences on Web Intelligence and Intelligent Agent Technologies" (November 2013), http://works.bepress.com/amit_sheth/62 (accessed 5 December 2018).

42 Eetu Mäkelä, Eero Hyvönen and Tuukka Ruotsalo, "How to Deal with Massively Heterogeneous Cultural Heritage Data-Lessons Learned in CultureSampo," *Semantic Web* 3, no. 1 (2012): pp. 85–109.

43 Herbert van de Sompel and Michael Nelson, "Memento: Time Travel for the Web. Arxiv preprint," (2009), http://arxiv.org/abs/0911.1112 (accessed 5 December 2018).

44 Bodenhamer, "The Potential of Spatial Humanities," p. 28.

45 Gabriele Bammer and Michael Smithson, eds., *Uncertainty and Risk: Multidisciplinary Perspectives* (London: Routledge, 2009); Matthijs Kouw, Charles van den Heuvel and Andrea Scharnhorst, "Exploring Uncertainty in Knowledge Representations: Classifications, Simulations, and Models of the World," in *Virtual Knowledge: Experimenting in the Humanities and the Social Sciences*, ed. P. Wouters et al., pp. 90–125 (Cambridge, MA: The MIT Press, 2012).
46 See for instance the visualization of Kevin Boyack, Katy Börner and Richard Klavans, "Desiderata for Visualizing Uncertainty and Contextuality in the Digital Humanities: Experiences with the Digital Republic of Letters," in Heuvel; Weingart, *Visualizing Data Resources*, slides 20 and 21, http://ckcc.huygens.knaw.nl/wp-content/bestanden/2012/05/CHeuvel_Scott_Gotha.pdf (accessed 5 December 2018).
47 Nicole Coleman and Charles van den Heuvel, "Visualizing Uncertainty and Complexity: Humanistic Methods for Mapping the Intellectual Geography of The Early Modern World.: Submitted paper for the publication of a volume after the conference Intellectual Geography: Comparative Studies 1550–1700, held from 5–7 September 2011 at St' Anne's College, University of Oxford."
48 The authors wish to thank Scott Weingart for sharing his article The Networked Structure of Scientific Growth (Scott Weingart, "The Networked Structure of Scientific Growth." *The Scottbot Irregular* (22 February 2012). Unpublished. http://www.scottbot.net/HIAL/index.html@p=12050.html (accessed 5 February 2020)). On page 3 in the manuscript he refers to J. Drucker (2011).
49 Scott Weingart, "The Networked Structure of Scientific Growth." *The Scottbot Irregular* (22 February 2012). Unpublished. http://www.scottbot.net/HIAL/index.html@p=12050.html (accessed 5 February 2020), p. 10. refers to Joshua Epstein, "Why Model?," *Journal of Artificial Societies and Social Simulation* 11, no. 4 (2008): p. 12.
50 See for instance the visualizations of Kruscke and Underwood in Charles van den Heuvel, Scott Weingart, "Desiderata for Visualizing Uncertainty and Contextuality in the Digital Humanities. Experiences with the Digital Republic of Letters," in ibid. *Visualizing Data Resources*.
51 A similar plea for mixed methods of quantitative and qualitative approaches in network analysis we can observe in the social sciences. Compare: Silvia Domínguez and Betina Hollstein, ed., *Mixed Methods Social Networks Research: Designs and Applications* (Cambridge: Cambridge University Press, 2014).

Bibliography

Akker, Chiel, Susan van den Legêne, Mareike van Erp et al. "Agora Digital Hermeneutics: Online Understanding of Cultural Heritage." *Proceedings of the 3rd International Conference on Web Science (WebSci' 11)*, Koblenz, Germany, 14 June 2011. www.cs.vu.nl/~guus/papers/Akker11a.pdf (accessed 5 December 2018).

Alcantara, Marlon Cesar, Marco Braga and Charles van den Heuvel, "Historical Networks in Science Education. *A Case Study of an Experiment with Network Analysis by High School Students*," Science and Education 29, pp. 101–121, Dordrecht: Springer Netherlands 2020 https://doi.org/10.1007/s11191-019-00096-4 ISSN: 1573-1901 (Online)

Álvares Francés, L., and Charles van den Heuvel. "End Report Mapping Notes and Nodes in Networks." (2014). www.huygens.knaw.nl/wp-content/uploads/2015/05/EndReportMNN.pdf (accessed 5 December 2018).

Bammer, Gabriele, and Michael Smithson, eds. *Uncertainty and Risk: Multidisciplinary Perspectives*. London: Routledge, 2009.

Blei, David M., and John D. Lafferty. "Dynamic Topic Models." In *Dynamic Topic Models: Proceedings of the 23rd international Conference on Machine Learning.* Edited by William Cohen and Andrew Moore, pp. 113–20, 2006. doi:10.1145/1143844.1143859 (accessed 5 December 2018).

Bodenhamer, David J. "The Potential of Spatial Humanities." In *The Spatial Humanities: GIS and the Future of Humanities Scholarship.* Edited by David J. Bodenhamer. Spatial Humanities. Bloomington: Indiana University Press, Ch. 2, 2010.

Boyack, Kevin W., Katy Börner, and Richard Klavans. "Desiderata for Visualizing Uncertainty and Contextuality in the Digital Humanities: Experiences with the Digital Republic of Letters." In Heuvel; Weingart, *Visualizing Data Resources.* http://ckcc.huygens.knaw.nl/wp-content/bestanden/2012/05/CHeuvel_Scott_Gotha.pdf (accessed 5 December 2018).

Capurro, Rafael. "Digital Hermeneutics: An Outline." (2010). www.capurro.de/digitalhermeneutics.html (accessed 5 December 2018).

Coleman, Nicole, and Charles van den Heuvel. "Visualizing Uncertainty and Complexity: Humanistic Methods for Mapping the Intellectual Geography of the Early Modern World." *Submitted Paper for the Publication of a Volume after the Conference Intellectual Geography: Comparative Studies 1550–1700*, held from 5–7 September 2011, St' Anne's College, Oxford, United Kingdom, University of Oxford.

Domínguez, Silvia, and Betina Hollstein, eds. *Mixed Methods Social Networks Research: Designs and Applications.* Cambridge: Cambridge University Press, 2014.

Epstein, Joshua M. "Why Model?" *Journal of Artificial Societies and Social Simulation* 11, no. 4 (2008): p. 12.

Heuvel, Charles van den. "Netwerken van vertrouwelijkheid en geheimhouding in brieven en tekeningen." In *In vriendschap en vertrouwen. Cultuurhistorische essays over confidentialiteit.* Edited by J. Gabriëls et al. Hilversum: Verloren, 2014.

Heuvel, Charles van den. "Mapping Knowledge Exchange in Early Modern Europe: Intellectual and Technological Geographies and Network Representations." *International Journal of Humanities and Arts Computing* 9, no. 1 (2015): pp. 95–114. doi:10.3366/ijhac.2015.0140 (accessed 5 December 2018).

———. "Mapping Notes and Nodes: Building a Multi-Layered Network for a History of the Cultural Industry." (2015). http://dh2015.org/abstracts/xml/HEUVEL_Charles_van_den_Mapping_Notes_And_Nodes__B/HEUVEL_Charles_van_den_Mapping_Notes_And_Nodes__Buildin.html (accessed 5 December 2018).

Heuvel, Charles van den, and Scott Weingart. "Desiderata for Visualizing Uncertainty and Contextuality in the Digital Humanities: Experiences with the Digital Republic of Letters." In Heuvel; Weingart, *Visualizing Data Resources.*

———, eds. *Visualizing Data Resources: The Potential of the Wikimedia Platform for the Digital Humanities.* Forschungszentrum Gotha, 2012. www.scottbot.net/HIAL/wp-content/uploads/2012/05/CHeuvel_Weingart.pdf (accessed 5 December 2018).

Heuvel, Charles van den et al. "Modeling Confidentiality and Secrecy in Knowledge Exchange Networks of Letters and Drawings in the Early Modern Europe." *Nuncius*, no. 31 (2016): pp. 78–106.

Heuvel, Charles van den, "Modelling Texts and topics", In: Howard Hotson and Thomas Wallnig (Eds.) *Reassembling the Republic of Letters in the Digital Age. Standards, Systems, Scholarship,* Göttingen: University Press, pp. 265–280, 2019. Doi:10.17875/gup2019-1146

Hinton, Geoffrey, Simon Osidero, and Yee Whye Teh. "A Fast Learning Algorithm for Deep Belief Nets." *Neural Computation* 18, no. 7 (July 2006): pp. 1527–54.

Hoogewerff, Godefridus Joannes, ed. "De twee reizen van Cosimo de' Medici, prins van Toscane door de Nederlanden (1667–1669)." In *Journalen en Documenten*, Amsterdam: J. Müller, 1919.

Hotson, Howard and Thomas Wallnig (Eds.) *Reassembling the Republic of Letters in the Digital Age. Standards, Systems, Scholarship*, Göttingen University Press 2019, pp. 237–264. Doi:10.17875/gup2019-1146.

Kim, Yoon, Yi-I Chiu, Kentaro Hanaki, et al. "Temporal Analysis of Language through Neural Language Models." *Proceedings of the ACL 2014 Workshop on Language Technologies and Computational Social Science*, Baltimore, Maryland, USA, 2014, pp. 61–5.

Kouw, Matthijs, Charles van den Heuvel, and Andrea Scharnhorst. "Exploring Uncertainty in Knowledge Representations: Classifications, Simulations, and Models of the World." In *Virtual Knowledge: Experimenting in the Humanities and the Social Sciences*. Edited by P. Wouters et al., pp. 90–125. Cambridge, MA: The MIT Press, 2012.

Lathisena, Sarasi, Pascal Hitzler, Amit Sheth, and Pratek Jain. "Automatic Domain Identification for Linked Open Data." *Proceedings of the 2013 IEEE/WIC/ACM International Joint Conferences on Web Intelligence and Intelligent Agent Technologies*, 2013, p. 11. http://works.bepress.com/amit_sheth/62 (accessed 5 December 2018).

Latour, Bruno. *Reassembling the Social: An Introduction to Actor-Network-Theory*. Oxford: University Press, 2005.

Lux, David S., and Harold J. Cook. "Closed Circles or Open Networks? Communicating at a Distance during the Scientific Revolution." *History of Science*, no. 36 (1998): pp. 179–211.

Mäkelä, Eetu, Eero Hyvönen, and Tuukka Ruotsalo. "How to Deal with Massively Heterogeneous Cultural Heritage Data-Lessons Learned in CultureSampo." *Semantic Web* 3, no. 1 (2012): pp. 85–109.

Martelli, Francesco, ed. *Il viaggio in Europa di Pietro Guerrini (1682–1686): Edizione della corrispondenza e dei disegni di un inviato di Cosimo III dei Medici*. Documenti di storia italiana Serie II 11. Florence: Leo S. Olschki, 2005.

Mihalcea, Rada, and Vivi Nastase. "Word Epoch Disambiguation: Finding How Words Change Over Time." *Proceedings of the 50th Annual Meeting of the Association for Computational Linguistics*, pp. 259–263, Jeju, Republic of Korea, 8–14 July 2012.

Mirto, Alfonso, and Henk Th. van Veen. *Pieter Blaeu: lettere ai Fiorentini, Antonio Magliabechi, Leopoldo e Cosimo III de' Medici e altri*. Firenze, Amsterdam, Maarssen: Centro Di, 1993.

Molhuysen, P.C., Meulenbroek, B.L., Witkam, P.P., Nellen, H.J.M., and Ridderikhoff, C. M., eds. *Briefwisseling van Hugo Grotius*. Vol. 1–17. Rijksgeschiedkundige Publicatiën, Grote Serie. The Hague: Martinus Nijhoff, later Instituut voor Nederlandse Geschiedenis, 1928–2001.

Nellen, Henk. "'In Strict Confidence': Grotius' Correspondence with his Socinian Friends." In *Self-Presentation and Social Identification: The Rhetoric and Pragmatics of Letter Writing in Early Modern Times*. Edited by Toon van Houdt, G. Tournoy and C. Matheeusen, pp. 227–45. Supplementa Humanistica Lovaniensia 18. Leuven: University Press, 2002.

———. "The Correspondence of Hugo Grotius." In *Les grands intermédiaires culturels de la République des Lettres: Études de réseaux de correspondances du XVIe au XVIIIe siècles*. Edited by C. Berkvens-Stevelinck, pp. 127–64. Les dix-huitièmes siècles no 91. Paris: Honoré Champion, 2005.

Ravenek, Walter, Charles van den Heuvel, Guido Gerritsen, "The ePistolarium: origins and techniques." In Jan Odijk, and Arjan van Hessen (eds.) *CLARIN in the Low Countries*. London: Ubiquity Press. pp. 317–323. 2017.

Sagi, Eyal, Stefan Kaufmann, and Brady Clark. "Semantic Density Analysis Comparing Word Meaning across Time and Phonetic Space." *Proceedings of the EACL 2009 Workshop on GEMS: Geometrical Models of Natural Language Semantics, Athens, Greece, 31 March*, 2009, pp. 104–11.

Underwood, Ted. "The Stone and the Shell." (2011). http://tedunderwood.wordpress.com/2011/09/16/topics-are-also-trends (accessed 5 December 2018).

Valleriani, Matteo, Florian Kräutli, Maryam Zamani, Alejando Tejedor, Christoph Sander, Malte Vogl, Sabine Bertram, Gesa Funke and Holger Kantz. "The Emergence of Epistemic Communities in the *Sphaera* Corpus: Mechanisms of Knowledge Evolution," *Journal of Historical Network Research* 3 (2019): pp. 50–91. https://doi.org/10.25517/jhnr.v3i1.63 (accessed 2 March 2020).

van de Sompel, Herbert, and Michael Nelson. "Memento: Time Travel for the Web." *Arxiv Preprint* (2009). http://arxiv.org/abs/0911.1112 (accessed 5 December 2018).

van Houdt, Toon, Papy Jan, Gilbert Tournoy, and Constant Matheeusen, eds. *Self-Presentation and Social Identification: The Rhetoric and Pragmatics of Letter Writing in Early Modern Times*. Supplementa Humanistica Lovaniensia 18. Leuven: University Press, 2002. www.loc.gov/catdir/enhancements/fy1608/2002494848-d.html (accessed 5 December 2018).

Vugt, Ingeborg van. "Storia e Geografia di una rete epistolare." In *Antonio Magliabechi nell'Europa dei Saperi*. Edited by P. Boutier, M.P. Paoli and C. Viola, pp. 259–292. Pisa: Edizioni della Normale, 2017.

Vugt, Ingeborg van. *The structure and dynamics of scholarly networks between the Dutch Republic and the Grand Duchy of Tuscany in the 17th century*. Unpublished Dissertation. Scuola Normale Superiore di Pisa/University of Amsterdam, 2019. Https://hdl.handle.net/11245.1/94502a28-e642-4ecc-81e2-100fda93ecba (accessed 2 March 2020).

Vugt, Ingeborg van. "Using Multi-layered Networks to Disclose Books in the Republic of Letters," *Journal of Historical Network Research* 1 (2017): pp. 25–51. Retrieved from //jhnr.uni.lu/index.php/jhnr/article/view/7 (accessed 2 March 2020).

Wagenaar, Lodewijk, ed. *Een Toscaanse prins bezoekt Nederland. De twee reizen van Cosimo de' Medici, 1667–1669*. Amsterdam: Uitgeverij Bas Lubberhuizen, 2014.

Weingart, Scott. "The Networked Structure of Scientific Growth." *The Scottbot Irregular* (22 February 2012). Unpublished. http://www.scottbot.net/HIAL/index.html@p=12050.html (accessed 5 February 2020).

Windolf, Paul H. "Social Capital and Social Inequality: Corporate Networks in Germany and the United States (1896–1938)." *SSRN Electronic Journal* (2014). doi:10.2139/ssrn.2529698 (accessed 5 December 2018).

Wittek, Peter, and Walter Ravenek. "Supporting the Exploration of a Corpus of 17th Century Scholarly Correspondences by Topic Modeling." In *Proceedings of Supporting Digital Humanities: Answering the Unaskable*. Edited by B. Maegaard. Copenhagen, 2011.

Wouters, Paul, Anne Beaulieu, Andrea Scharnhorst, and Sally Wyatt, eds. *Virtual Knowledge: Experimenting in the Humanities and the Social Sciences*. Cambridge, MA: The MIT Press, 2012.

Zuidervaart, Huib. "The ePistolarium Tool: Experiences in the Development of a Digital Tool for the Research of 17th Century Scholarly Correspondences." www.culturesofknowledge.org/?page_id=4356 (accessed 5 December 2018).

4.2 Networks as gateways. Gleanings from applications for the exploration of historical data[1]

Marten Düring

Introduction

The mass availability of digitised sources, powerful information retrieval systems, knowledge bases with relevant information for historians and user-friendly data visualisation techniques open up new opportunities for historians. In and around the digital humanities, applications emerged that use network visualisations – node link diagrams, for the most part – as gateways for the exploration and curation of historical and cultural heritage data.[2] Most of the research that is discussed in this book focuses on question-driven datasets, i.e., data that was created or curated with the sole purpose of answering specific research questions. In this chapter we discuss data that was created for general-purpose content discovery or linking. The surveyed applications offer ambitious and inspiring conceptualisations of historical networks and experiment with new ways of collecting and studying historical relational data. All of them share the aim to allow non-experts to benefit from the strengths of network visualisations and relational data without burdening them with the steep learning curve associated with network analysis. This learning curve together with high costs for data collection, restrictive data models and far from certain prospects of success in this author's experience constitute the most common reasons why historians choose not to apply network analysis methods in their research. But which added value do these applications have for historians? How do they conceptualise networks? How do they facilitate the exploration of relational data? And what is the price to pay for gentle learning curves and automatically generated data?

Lauren Klein et al.'s definition of exploratory data exploration[3] captures what contribution the surveyed applications could make to historical research workflows: with Tukey (1977) they stress that exploratory data analysis "is intended to help the researcher develop a general sense of the properties of the data set before embarking on more specific inquiries." These preliminary, lightweight analyses help humanists assess "What is there." The surveyed applications encourage users to create and explore paths guided by their interests and with the goal to discover relevant links along the way but also to develop a sense of the composition of a corpus, its inherent biases and its value for research.

These paths ideally lead to new perspectives, confirm hypotheses, refine research questions and trigger new questions.[4] Dörk et al. capture this idea through their "information flaneur" persona who navigates information landscapes guided by curiosity, reflection and imagination and makes serendipitous discoveries.[5]

Related work

It is a truism that network visualisations can be beautiful, that they are popular with audiences but are also difficult to use as meaningful carriers of information. We have a good overview of the wide range of challenges associated with the conceptualisation, representation, interpretation, documentation, communication and teaching of network analysis and visualisation in the humanities.[6] At present there are, to our knowledge, no systematic evaluations of the surveyed applications in particular and for exploratory network data visualisations in the humanities in general.

With regard to humanities user interfaces in general, Dörk et al. have identified the following additional challenges: the representation of large datasets, the need for multiple perspectives to represent the diversity of content, bridging the gap between abstract visualisations and objects and the need for an interface that appeals to broad audiences.[7] For interface design more generally, Ben Shneiderman's paradigm "overview first, zoom and filter, then details-on-demand"[8] continues to be influential. Much current work favours user-centred design,[9] focusing on zoomable interfaces,[10] coordinated views that enable the combination of different forms of data visualisation, content organisation based on timelines or topics[11] and "generous interfaces"[12] that provide access to content independently of a search query and without requiring a pre-existing interest in specific parts of the collection.

Quality standards for network-based interfaces developed in visual analytics, an applied subfield in computer science, offer additional indicators to assess the quality of the surveyed visualisations. In their survey article, Herman et al. point to the major challenges in the design of graph-based applications in information visualisation, with the goal of reducing visual complexity so as not to overwhelm users: many algorithms don't scale very well, and the size of a graph often needs to be reduced by means of rankings or other forms of selection to allow comprehension and detailed analysis. Incremental exploration techniques focus users' attention on a subset of the overall visualisation so as to not overwhelm them. Bundles of edges have emerged as an effective means to address the problem of planarity, i.e., the avoidance of edge crossings.[13] There may also be orientation issues: two runs of a layout algorithm often yield a very different organisation of nodes, meaning that users lose their mental maps of graphs and have to reorient themselves after every run.[14] Interactive visualisations typically allow users to pan and zoom but again need to ensure that the observed movements are not experienced as overwhelming. Semantic zoom techniques link the display of selected information to appropriate zoom levels; a common yet basic example is node labels, which appear as users zoom into a network.

Focus and context problems emerge when a certain area of a visualisation is selected for closer inspection and the surrounding context disappears as a consequence.[15]

Survey

This chapter reviews the relationship between extraction and exploratory visualisation techniques of nine digital humanities applications and discusses their possible contribution to the study of historical networks. The surveyed applications were selected on the basis of three criteria: 1) usage of interactive graph visualisations as part of an interface, 2) use of data with immediate relevance for historical research and 3) public accessibility.

During the survey we pay particular attention to the context in which they were created, their overall objective, the type of data they handle and how it was created, target audiences and – most interesting for the purpose of this chapter – how they utilise network visualisations to help with the exploration of their corpora. Each description highlights features which are immediately relevant for historical research and are unique to or particularly well implemented in the respective application.[16]

In addition, we have sought to identify key features of the applications organised in three groups: *Types of visualisations* characterises the chosen representation of relational data (multimodal/dynamic network, maps, timelines) as well as representations of node centrality and edge weight. *Visualisation features* is based on the recommendations compiled by Herman et al., which we outlined earlier and include graph subsets, panning, zooming, static graph layouts, semantic zoom and edge bundles. *User interactions* collects information on the interactivity of the visualised networks, whether users can move between data visualisations and underlying sources, whether (shortest) path queries between selected nodes are possible, whether nodes or edges can be created by users, whether the application combines data visualisation with narratives about the content and the ability to calculate network centralities or clustering. The group *Data creation* collects information on the disambiguation of named entities and events to external identifiers such as VIAF or Wikidata, whether the relational data stems from on a pre-existing database, whether nodes and edges were automatically extracted from unstructured text, whether named entities have been detected automatically and whether term frequencies have been calculated.

SDFB – Six Degrees of Francis Bacon

Six Degrees of Francis Bacon (SDFB)[17] provides information about relationships between people in early modern Britain, creating a workable infrastructure to identify links and potentially influential figures. SDFB offers an exploratory interface and data entry system for the retrieval and continued curation of data on personal relationships between historical figures. Its core service for a mainly scholarly community is to provide a bird's-eye view on who interacted

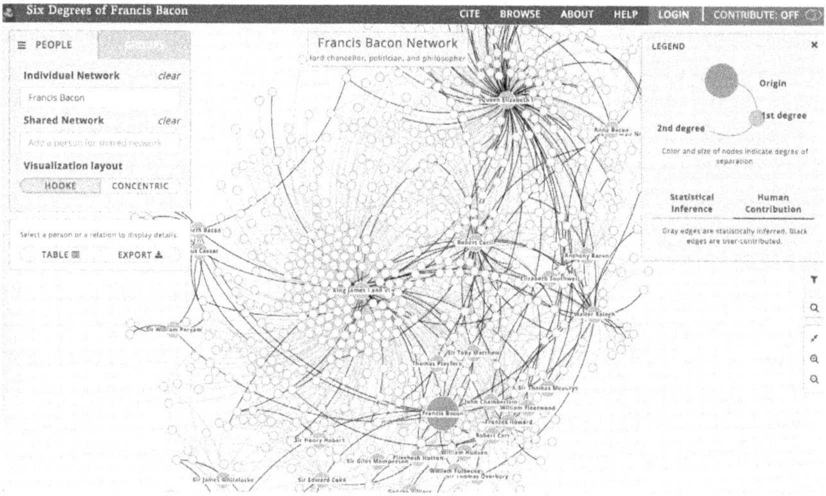

Figure 4.2.1 Six Degrees of Francis Bacon seeks to reconstruct who knew whom in Early Modern England (Screenshot)

with whom: links observed in SDFB raise questions and prompt further investigation outside the application. SDFB has been in development since 2010. Based on the Oxford Dictionary of Biographies,[18] SDFB uses text mining to extract relationships between more than 15,000 individuals. At the time of writing, it lists 113 distinct relationship types, which were developed by a community of users and range from the generic "acquaintance of" to the more specific "attracted to" and "step-sibling of." The list of types covers kinship, education and also financial interactions. Users can operate with qualifying statements such as "before", "in", "after", "circa" and "after or in" to address uncertainties in dates.

Users can choose to explore a person's ego network with the option to filter by date range. An additional filter separates human annotations from automatically generated ties. A confidence score indicates the likelihood that an inferred edge is valid.[19] To find out who connects two people, SDFB users can run a shortest path query, which lists all shared connections. Groups can be searched by their name, and people who are members of several groups can be identified. A click on a node reveals a timeline with life dates, a brief description of the person's occupation or other characterisation of their historical significance, group memberships if available, external links to the person's identifier in the Oxford Dictionary of Biographies and finally also links to search queries in JSTOR[20] and Google. A click on an edge reveals all of the previously mentioned together with computed confidence scores regarding the type of relationship and indications of the time period it covered.

To address shortcomings in data quality and availability, SDFB is designed so as to be extended by community members who are encouraged to add and revise relationships pending the approval of moderators. The system allows users to qualify relationship types and to add start and end dates for relationships together with a compulsory justification and optional references to external sources and notes. In the past, SDFB has successfully launched crowdsourcing campaigns, for example, to enrich the corpus with data on women. The project blog mentions future visualisations of the data, but these were not yet available at the time of writing.[21]

What makes it interesting: SDFB is a good example of the way computational text analysis can be used to extract actors and relations from unstructured text and enrich them with persistent identifiers. Inevitable imperfections during this process are partially compensated by the computed confidence score and the creation and maintenance of a community of users. Like other applications, SDFB gives great freedom regarding the creation of relationship types and by supporting fuzzy dates whilst still maintaining control by means of moderators. Also not to be taken for granted, SDFB lets registered users download all their data.

Kindred Britain

Kindred Britain (published in 2013) is an enhanced representation of a database of kinship relationships between some 30,000, mostly 19th-century British personalities with a rich interface for the exploration of family trees. A combination of network visualisations, timelines and maps supports the coordinated views of the data. Networks represent genealogical ties and nodes are positioned following a linear time scale on the y-axis. The timeline displays life spans of related people horizontally and highlights events that are shared by two or more people by means of vertical lines connecting them. The map shows the locations of selected people at a given time. All three visualisations react to user selections simultaneously: a node selected in the graph will highlight corresponding items in the timeline as well as in the map below and vice versa.

A click on a node reveals attribute data based on the extent to which a person is linked to others, for example, with regard to their centrality in the graph and the number of family members. Close links to other people are provided as well as a short biographical description. Another noteworthy feature is the tongue-in-cheek "Tragedy Index", which assigns a score to the frequency of unfortunate events people encountered throughout their lives: deaths of spouses, divorces and other negative events are all included. Its strongest feature is perhaps the option to display shortest paths between individuals, revealing shared family ties across time. A selection of these paths are listed in the Connections segment, which encourages users to explore them. Kindred Britain also attempts to further bridge the gap between explorative data visualisation and storytelling: "stories" combine traditional narratives with views of the data as a means of contextualisation.

What makes it interesting: Kindred Britain excels at the smooth integration of different data visualisation techniques; kinship relationships go hand in hand

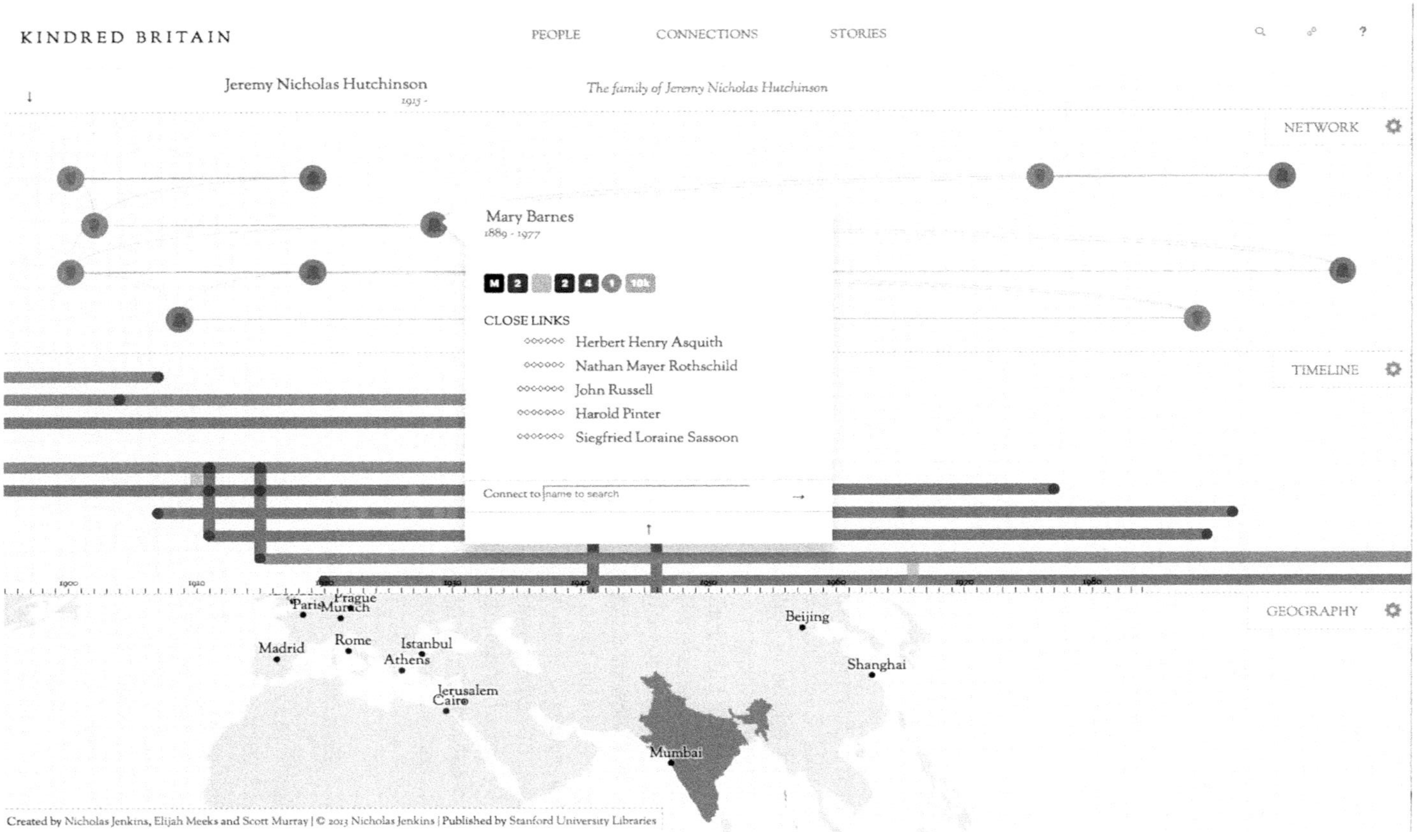

Figure 4.2.2 Kindred Britain uses a variety of visualisations to open up a pre-existing database on 19th-century British kinship relationships for exploration (screenshot)

with a timeline and geographical map. Stories enriched by network visualisations bring data to life and create a meaningful relationship between data visualisation and narrative.

ALCIDE – Analysis of Language and Content in a Digital Environment

ALCIDE,[22] developed by Fondazione Bruno Kessler in 2015, is a tool suite that combines computational text analysis with data visualisation. The system was developed for the study of historical text corpora and uses co-occurrence networks of people (a link is created if two people are mentioned within a text) together with term frequencies, maps, keyword search and keyword extraction as a means to enable scholars to ask "who, where, when, what and how" questions about their corpora. Co-occurrence networks are based on the assumption that co-occurring entities are likely to have something to do with each other.[23] Graph visualisations and some of the other visualisations are interactive and reveal links to underlying documents. People mentioned in a given corpus are represented as nodes, which are linked to each other either if they appear together in a sentence or if they appear within a set number of words from each other. Users can choose which approach they prefer and also set the number of tokens to control the type and the number of relationships that will be displayed in the graph. A greater distance results in a more densely connected network with a higher chance that there are no immediate relationships between two people, whereas a lower number will produce the opposite. ALCIDE is designed to use text analysis and data visualisations as a way of offering a complementary view of a corpus. The goal is to enrich users' close reading experience and to enable them to switch between data visualisation and close reading.

What makes it interesting: ALCIDE offers a complete workflow from the detection of named entities in unstructured text to the exploration of annotated texts by means of multiple data visualisation tools. Particularly noteworthy features are the ability to go back and forth between visualisations and source documents and the data management facility.

APIS – Austrian Prosopographical Information System

The Austrian Prosopographical Information System (APIS) has the ambition to "collect, curate, enrich, analyze, visualize, and export prosopographical and biographical data."[24] Created by the Austrian Academy of Sciences, APIS is based on the Austrian Biographical Dictionary, which covers approximately 18,000 people who lived in the former Austro-Hungarian monarchy as well as the First and Second Austrian Republics. All data will be published as part of the Linked Open Data Cloud and thereby made accessible for future reuse. APIS will provide users with tools for the annotation of texts, which are based on the entity types Persons, Places, Institutions, Events and Works. Like ERNiE (see later), APIS combines networks and maps for the exploration of data but

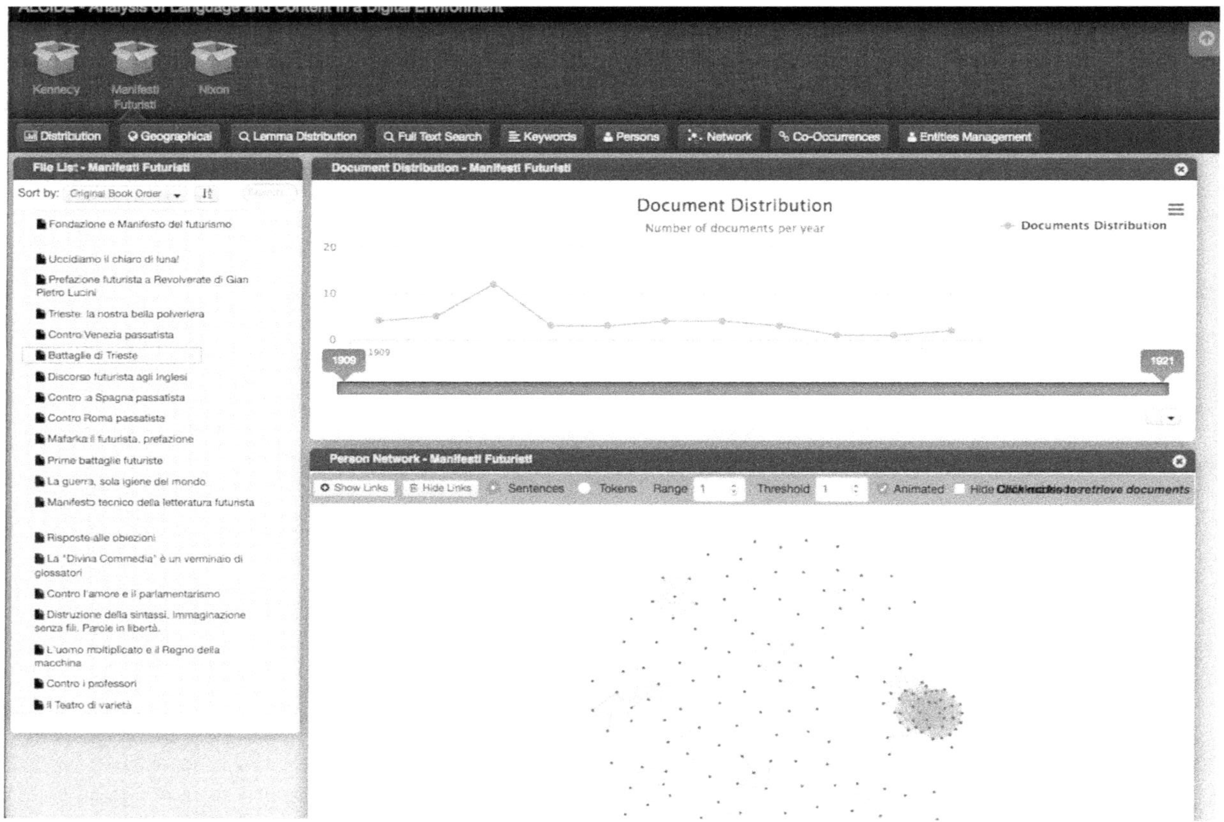

Figure 4.2.3 ALCIDE supports close reading with text analysis and a variety of visualisation tools (screenshot)

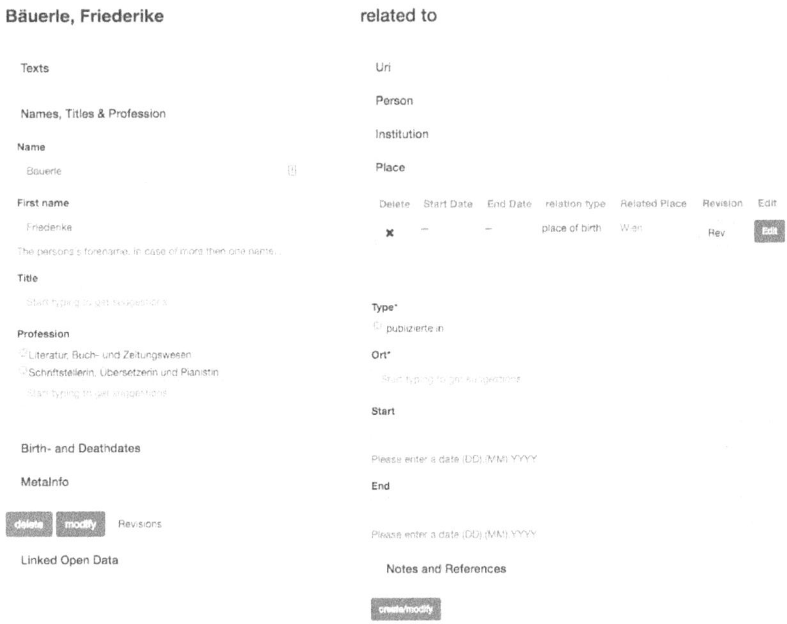

Figure 4.2.4 APIS supports fine-grained collection of prosopographical data including relationships between people, institutions, events and places (screenshot)

concentrates on the interlinking of annotations via free-to-choose relationship types.

What makes it interesting: even at this early stage, APIS' dedication to the Linked Open Data Cloud appears particularly promising as a way of breaking up information silos created by individual corpora.

histograph

histograph[25] was designed for the exploration and crowd-based annotation of multimedia document collections. The application combines the graph-based exploration of multimedia collections with crowd-based annotations and a recommender system.[26] A first demo (to which this author contributed as well) provides a new perspective on the collections of the former Centre virtuel de la connaissance sur l'Europe (CVCE),[27] which document the history of European integration based on some 20,000 digitised and multilingual text documents, photos, audio recordings and videos.

Like ALCIDE, histograph operates based on co-occurrence networks of entities such as people who are mentioned together in a variety of text documents

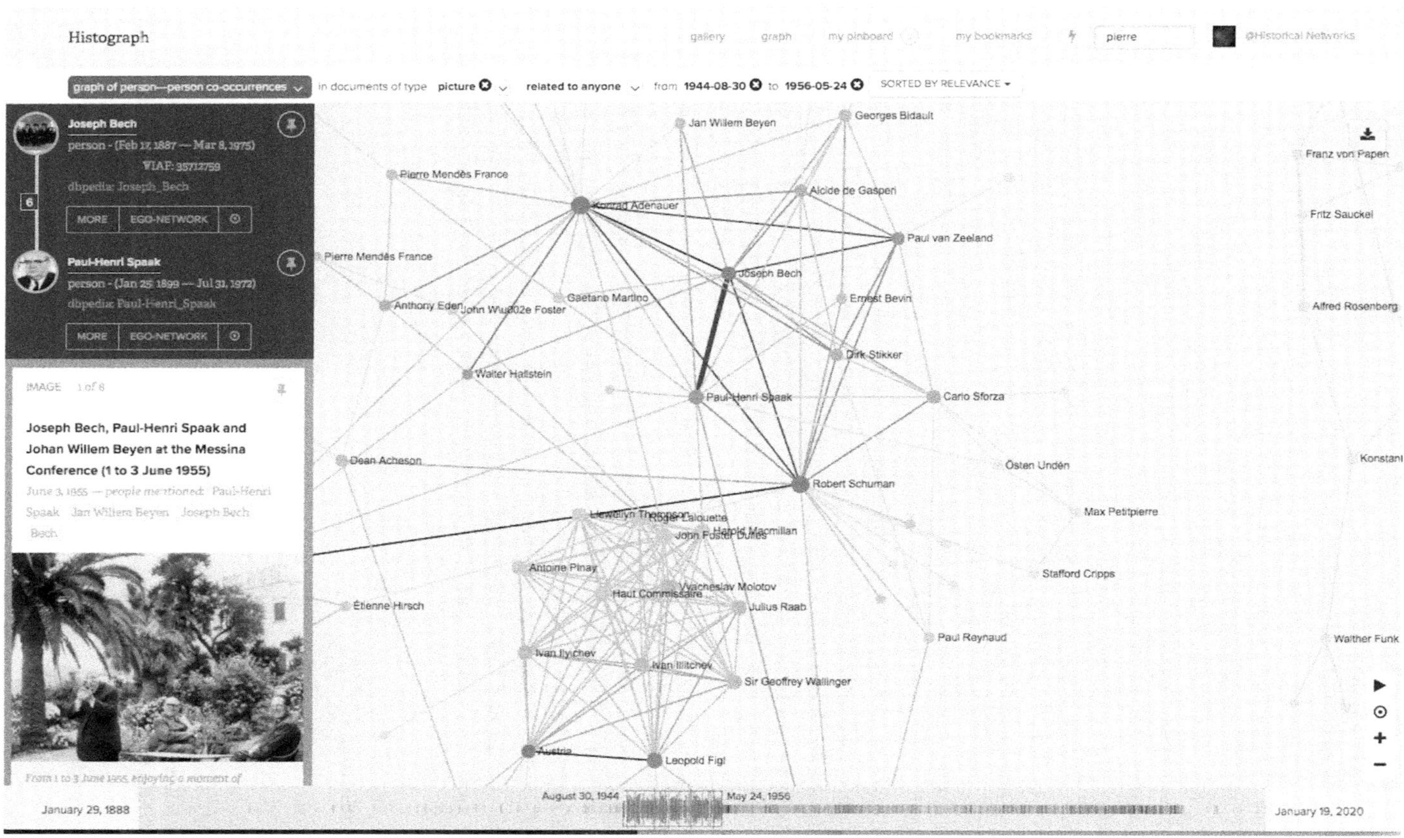

Figure 4.2.5 histograph uses co-occurrence relationships, annotations, recommendations and filters for the exploration of multimedia corpora (screenshot)

and images. Filters allow users to narrow down their search, for example by limiting it to certain document types, time periods or mentioned named entities such as people, institutions or places. The graph view provides a bird's-eye view of the people who co-occur with a given person and also supports shortest path queries. histograph displays unipartite person-to-person co-occurrences but also bipartite visualisations (e.g. person to document). The graph responds to filters and is closely connected to the underlying documents: a click reveals the documents that constitute a co-occurrence relationship. The combination of abstract data visualisation and immediate access to the underlying primary sources allows users to decide whether or not a relationship is of interest to them.[28] Where there are too many edges to display, histograph shows only a subset of highest ranking edges: each co-occurrence relationship receives a score, which depends on the distance between entities in the text and their overall distribution within the corpus. The Jaccard similarity coefficient[29] takes into account the number of documents in which two entities appear divided by the total number of documents in which either appear.

histograph uses methods for the automatic annotation of named entities such as people, institutions and places.[30] These named entities can be further enriched by knowledge bases such as DBpedia and VIAF.[31] Generic crowds or dedicated teams can fix mistakes in the automated annotations by means of four options that correspond with frequently observed error types: fragmentary detection of entities, homonyms, wrong entity types and false detection of an entity. Rather than relying on centralised quality control, annotations can be confirmed by users and voted up or down.

What makes it interesting: histograph was designed to manage multimedia collections and is not limited to texts alone. Snippet previews within the network visualisation help to bridge the gap between data visualisation and content. An effective search and filter functionality makes it easy to find relevant material in the corpus, and this is also supported by the uni- and bipartite network visualisations. Annotations and error correction functions are kept lightweight and easy to use.

ePistolarium

ePistolarium,[32] which is also referred to in the contribution by Charles van den Heuvel et al. in this volume, was first published in 2013 as part of the project "Circulation of Knowledge and Learned Practices in the 17th-century Dutch Republic". The application was designed around a corpus of some 20,000 scholarly letters with the aim of facilitating exploration of the corpus and investigating knowledge networks and the spread of ideas.[33] Users can search for keywords (which can be enriched by means of associated words based on topic modelling) in similar text passages and use multi-faceted search options to identify letters for closer inspection. Alternative entry points into the corpus are person roles (senders, recipients or simply mentioned) and locations of senders and recipients. Finally, the Correspondences section lists a sender and

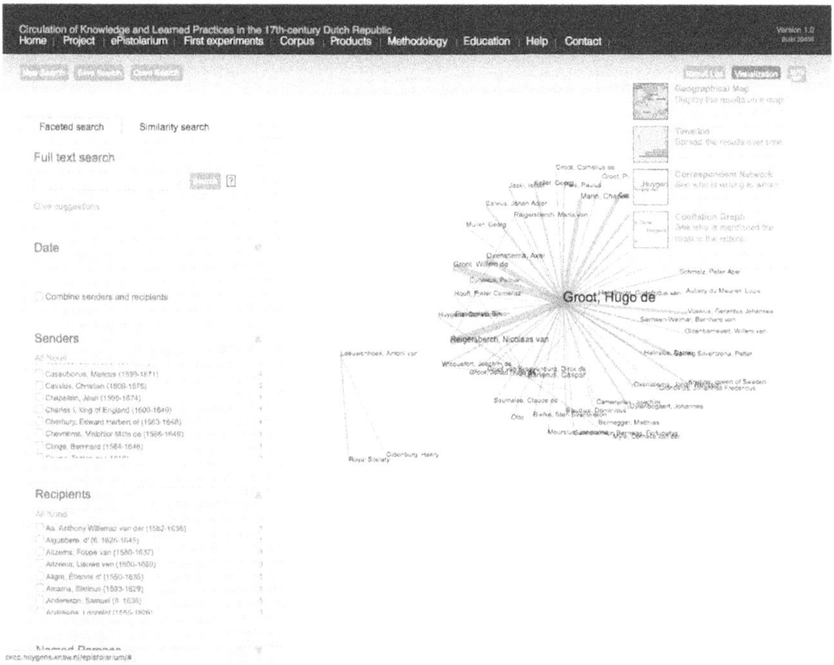

Figure 4.2.6 ePistolarium uses networks, maps and timelines for visualisation and a variety of text analysis tools for the interlinking of 17th-century correspondence (screenshot)

all recipients in the form of a star network, which can be enriched by adding additional senders in order to explore the overlap in the correspondence between two senders. Query results are visualised in the form of 1) lists of letters, 2) network visualisations that display relationships between senders and recipients or co-citations, 3) geographical maps that show the locations of selected nodes and 4) a timeline view. Edge types are based on letter exchanges and can be filtered further using time periods and sender/recipient locations. In the current version[34] of ePistolarium, graphs are not interactive, and the content and metadata of letters need to be retrieved via the list view. Users are encouraged to make searches and selections and to explore the results; additional explorations (keywords, filters) on top of query results were not a priority in this version.

What makes it interesting: the application excels at combining accessible (faceted) search interfaces and result lists with more advanced techniques such as graph visualisations, maps, topic modelling and keyword analysis.

ERNiE – Encyclopedia of Romantic Nationalism

The Encyclopedia of Romantic Nationalism (ERNiE)[35] is an online platform that combines the display of a collection of digitised primary sources on nationalism

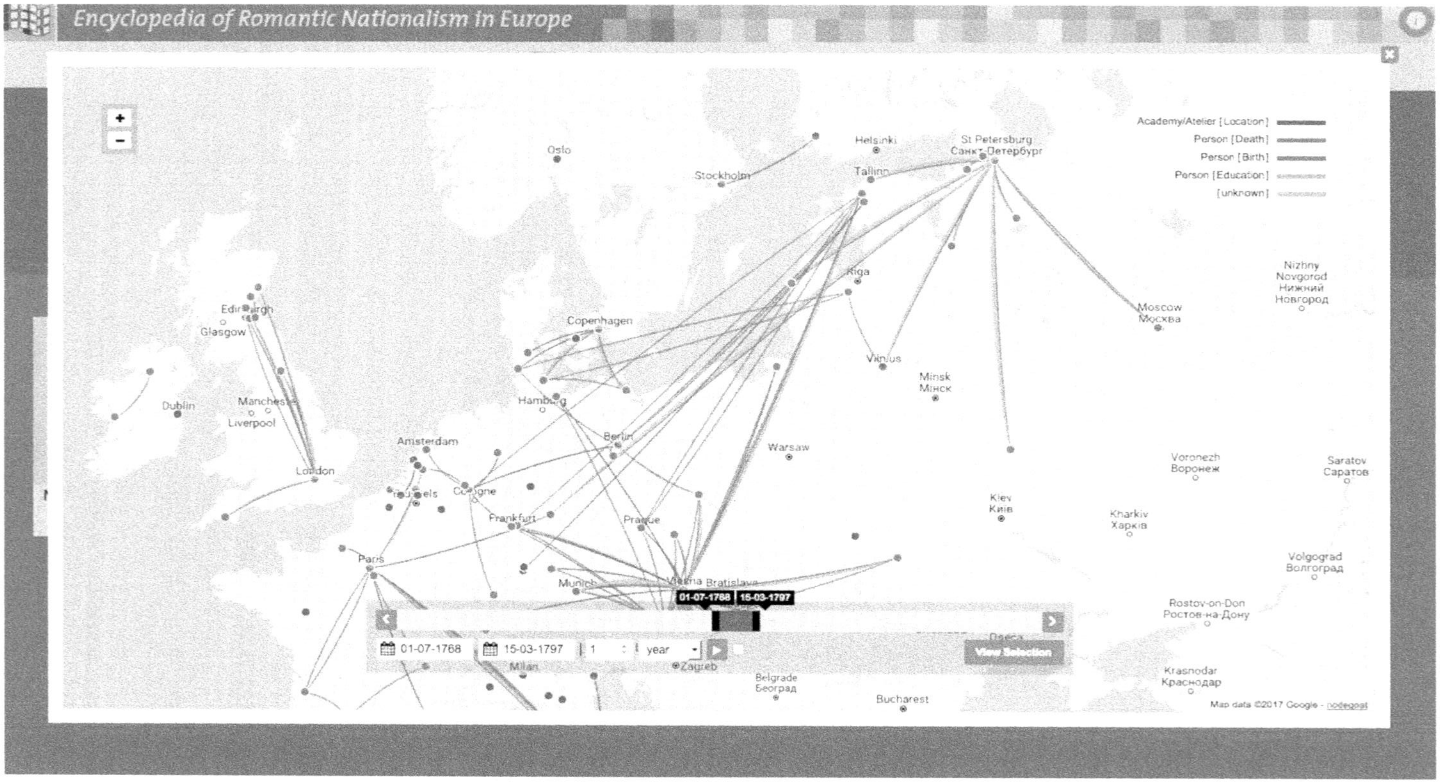

Figure 4.2.7 ERNiE uses combinations of networks, maps, galleries and lists to open up a multimedia collection on romantic nationalism based on Nodegoat (screenshot)

with networks, timelines and maps to support its exploration. The application is based on Nodegoat,[36] a general-purpose tool for the creation of complex data structures and the storage of digital media developed by Lab1100 (see the aforementioned contribution by Charles van den Heuvel in this volume). This includes the creation of complex data structures that reflect complicated relationships between entities such as people, objects, institutions and places. Nodegoat lets users enrich their data, for example by geo-tagging city names. ERNiE is a co-created public-facing interface that allows users to enrich texts with visualisations and to explore links between different types of digitised objects such as letters and texts, data, monuments and the changing locations of people. ERNiE uses maps, networks, networks on top of maps and lists to explore the data. All networks and maps can be filtered and animated via a timeline that lets users decide on the length of the time period (days, weeks, months, years). Nodes are interactive, and a click yields an overlay window that provides data about the node for further inspection.

What makes it interesting: the combination of a highly flexible database, tools for the exploration of network and spatial data and an integrated presentation platform is unique among the surveyed applications.

HuNI – Humanities Networked Infrastructure

The Humanities Network Infrastructure (HuNI)[37] has been designed to encourage users to create links between digitised cultural heritage objects. Published in 2014 by a group of 13 Australian institutions and led by Deakin University, HuNI's ambition is to be both a "data warehouse" that pre-processes data from different institutions and a "data lake" in which a grassroots strategy is used to link objects and create collections.[38] HuNI refrains from adding any automatically generated links and relies solely on its users to create them by means of six node types (Concept, Event, Organisation, Person, Place or Work) and 18 relationship types such as "editor of", "organised by" or "visited", also giving them the possibility to create their own types. It does not currently let users search for paths within this dataset, but multimodal graph visualisations enable users to jump from one node to the next and thereby explore links between collections. HuNI seeks to make collaboration and serendipitous discoveries possible and to respect and support humanist interpretation by keeping authoritative taxonomies to a minimum.[39]

What makes it interesting: HuNI is designed to interlink different collections from different institutions in a highly flexible yet lightweight manner. The ability to go beyond the limitations of a single corpus is crucial for historical research.

RoSE – Research-oriented Social Environment

The Research-oriented Social Environment (RoSE)[40] was first conceived of as part of the Transliteracies project[41] funded by the University of California, led by Alan Liu and completed in 2012. Not bound by a specific group of people or a period, RoSE is presented as a crowdsourced "knowledge-exploration

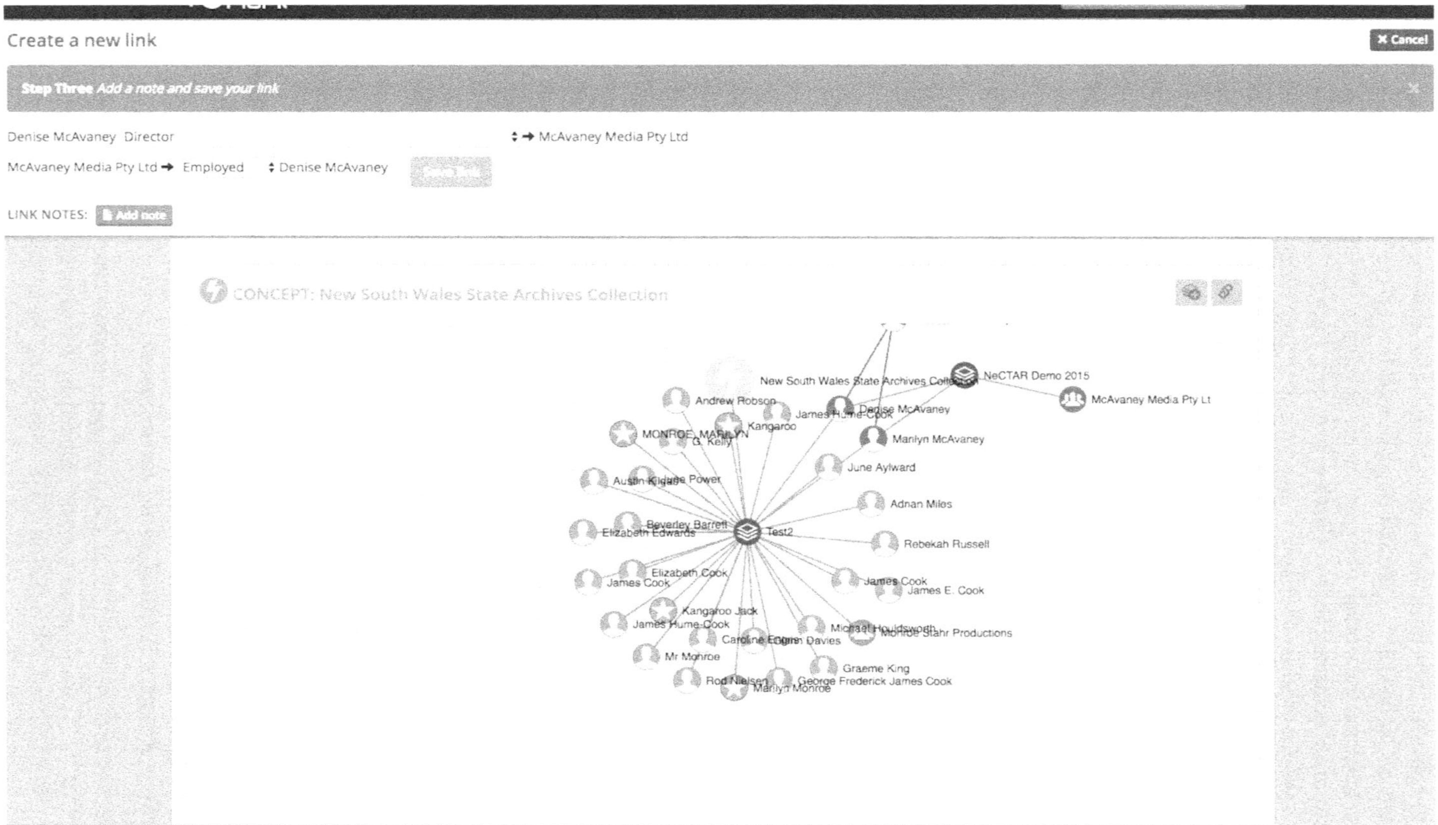

Figure 4.2.8 HuNI seeks to interlink Australian cultural heritage catalogues by means of crowdsourcing (screenshot)

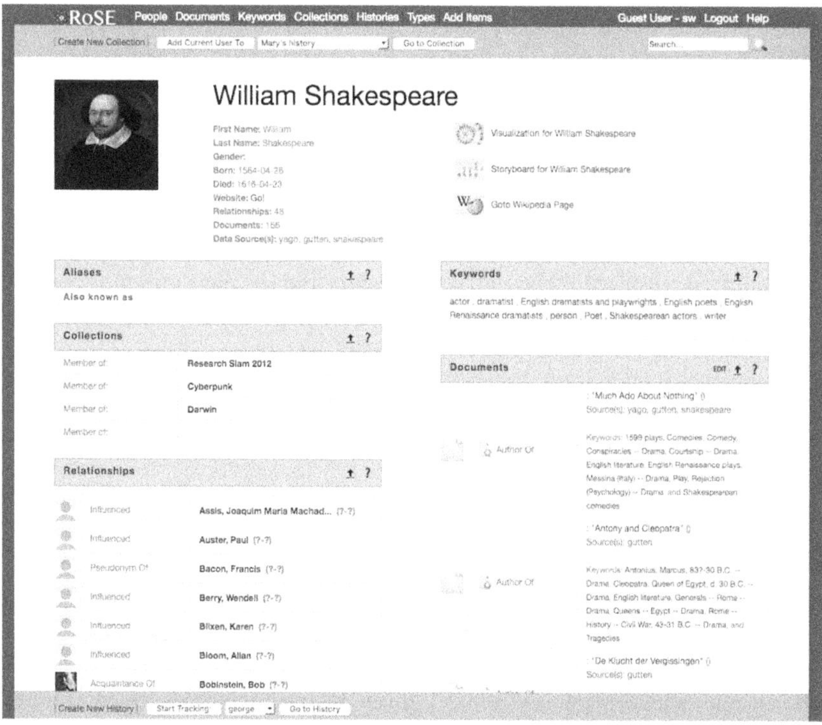

Figure 4.2.9 RoSE lets users explore, enrich and interlink bibliographical records (screenshot)

system that fuses a social-computing model to humanities bibliographical resources to allow users to explore the present and past of the human record as one 'social network.'"[42] RoSE has the ambition to become "a contextual discovery tool for the formative stages of learning about a topic."[43] The system currently contains data on some 21,000 people, 48,000 documents and 20,000 keywords that were added by users or imported from the knowledge base YAGO,[44] the open book repository Project Gutenberg[45] and the SNAC – Social Networks & Archival Context project,[46] which links person information across archives. Like SDFB, RoSE enables users to design their own relationship types in order to embrace the complexity of observed relationships and add to existing bibliographical metadata. RoSE seeks to mimic social networking sites and operates with simple list formats such as aliases, collections (in which a person is listed), relationships, associated keywords and documents.

For the exploration of its content, RoSE relies on interactive network visualisations, which allow users to explore dyads between nodes, the network of associated documents and collections as well as a timeline. All network visualisations can be filtered by the number of people and documents, and the spatial layout can be improved by increasing the distances between nodes. RoSE incorporates interesting features that facilitate the exploration of its content: back and forward buttons and a list of previously visited items give users a sense of their path through the data. Much like Kindred Britain, RoSE also attempts to use its data to enrich narratives. The Storyboard feature lets users draw edges and nodes, aiming to offer them the "visual equivalent of writing a sentence."[47]

Where SDFB focuses on personal links only, RoSE is driven by the idea of a crowdsourced enhanced bibliography, using data and visualisations to reflect the idea that everything is linked to everything else and serving as a starting point for the scholarly exploration of any research topic.

What makes it interesting: RoSE is driven by the idea of open collaboration in the form of a crowdsourced bibliography and a shared knowledge repository that serves as an entry point for researchers.

Discussion

The previous section introduced nine applications that were built based on the assumption that the exploration of relationships between people and objects can effectively guide scholarly research and lead to the discovery of relevant relationships and content. Table 4.2.1 gives an overview of selected features that were integrated into the surveyed applications. Among the visualisation features, all surveyed networks are interactive (9/9) and multimodal networks are common (6/9) as well as queries for paths between nodes (5/9) and geographic maps and timelines (both 5/9). Links between data visualisation and underlying source documents are also frequent (6/9). With regard to the recommendations for effective graph visualisations as laid out by Herman et al., the options to display a subset of the graph (8/9) are well-supported as well as panning (8/9) and zooming (6/9). Static graph layouts (3/9), semantic zoom (2/9) and edge bundling (0/9) are rarely or not at all implemented. Features that are suitable for more quantitative analysis such as network centralities and clustering (0/9) or term frequencies (1/9) are clearly less relevant. Even easy to implement features such as data export (3/9), representations of node centrality (4/9) and edge weight (3/9) are not very frequent. This suggests that the discovery of links across different types of entities across time and space combined with close reading of source documents is considered more valuable than the analysis of the network topographies in which these entities can be found.

Almost all applications combine data from multiple sources such as pre-existing databases, named entity recognition from unstructured text or manual annotation. Identifiers from third party providers such as GND,[48] VIAF[49] or Wikidata[50] are relatively popular (6/9) and allow the enrichment of data with external information and lower the bar for data linking and reuse significantly.

Table 4.2.1 Overview of key features in the surveyed applications (sorted by frequency)

		SDFB	*Kindred Britain*	*ALCIDE*	*APIS*	*Histo-graph*	*Episto-larium*	*ERNiE*	*HuNI*	*RoSE*	*Count*
Types of vis.	Multimodal vis.	-	-	-	x	x	x	x	x	x	6
	Timeline	-	x	x	-	x	x	x	-	-	5
	Geographical maps	-	x	x	-	-	x	x	-	-	4
	Node centrality	x	-	-	-	x	x	x	-	-	4
	Edge weight	x	-	-	-	x	x	-	-	-	3
	Dynamic graph	-	-	-	-	x	-	-	-	-	1
Vis. Features	Graph subsets	x	x	-	x	x	x	x	x	x	8
	Panning	x	x	x	x	x	-	x	-	x	7
	Zooming	x	-	x	x	x	-	x	-	x	6
	Static graph layout	x	x	-	-	-	-	-	-	x	3
	Semantic zoom	-	-	x	-	x	-	-	-	-	2
	Edge bundles	-	-	-	-	-	-	-	-	-	0
Interactions	Interactive nodes/edges	x	x	x	x	x	x	x	x	x	9
	Direct link to sources	-	-	x	x	x	x	x	-	x	6
	Query (shortest) paths	x	x	-	x	x	x	-	-	-	5
	Network data export	x	-	-	x	-	x	-	-	-	3
	Datavis narratives	-	x	-	-	-	-	-	-	x	2
	Clustering/centralities	-	-	-	-	-	-	-	-	-	0
Data creation	External Ids	x	x	-	x	x	-	x	-	x	6
	Based on pre-existing data	-	x	-	x	-	x	-	x	x	5
	Autom. extracted data	x	-	x	x	x	x	-	-	-	5
	NER	x	-	x	x	x	x	-	-	-	5
	Users add Nodes/Edges	x	-	-	x	x	-	-	x	x	5
	Term frequencies	-	-	x	-	-	-	-	-	-	1

Conclusion

Relational data in the surveyed applications differ significantly from the data that is typically used in SNA-like research: carefully crafted models of social structures stand in stark contrast to an automatically aggregated set of relationships between persons, places, institutions or objects within a corpus of documents or database. The following aspects make them particularly relevant for historical research:

(1) Aggregation effects ("I discovered that X was connected to Y") e.g. via path queries are notable examples of the strengths of the surveyed applications. Broad definitions of what counts as a relation make these discoveries more likely.
(2) Links between data visualisation and the digitised documents from which they were extracted help to bridge the gap between abstract data exploration and close reading. It is essential that the discovery of promising links is followed by a closer inspection of the underlying sources. This is an invitation to go beyond the restrictive network models typically used in SNA and is actually closer to the needs of many historians who seek to find a way to create and explore a large number of complex relationships between very different types of entities such as people, institutions, events, ideas or objects.
(3) Multiple approaches for data collection: combinations of pre-existing, automatically generated and (crowd-based) manually created data can be an effective strategy to lower the cost of data collection, whilst obtaining high quality and project-specific data.
(4) Entity disambiguation and linking to existing repositories allows scholars to tap into a wealth of available information, to conceptualise networks in broader terms and to easily share their data with others. APIS' dedication to Linked Open Data is another step in this direction.
(5) Simplified, interactive multimodal network visualisations lower the learning curve and serve as gateways to the underlying content but also allow the critical assessment of its composition e.g. by revealing biases or gaps.
(6) Multimodality: while node-link diagrams are the dominant form for representing networks, most of the surveyed applications also offer alternatives such as timelines, maps, lists and faceted search to visualise and explore their data and thereby open up room for different angles and observations.
(7) Data visualisation can effectively enrich narratives, as demonstrated by Kindred Britain. There is no reason not to apply a similar visual storytelling approach for dedicated historical network analysis projects.

While the surveyed applications offer promising outlooks for future, network-enhanced historical research workflows, a number of challenges remain: not all research-relevant primary sources are available in digital form or can be digitised with reasonable effort. With the exception of ERNiE, which is based on Nodegoat, the surveyed applications are more or less advanced prototypes and often custom-built for specific projects. In contrast to most SNA software, which is suitable for self-study, the majority of the applications investigated in this chapter require support from a developer and need to be customised. Many of the techniques for the automated extraction of network data also require experts, and their performance depends heavily on language and OCR quality. Finally, close collaborations with visualisation experts can help to improve the value of the visualisations for exploration.

These limitations and challenges notwithstanding, the surveyed applications demonstrate that novel techniques in the creation and enrichment of research data paired with accessible albeit versatile visualisation techniques can offer new perspectives on the notions of historical networks, which integrate well with traditional historical research workflows.

Notes

1 This work was partially funded by the FNR grant BLIZAAR INTER/ANR/14/9909176. I would like to thank Martin Stark, Linda von Keyserlingk and Catherine Jones for their helpful feedback on earlier versions of this chapter.
2 For a continuously updated list of applications, see: Marten Düring, "Bibliography: Interactive Graph Exploration," *Historical Network Research*, http://historicalnetworkresearch.org/resources/bibliography/ (accessed 8 February 2019).
3 Lauren F. Klein, Jacob Eisenstein and Iris Sun, "Exploratory Thematic Analysis for Digitized Archival Collections," *Digital Scholarship in the Humanities* 30, no. 1 (2015), https://doi.org/10.1093/llc/fqv052: pp. i130–i41 (accessed 8 February 2019).
4 Uta Hinrichs, Mennatallah El-Assady, Adam James Bradely, Stefania Forlini and Christopher Collins, "Risk the Drift! Stretching Disciplinary Boundaries through Critical Collaborations between the Humanities and Visualization," *Paper Presented at the 2nd Workshop on Visualization for the Digital Humanities*, 2017, https://research-repository.st-andrews.ac.uk/bitstream/handle/10023/11778/vis4dh17_position_pa per_cameraReady_2.pdf?sequence=1&isAllowed=y (accessed 8 February 2019); Jänicke, "Valuable Research for Visualization and Digital Humanities: A Balancing Act," *Workshop on Visualization in the Digital Humanities Vis4DH 17*, 2017, www.informatik.uni-leipzig.de/~stjaenicke/balancing.pdf (accessed 8 February 2019).
5 Marian Dörk, Sheelagh Carpendale and Carey Williamson, "The Information Flaneur: A Fresh Look at Information Seeking," in *Proceedings of the SIGCHI Conference on Human Factors in Computing Systems*, pp. 1215–24, http://dl.acm.org/citation.cfm?id=1979124 (accessed 8 February 2019). This term can also be interpreted as a response to Joanna Drucker's call for interfaces which represent the critical and interpreting nature of the humanities, see: Johanna Drucker, "Performative Materiality and Theoretical Approaches to Interface," *Digital Humanities Quarterly* 7, no. 1 (2013), www.digitalhumanities.org/dhq/vol/7/1/000143/000143.html (accessed 26 November 2016).
6 Micki Kaufman, Zoe LeBlanc, Matthew Lincoln, Yannick Rochat and Scott B. Weingart, "Visualizing Futures of Networks in Digital Humanities Research," *Abstract for DH2017* (2017), https://dh2017.adho.org/abstracts/428/428.pdf (accessed 8 February 2019); Mia Ridge, "Network Visualisations and the 'So What?' Problem," (2016), www.openobjects.org.uk/2016/06/network-visualisations-problem/ (accessed 8 February 2019); Hein van den Berg, Arianna Betti, Thom Castermans, Rob Koopman and Bettina Speckmann, "A Philosophical Perspective on Visualization for Digital Humanities," *3rd Workshop on Visualization for the Digital Humanities* (2018), http://vis4dh.dbvis.de/papers/2018/A%20Philosophical%20Perspective%20on%20Visualization%20for%20Digital%20Humanities.pdf (accessed 8 February 2019);

Hinrichs et al., "Risk the Drift!"; Martin Grandjean, "Introduction à la Visualisation de Données: L'analyse de Réseau en Histoire," *Geschichte und Informatik* 18, no. 19 (2015): pp. 109–28, https://halshs.archives-ouvertes.fr/halshs-01525543/document (accessed 8 February 2019).

7 Marian Dörk, Christopher Pietsch and Gabriel Credico, "One View Is Not Enough: High-Level Visualizations of a Large Cultural Collection," *Information Design Journal* 23, no. 1 (2017), doi:10.1075/idj.23.1.06dor: pp. 39–47 (accessed 8 February 2019).

8 Ben Shneiderman, "The Eyes Have It: A Task by Data Type Taxonomy for Information Visualizations," *Proceedings 1996 IEEE Symposium on Visual Languages*, pp. 336–43, https://ieeexplore.ieee.org/document/545307 (accessed 8 February 2019).

9 A point also highlighted by Fred Gibbs and Trevor Owens, "Building Better Digital Humanities Tools: Toward Broader Audiences and User-Centered Designs," *Digital Humanities Quarterly* 6, no. 2 (12 October 2012), www.digitalhumanities.org/dhq/vol/6/2/000136/000136.html (accessed 8 February 2019); Max Kemman and Martijn Kleppe, "User Required? On the Value of User Research in the Digital Humanities," *Linköping Electronic Conference Proceedings*, no. 116 (26 August 2015), http://orbilu.uni.lu/handle/10993/21812: pp. 63–74 (accessed 8 February 2019).

10 Katrin Glinka, Christopher Pietsch and Marian Dörk, "Past Visions and Reconciling Views: Visualizing Time, Texture and Themes in Cultural Collections," *Digital Humanities Quarterly* 11, no. 2, www.digitalhumanities.org/dhq/vol/11/2/000290/000290.html (accessed 1 January 2017); Florentina Armaselu, "The Text as a Scalable Structure," *International Journal of the Book* 4, no. 4 (2007), https://cgscholar.com/bookstore/works/the-international-journal-of-the-book-vol-4-issue-4-2007 (accessed 8 February 2019); Benjamin B. Bederson, "The Promise of Zoomable User Interfaces," *Behaviour & Information Technology* 30, no. 6 (November 2011), doi:10.1080/0144929X.2011.586724: pp. 853–66 (accessed 8 February 2019).

11 Andreas Girgensohn et al., "Flexible Access to Photo Libraries via Time, Place, Tags, and Visual Features," *JCDL '10 Proceedings of the 10th Annual Joint Conference on Digital Libraries*, pp. 187–96, https://dl.acm.org/citation.cfm?id=1816151&dl=ACM&coll=DL (accessed 8 February 2019).

12 Mitchell Whitelaw, "Generous Interfaces for Digital Cultural Collections," *Digital Humanities Quarterly* 9, no. 1 (2015), www.digitalhumanities.org/dhq/vol/9/1/000205/000205.html (accessed 26 November 2016).

13 Helen Purchase, "Which Aesthetic Has the Greatest Effect on Human Understanding?," in *Graph Drawing: 5th International Symposium, GD '97, Rome, Italy, September 18–20, 1997: Proceedings*, ed. Giuseppe DiBattista 1353, pp. 248–61 (Berlin, Heidelberg: Springer, 1997), https://link.springer.com/chapter/10.1007/3-540-63938-1_67 (accessed 8 February 2019); Danny Holten, "Hierarchical Edge Bundles: Visualization of Adjacency Relations in Hierarchical Data," *IEEE Transactions on Visualization and Computer Graphics* 12, no. 5 (October 2006), doi:10.1109/TVCG.2006.147: pp. 741–8 (accessed 8 February 2019).

14 Helen Purchase, Eve Hoggan and Carsten Görg, "How Important Is the "Mental Map"? An Empirical Investigation of a Dynamic Graph Layout Algorithm," in *Graph Drawing: 14th International Symposium, GD 2006, Karlsruhe, Germany, September 18–20, 2006, Revised Papers*, ed. Michael Kaufmann and Dorothea Wagner, pp. 184–95 (Berlin, Heidelberg: Springer, 2006), https://link.springer.com/chapter/10.1007/978-3-540-70904-6_19 (accessed 8 February 2019).

15 Ivan Herman, Guy Melancon and M. S. Marshall, "Graph Visualization and Navigation in Information Visualization: A Survey," *IEEE Transactions on Visualization and Computer Graphics* 6, no. 1 (2000), doi:10.1109/2945.841119: pp. 24–43 (accessed 8 February 2019).

16 All of the surveyed applications use automated graph layouts and data visualisation libraries such as D3 (https://d3js.org/) or Sigma (http://sigmajs.org/) to facilitate the

interactive exploration of the graphs. Minor flaws such as overlapping node labels would require manual correction and are therefore considered tolerable in these contexts.

17 "Six Degrees of Francis Bacon," http://sixdegreesoffrancisbacon.com/?og=1 (accessed 8 February 2019).

18 "Oxford Dictionary of National Biography," www.oxforddnb.com/public/index.html (accessed 8 February 2019).

19 For a detailed description, see the annex in Christopher Warren et al., "Six Degrees of Francis Bacon: A Statistical Method for Reconstructing Large Historical Social Networks," *Digital Humanities Quarterly* 10, no. 3 (2016), http://digitalhumanities.org/dhq/vol/10/3/000244/000244.html (accessed 8 February 2019).

20 "JSTOR," www.jstor.org/ (accessed 8 February 2019).

21 "Six Degrees of Francis Bacon," http://6dfb.tumblr.com/?og=1 (accessed 8 February 2019).

22 "ALCIDE (Analysis of Language and Content In a Digital Environment)," https://dh.fbk.eu/projects/alcide-analysis-language-and-content-digital-environment (accessed 8 February 2019); Giovanni Moretti et al., "Knowledge-Based Systems ALCIDE: Extracting and Visualising Content from Large Document Collections to Support Humanities Studies," *Knowledge-Based Systems* 111 (2016), doi:10.1016/j.knosys.2016.08.003: pp. 100–12 (accessed 8 February 2019).

23 For a very thorough review on this topic see Stefan Evert, "The Statistics of Word Cooccurrences: Word Pairs and Collocations" (Dissertation, Universität Stuttgart, 2005), https://elib.uni-stuttgart.de/handle/11682/2573 (accessed 8 February 2019).

24 "Apis-WebApp," https://apisdev.eos.arz.oeaw.ac.at/ (accessed 8 February 2019).

25 Daniele Guido, Lars Wieneke and Marten Düring, *Histograph: Graph-Based Exploration, Crowdsourced Indexation, Version 0.7* (Luxembourg: CVCE, 2016), http://histograph.eu (accessed 8 February 2019).

26 histograph has its roots in the FP7-funded project CUbRIK, which explored the potential of human annotation in interaction with machine learning algorithms in multimedia search. See: Lars Wieneke et al., "HistoGraph as a Demonstrator for Domain Specific Challenges to Crowd-Sourcing," in *Social Informatics: SocInfo 2014 International Workshops, Barcelona, Spain, November 11, 2014, Revised Selected Papers*, ed. Luca M. Aiello and Daniel McFarland, pp. 469–76 (Basel: Springer International Publishing, 2014).

27 "CVCE Website," www.cvce.eu/ (accessed 8 February 2019).

28 Uboldi et al. formulated very similar requirements as a consequence of a workshop together with humanities users. Giorgio Uboldi et al., "Knot: An Interface for the Study of Social Networks in the Humanities," *Proceedings of the Biannual Conference of the Italian Chapter of SIGCHI*, CHItaly' 13, pp. 15:1–15:9 (New York: ACM, 2013), www.semanticscholar.org/paper/Knot%3A-an-interface-for-the-study-of-social-networks-Uboldi-Caviglia/6b39abd591441774a9203bd003d583996c31f0b7 (accessed 8 February 2019).

29 "Jaccard index," *Wikipedia*, 2016, https://en.wikipedia.org/w/index.php?title=Jaccard_index&oldid=754661103 (accessed 8 February 2019).

30 "TextRazor: The Natural Language Processing API," www.textrazor.com/ (accessed 8 February 2019).

31 "DBpedia," https://dbpedia.org (accessed 8 February 2019).

32 "EPistolarium," http://ckcc.huygens.knaw.nl/epistolarium/ (accessed 8 February 2019).

33 A revised version is in the making but was not yet available for testing at the time of writing.

34 A redeveloped and enhanced version is expected to be published in the future.

35 "Encyclopedia of Romantic Nationalism in Europe," http://romanticnationalism.net/viewer.p/21 (accessed 8 February 2019).

36 "Nodegoat," http://nodegoat.net/ (accessed 8 February 2019).

37 "Humanities Networked Infrastructure (HuNI)," https://huni.net.au/#/search (accessed 8 February 2019).

38 Deb Verhoeven and Toby Burrows, “Aggregating Cultural Heritage Data for Research Use: The Humanities Networked Infrastructure (HuNI),” in *Metadata and Semantics Research Research*, pp. 417–23 (Cham: Springer, 2015).
39 Ibid.
40 “RoSE,” http://rose.english.ucsb.edu/ (accessed 8 February 2019).
41 “Transliteracies-Research Project,” http://transliteracies.english.ucsb.edu/category/research-project (accessed 8 February 2019).
42 Alan Liu, “Funded Collaborative Projects,” http://liu.english.ucsb.edu/category/new-media-projects/funded-collaborative-projects (accessed 8 February 2019).
43 Alan Liu et al., “Friending the Humanities Knowledge Base: Exploring Bibliography as Social Network in RoSE” (White Paper for the NEH Office of Digital Humanities: Rose Digital Humanities Start-up Grant (Level 2) HD-51433-11, University of California, Santa Barbara, 21 December 2012), www.academia.edu/4154611/Friending_the_Humanities_Knowledge_Base_Exploring_Bibliography_as_Social_Network_in_RoSE (accessed 8 February 2019).
44 “YAGO,” www.mpi-inf.mpg.de/departments/databases-and-information-systems/research/ambiverse-nlu/aida/ (accessed 8 February 2019).
45 “Project Gutenberg,” www.gutenberg.org/ (accessed 8 February 2019).
46 “SNAC-Social Networks and Archival Context,” http://socialarchive.iath.virginia.edu/ (accessed 8 February 2019).
47 Liu et al., “Friending the Humanities Knowledge Base: Exploring Bibliography as Social Network in RoSE”.
48 “Gemeinsame Normdatei (GND),” www.dnb.de/DE/Professionell/Standardisierung/GND/gnd_node.html (accessed 8 February 2019)
49 “VIAF,” www.viaf.org/ (accessed 8 February 2019).
50 “Wikidata: A Free Collaborative Knowledgebase,” www.wikidata.org/wiki/Wikidata:Main_Page (accessed 8 February 2019).

Bibliography

“ALCIDE (Analysis of Language and Content in a Digital Environment).” https://dh.fbk.eu/projects/alcide-analysis-language-and-content-digital-environment (accessed 8 February 2019).

Apis. “Apis-WebApp.” https://apisdev.eos.arz.oeaw.ac.at/ (accessed 8 February 2019).

Armaselu, Florentina. “The Text as a Scalable Structure.” *International Journal of the Book* 4, no. 4 (2007): pp. 53–61.

Bederson, Benjamin B. “The Promise of Zoomable User Interfaces.” *Behaviour & Information Technology* 30, no. 6 (2011): pp. 853–66. doi:10.1080/0144929X.2011.586724.

Briatte, François. “Awesome Network Analysis.” https://github.com/briatte/awesome-network-analysis (accessed 8 February 2019).

CVCE. “CVCE Website.” www.cvce.eu/ (accessed 8 February 2019).

DBpedia. “DBpedia.” http://wiki.dbpedia.org/ (accessed 8 February 2019).

Dörk, Marian, Sheelagh Carpendale, and Carey Williamson. “The Information Flaneur: A Fresh Look at Information Seeking.” *Proceedings of the SIGCHI Conference on Human Factors in Computing Systems*, 2011, pp. 1215–24. http://dl.acm.org/citation.cfm?id=1979124 (accessed 8 February 2019).

Dörk, Marian, Christopher Pietsch, and Gabriel Credico. “One View Is Not Enough: High-Level Visualizations of a Large Cultural Collection.” *Information Design Journal* 23, no. 1 (2017): pp. 39–47. doi:10.1075/idj.23.1.06dor.

Drucker, Johanna. "Performative Materiality and Theoretical Approaches to Interface." *Digital Humanities Quarterly* 7, no. 1 (2013). www.digitalhumanities.org/dhq/vol/7/1/000143/000143.html (accessed 8 February 2019).

Düring, Marten. "Bibliography: Interactive Graph Exploration." http://historicalnetwork-research.org/resources/bibliography/ (accessed 8 February 2019).

Edelstein, Jeff, Lola Galla, Carolyn Li-Madeo, Julia Marden, Alison Rhonemus, and Noreen Y. Whysel. "Linked Open Data for Cultural Heritage." *Columbia University Academic Commons* (2013). https://academiccommons.columbia.edu/catalog/ac:168437 (accessed 8 February 2019).

"EPistolarium." http://ckcc.huygens.knaw.nl/epistolarium/ (accessed 8 February 2019).

Encyclopedia of Romantic Nationalism. "Encyclopedia of Romantic Nationalism in Europe." http://romanticnationalism.net/viewer.p/21 (accessed 8 February 2019).

Evert, Stefan. "The Statistics of Word Cooccurrences: Word Pairs and Collocations." Dissertation, Universität Stuttgart (2005). https://elib.uni-stuttgart.de/handle/11682/2573 (accessed 8 February 2019).

Gibbs, Fred, and Trevor Owens. "Building Better Digital Humanities Tools: Toward Broader Audiences and User-Centered Designs." *Digital Humanities Quarterly* 6, no. 2 (12 October 2012).

Girgensohn, Andreas, Frank Shipman, Jane Hunter, Thea Turner, and Lynn Wilcox. "Flexible Access to Photo Libraries via Time, Place, Tags, and Visual Features." In *JCDL '10 Proceedings of the 10th Annual Joint Conference on Digital Libraries*, pp. 187–96. New York: ACM, 2010. doi:10.1145/1816123.1816151.

Glinka, Katrin, Christopher Pietsch, and Marian Dörk. "Past Visions and Reconciling Views: Visualizing Time, Texture and Themes in Cultural Collections." *Digital Humanities Quarterly* 11, no. 2 (1 January 2017). www.digitalhumanities.org/dhq/vol/11/2/000290/000290.html (accessed 8 February 2019).

Guido, Daniele, Lars Wieneke, and Marten Düring. *Histograph: Graph-Based Exploration, Crowdsourced Indexation, Version 0.7*. Luxembourg: CVCE, 2016. http://histograph.eu (accessed 8 February 2019).

Herman, Ivan, Guy Melancon, and M. Scott Marshall. "Graph Visualization and Navigation in Information Visualization: A Survey." *IEEE Transactions on Visualization and Computer Graphics* 6, no. 1 (2000): pp. 24–43. doi:10.1109/2945.841119.

Holten, Danny. "Hierarchical Edge Bundles: Visualization of Adjacency Relations in Hierarchical Data." *IEEE Transactions on Visualization and Computer Graphics* 12, no. 5 (October 2006): pp. 741–8. doi:10.1109/TVCG.2006.147.

"Humanities Networked Infrastructure (HuNI)." https://huni.net.au/#/search (accessed 8 February 2019).

"Jaccard index." *Wikipedia* (2016). https://en.wikipedia.org/w/index.php?title=Jaccard_index&oldid=754661103 (accessed 8 February 2019).

"JSTOR." www.jstor.org/ (accessed 8 February 2019).

Kemman, Max, and Martijn Kleppe. "User Required? On the Value of User Research in the Digital Humanities." *Linköping Electronic Conference Proceedings*, no. 116 (26 August 2015): pp. 63–74. http://orbilu.uni.lu/handle/10993/21812 (accessed 8 February 2019).

Koeser, Rebecca Sutton, Brian Croxall, Alice Hickcox, Elizabeth R. Roke, Kevin Glover, and Stephen Enniss. "Belfast Group Poetry | Networks." (2015). http://belfastgroup.digitalscholarship.emory.edu/ (accessed 8 February 2019).

Liu, Alan. "Funded Collaborative Projects." http://liu.english.ucsb.edu/category/new-media-projects/funded-collaborative-projects (accessed 8 February 2019).

Liu, Alan, Rama Hoetzlein, Rita Raley, Ivana Anjelkovic, Salman Bakht, Joshua Dickinson, and Michael Hetrick et al. "Friending the Humanities Knowledge Base: Exploring Bibliography as Social Network in RoSE." White Paper for the NEH Office of Digital Humanities: Rose Digital Humanities Start-Up Grant (Level 2) HD-51433-11, University of California, Santa Barbara (21 December 2012). www.academia.edu/4154611/Friending_the_Humanities_Knowledge_Base_Exploring_Bibliography_as_Social_Network_in_RoSE (accessed 8 February 2019). www.academia.edu/4154611/Friending_the_Humanities_Knowledge_Base_Exploring_Bibliography_as_Social_Network_ in_RoSE (accessed 8 February 2019).

Moretti, Giovanni, Rachele Sprugnoli, Stefano Menini, and Sara Tonelli. "Knowledge: Based Systems ALCIDE: Extracting and Visualising Content from Large Document Collections to Support Humanities Studies." *Knowledge-Based Systems* 111 (2016): pp. 100–12. doi:10.1016/j.knosys.2016.08.003.

"Nodegoat." http://nodegoat.net/ (accessed 8 February 2019).

"Oxford Dictionary of National Biography." www.oxforddnb.com/public/index.html (accessed 8 February 2019).

"Project Gutenberg." www.gutenberg.org/ (accessed 8 February 2019).

Purchase, Helen. "Which Aesthetic Has the Greatest Effect on Human Understanding?" In *Proceedings of the Graph Drawing: 5th International Symposium, GD '97, Rome, Italy, September 18–20, 1997*. Edited by Giuseppe DiBattista, pp. 248–61, 1353. Berlin, Heidelberg: Springer, 1997. doi:10.1007/3-540-63938-1_67.

Purchase, Helen, Eve Hoggan, and Carsten Görg. "How Important Is the 'Mental Map'?: An Empirical Investigation of a Dynamic Graph Layout Algorithm." In *Graph Drawing: 14th International Symposium, GD 2006, Karlsruhe, Germany, September 18–20, 2006, Revised Papers*. Edited by Michael Kaufmann and Dorothea Wagner, pp. 184–95, 4372. Berlin, Heidelberg: Springer, 2006. doi:10.1007/978-3-540-70904-6_19.

"RoSE." http://rose.english.ucsb.edu/ (accessed 8 February 2019).

Shneiderman, Ben. "The Eyes Have It: A Task By Data Type Taxonomy for Information Visualizations." In *Proceedings 1996 IEEE Symposium on Visual Languages*, pp. 336–43. doi:10.1109/VL.1996.545307.

"Six Degrees of Francis Bacon." http://6dfb.tumblr.com/?og=1 (accessed 8 February 2019).

"Six Degrees of Francis Bacon." http://sixdegreesoffrancisbacon.com/?og=1 (accessed 8 February 2019).

"SNAC: Social Networks and Archival Context." http://socialarchive.iath.virginia.edu/ (accessed 8 February 2019).

"TextRazor: The Natural Language Processing API." (2016). www.textrazor.com/ (accessed 8 February 2019).

Thomas III, William G., Kaci Nash, Laura Weakly, Karin Dalziel, and Jessica Dussault. "O Say Can You See: Early Washington, D.C., Law & Family." http://earlywashingtondc.org/ (accessed 8 February 2019).

"Transliteracies: Research Project." http://transliteracies.english.ucsb.edu/category/research-project (accessed 8 February 2019).

Uboldi, Giorgio, Giorgio Caviglia, Nicole Coleman, Sébastien Heymann, Glauco Mantegari, and Paolo Ciuccarelli. "Knot: An Interface for the Study of Social Networks in the Humanities." In *Proceedings of the Biannual Conference of the Italian*

Chapter of SIGCHI, pp. 15:1–15:9. CHItaly' 13. New York: ACM, 2013. doi:10.1145/2499149.2499174.

Verhoeven, Deb, and Toby Burrows. "Aggregating Cultural Heritage Data for Research Use: The Humanities Networked Infrastructure (HuNI)." In *Metadata and Semantics Research Research*, pp. 417–23. Cham: Springer, 2015. doi:10.1007/978-3-319-24129-6_36.

"VIAF." http://viaf.org/viaf/data/ (accessed 8 February 2019).

Warren, Christopher, Daniel Shore, Jessica Otis, Lawrence Wang, Mike Finegold, and Cosma Shalizi. "Six Degrees of Francis Bacon: A Statistical Method for Reconstructing Large Historical Social Networks." *Digital Humanities Quarterly* 10, no. 3 (2016).

Whitelaw, Mitchell. "Generous Interfaces for Digital Cultural Collections." *Digital Humanities Quarterly* 9, no. 1 (2015). www.digitalhumanities.org/dhq/vol/9/1/000205/000205.html (accessed 8 February 2019).

Wieneke, Lars, Marten Düring, Vincenzo Croce, and Jasminko Novak. "HistoGraph as a Demonstrator for Domain Specific Challenges to Crowd-Sourcing." In *Social Informatics: SocInfo 2014 International Workshops, Barcelona, Spain, November 11, 2014, Revised Selected Papers*. Edited by Luca M. Aiello and Daniel McFarland, pp. 469–76, 2014. doi:10.1007/978-3-319-15168-7_57.

"Wikidata." http://wikidata.org (accessed 8 February 2019).

"YAGO." www.mpi-inf.mpg.de/departments/databases-and-information-systems/research/yago-naga/yago (accessed 8 February 2019).

5

Outlook

5.1 Historical Network Research, Digital History, and Digital Humanities

Malte Rehbein

The scholarly landscape of Digital Humanities

This chapter is about three scholarly paradigms: Historical Network Research (HNR), Digital History, and Digital Humanities (DH). Albeit not new as grounding ideas,[1] all three have recently gained fresh qualities and are hence topical and currently widely discussed within the academic world and beyond. In particular, the expectations that are attached to them and the promises they offer for reaching new territory in research have led to broad and general attention. Regarding some of these promises, however, opinions differ. On the extreme sides of this dispute, technological ideologists and critical Luddites have hardly been able to find common ground.[2] A situation like this appears not to be untypical in academia when new methodologies, such as those proposed under the umbrella of these three paradigms, are introduced.

Looking forward beyond the dispute between enthusiasm and strict denial, there are several similarities among HNR and the two DHs. Moreover, they are interconnected in such a way that one might easily regard the first as the part of the other two: HNR is Digital History, Digital History is Digital Humanities,[3] and Digital Humanities embraces (historical) network research.

The overarching Digital Humanities[4] have been defined by Manfred Thaller as the sum of all attempts to apply information technology to the objects of Humanities scholarship. Further on, he points out that, first, the application of such technology serves the purpose of achieving results that would otherwise cause unreasonably high effort and second, that this process must be intersubjectively testable in order to have epistemic value.[5] Digital Humanities as a whole principally embraces many if not all disciplines within the Humanities such as language and literary studies, philosophy, archaeology, art history and the like, as well as, potentially, Social Sciences. Within this realm, Digital History[6] might be regarded as the subset of Digital Humanities that intersects with the study of history.

The sum of *all* applications of information technology to the Humanities is rather large, as not only many disciplines within the Humanities are integrated, but also a broad methodological variety is offered.[7] Various disciplines from sciences and engineering, not exclusively Computer Science, are included under this umbrella. Even if restricted to the subset of Digital History, the sum of applications ranges from simply accessing a digital surrogate of an archival entity at one's home desk to digitally editing a mediaeval manuscript in its entire tradition, to the development of a complex procedure in order to automatically classify historical photographs employing artificial neural networks. To better cope with this enormous range of Digital Humanities, I propose a twofold definition:

First, DH encompasses all kinds of research in the Humanities that partly gains its findings from applying computer-based procedures, practices, and tools. In this understanding, Digital Humanities is pure Humanities scholarship, as its objects and questions are those from the Humanities.

Second, DH encompasses the design, development, and generalization of these computer-based procedures, practices and tools, as well as the study of their underlying theories and models. In this understanding, Digital Humanities is rather an auxiliary science (Hilfswissenschaft)[8] located at the intersection between Humanities and Computer Science. Since its epistemological interest is particularly grounded in the functional question of this intersection, DH possesses, hence, its own objects and questions for study.

In case of the second viewpoint of DH, one might also speak of an "Information Science for the Humanities" or "Humanistic Informatics" (Aarseth). For the realm of historical research, I propose to use the term "Historical Information Science" (HIS), which has been suggested by others earlier on and in various translations.[9] Onno Boonstra, Leen Breure, and Peter Doorn define Historical Information Science as "the discipline that deals with specific information problems in historical research and in the sources that are used for historical research"[10] and characterize it as "a science of its own, with its own methodological framework".[11]

With the twofold definition of the overarching Digital Humanities in mind, Historical Information Science serves the second aspect of this definition while Digital History supports the first. Taking into account other disciplines of the Humanities which might or might not develop their own specific information sciences, the scholarly landscape of DH can be summarized as in Figure 5.1.1.[12] This model can be read from two directions. From the outer level, its starting point is a specific historical research problem that can be solved better or only (as in Thaller's definition) by applying computer technology (Digital History). Here, HIS serves as an auxiliary science as it provides suitable tools and practices. From the inner core level, however, the starting point is a methodological problem for which its applicability and generalization in historical research is to be found.[13] Digital Humanities in its generally accepted definitions encompasses the whole model including both viewpoints.

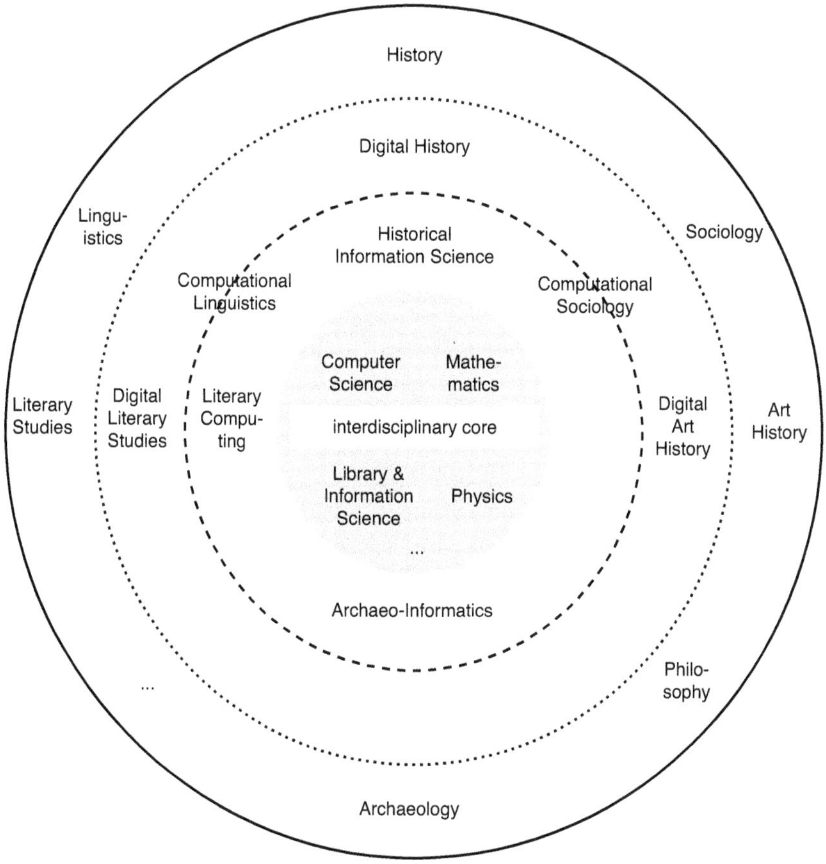

Figure 5.1.1 The scholarly landscape of Digital Humanities. An incomplete attempt

Digital Humanities and network research

It has often[14] been pointed out that the term "network" (as in HNR) is used in a twofold manner: as a widely used metaphor on the one hand (especially in the contexts of a network society,[15] "hyper connectivity",[16] media convergence, and the global internet)[17] and in the stricter sense, as formalisation of semantically annotated nodes and edges grounded on the mathematical graph theory on the other.[18] These semantics may describe communication and interaction among people, in which case a social network[19] is being referred to. However, the concept of network research can be applied to other semantics as well.

HNR[20] is understood as a transdisciplinary research paradigm in which a set of methods and techniques, adopted from other disciplines, is applied to study social (and other) networks of the past and embedded into larger contexts of historical research.[21] HNR can principally be applied without using computer technology (and has been since its beginnings).[22] However, in order to fully exploit its potential, computer-based procedures and tools, such as the management and visual analysis of large datasets, have become part of the HNR paradigm. It is here that HNR intersects with HIS, Digital History and DH.

An intersection among these, seen from the perspective of DH, also exists: networks and network research appear to be of growing interest in the international community of DH.[23] As an example, the widely received project Republic of Letters and its spin-offs visualize and analyse the scholarly exchange between erudite men during the age of Enlightenment by representing correspondence metadata as network graphs.[24] Furthermore, there are several cases in which the mathematical foundation of network graphs is employed for historical studies by applying a different semantic annotation of nodes and edges than that of a social network. In one example, Charlotte Schubert describes the attempt to build a taxonomy of classical textual sources supported by a network in which the longitudinal reception of texts by other authors is represented.[25] In a different setting, the author of this chapter examines the diachronic evolution of a continuously revised mediaeval text with the help of an interactive network graph. Here, nodes represent hypothetical stages of the text, edges represent hypothetical developments from one stage of the text to another, and a path from the (chronologically) first stage of the text to its last formulates a hypothesis of how the text evolved over time.[26] More examples in which network graphs are applied to historical research questions beyond HNR in its narrow definition can be easily found.

Apart from historical studies,[27] network research has been employed in Digital Humanities in a variety of other manners, for instance in empirical literary studies.[28] Literary scholars have also started to describe the configuration of characters in novels or in dramas to form hypotheses about typical patterns for a particular genre[29] by annotating characters as nodes and their co-occurrences or other interactions among them as edges within a network. In linguistics, graphs have long been used to describe relations between words (such as in WordNet).[30] In other works, networks of (co-) authorship or (co-) citation are studied.[31]

The scholarly landscape revisited

Besides the methodological and content-based relationship, similarities among HNR and DH can also be observed in other aspects of scholarship and academia. Current impact to HNR comes from case studies, often undertaken by early-stage scholars who might be motivated by the desire to leave trodden paths and to occupy a niche in which they can excel. Bringing together and transferring these start-up-like, often explorative and experimental studies into a larger scholarly context by attempting to consolidate and generalize their (methodological) findings, is, however, not pushed forward by an institution or otherwise motivated extrinsically. It rather stems from the community itself, informally and virtually.[32] This is comparable to the situation of the DH until the early 2000s.[33]

Back to the scholarly landscape mapped earlier on, case studies of HNR can be classified as Digital History because the starting point of their investigation is a historical research question. The generalization of their methodological findings, on the other hand, can be seen as Historical Information Science. Similar to earlier developments of DH, however, in an emerging research paradigm such as HNR, applying methods can hardly be separated from their critical assessment and contributions to their improvements. Many studies do both. HNR as a transdisciplinary research paradigm can, hence, be integrated into the model along the axis history – digital history – historical information science – interdisciplinary core as shown in Figure 5.1.2.

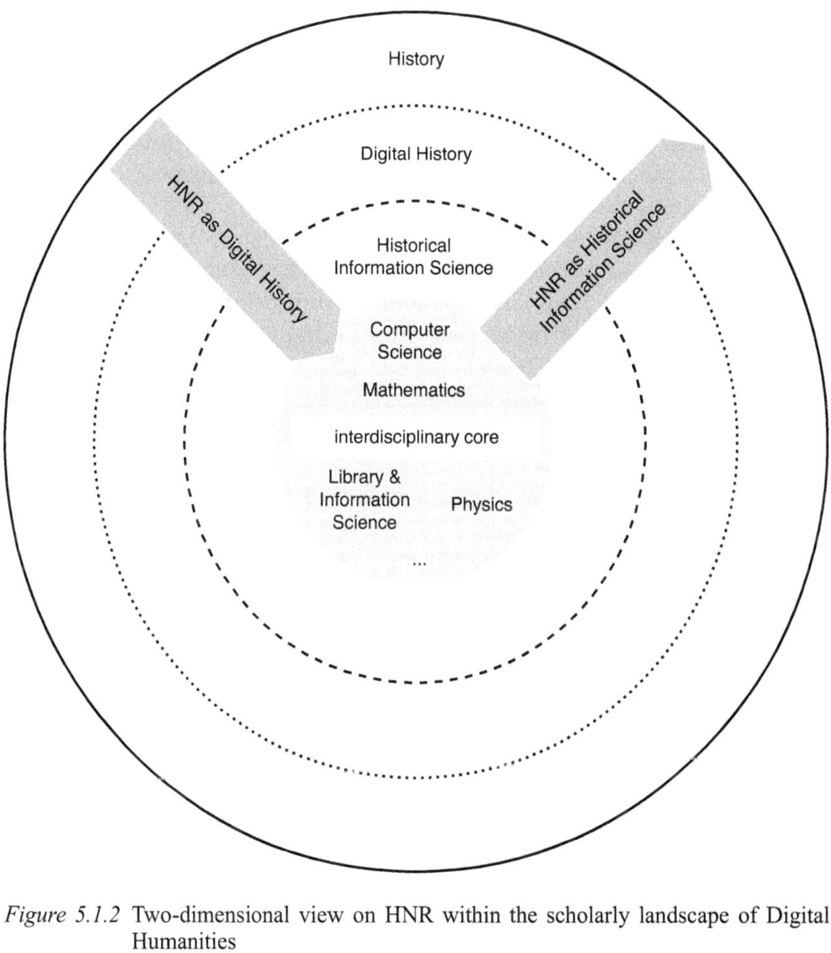

Figure 5.1.2 Two-dimensional view on HNR within the scholarly landscape of Digital Humanities

At the time of this writing, DH has been gaining wider recognition in academia. For this development, it might have been necessary to reach a critical mass of scholars and projects and especially to transfer their methods into the Humanities' canon,[34] but it was probably more the general, societal turn toward digitization at the beginning of this century that has drawn attention to this emerging area of research. As computational methods in the Humanities have been around for several decades, the availability of ready-to-use software will be a next major step forward toward their general acceptance. Only then could these methods be applied by the, as it were, masses of scholars who do not possess "arcane" knowledge such as computer programming, database development, or multimedia production.[35] A similar case in the academic past was that of applying statistical methods to larger sets of data and hence also bringing quantitative history to a new level. It started with the appearance of the Statistical Package for the Social Science (SPSS) software in 1968, which was marked, later on, by sociologist Barry Wellman as the "SPSS revolution".[36] Wellman illustrates its importance: "Now, we do not have to be giants. We can be ordinary people, using statistical packages to play with data and examine hundreds of analytic possibilities".[37] For DH, comparable developments have been under way for a while, such as web-based text analysis tools, off-the-shelf software for stylometric studies, or customized packages for digital editing. However, a wide break-through has not yet been reached. A thorough understanding of the principles of computer technology is often still required by its applicants and, I am tempted to say, at least partly for the good of scholarship.[38] Another aspect that needs to be discussed further in this context is that of the relation between method and software. Here, one has to be careful to insist on the directional dependency between the two. Matthias Bixler's and Daniel Reupke's argument that only using software transforms the metaphor of networks into a method[39] should be reversed: the rationale of the method does not lie within the software, but the software's rationale lies in the method: within the software, the method is operationalized.

Currently, HNR scholarship requires a large set of skills, most of which are not taught in regular History curricula. This, too, is paralleled in Digital Humanities and Digital History and seen as a problem.[40] Apart from the ability to critically interpret "results" from a network visualization within its historical context – which should stand above all – scholars working under the paradigm of HNR need to familiarize themselves, each to more or less extent, with theories of social networks, their mathematical, statistical and quantitative foundations, graph theory, visual literacy,[41] layout algorithms including their strengths and restrictions, human computer interaction, and, maybe most importantly, data modelling, formalization, and coding.

A key requirement to a wider acceptance of HNR methods, and this, too, can be observed equally in the broader field of DH, is the fact that generally all these skills are likewise required by the recipients (i.e., readers) of research results. As Katja Mayer points out: production and dissemination of knowledge are interconnected.[42] This fact in turn requires the applicants of new research paradigms to sensitively write in favour of audiences, the members of which, as stated earlier, do not necessarily possess such skills as long as they have not become part of the curricula in History (or other disciplines within the Humanities).[43] However, as scholarly "knowledge" exists only through a mutual agreement among peers,[44] addressing peers and allowing them to test hypotheses and results themselves is an essential, critical step in scholarship. This might also be a necessity if one aims at convincing a broader community of historians about the "added value" that HNR brings to historical research, which many might still deny.[45] Hence, to an extent, HNR, as well as Digital Humanities and Digital History, still exist only in particular niches and do not reach out as well as they could.[46]

Digital Humanities: an accelerator for Historical Network Research

As we have seen, various intersections between the research paradigm of Historical Network Research and the scholarly area commonly named Digital Humanities exist. In this final section, I will argue that current and past developments within DH can serve as an accelerator for HNR if applied thoughtfully. I will discuss four areas: digitization, automatic annotation, data linkage, and visualization and publication. These areas also partly mirror the four categories of "information problems" in historical research proposed by Boonstra, Breure, and Doorn.[47] A fifth and essential area, namely that of data modelling, is beyond the scope of this chapter and needs to be addressed separately.

Digitization

In HNR, simplified, data and information about historical actors, their behaviour, and especially interaction among them is collected, annotated, and modelled, then visually and/or statistically analysed. The data and information required is taken directly or indirectly (mediated through editions or research literature) from historical sources, the key role of which can hardly be denied for the study of History. Hence, obtaining access to sources is often a restricting factor in present day historical research.

The 19th and 20th centuries in particular have provided large amounts of edited historical records. But these are still only an extremely small portion of all surviving sources from the past, and, what is more, as a historical legacy, they often embrace only those records that represent the "big players" of the past. Yet, it is precisely such types of historical records that are interesting for HNR; serial sources in particular and heterogeneous sources beyond the scope of "big history" are, to date, hardly transcribed, let alone edited. Tedious visits to archives, finding, reading, and annotating relevant information is thus required in order for important questions to be answered. For HNR, too, it has been observed that pragmatic considerations – and not those of research interests or relevance – prevail. Various authors point out that only in cases in which sources are easily accessible and data, as well as information from these sources, can be extracted within reasonable time, is historical network research possible or sensible.[48]

The digitization of historical records, that is the process of transforming them into digital representations or surrogates, is one of the cornerstones of Digital Humanities.[49] It has been a research effort for a long time to put historical records efficiently and sustainably into digital format, make them openly accessible as well as analysable, and to preserve them for the future. The digitization of archival records follows a new paradigm and thus offers fresh quality: the availability of such records no longer depends on whether they have been edited in depth or made available in print (usually with the significant investment of time and money) or on a personal visit to the archives. Instead, the continuous digitization of archival records[50] and their availability in the virtual realm certainly brings new impact to HNR and might help in overcoming the aforementioned restrictions in research so far seen as necessary due to pragmatic considerations.

Naturally, digitization cannot solve challenges arising from the characteristics of historical research such as incomplete or missing sources: records that have not survived from the past cannot be digitized in the present and probably not in the future either. But digitization is beginning to play an ever-increasing role in cultural heritage in general and historical records in particular. After the last "wave" of digitization projects and programmes focused on printed material and rare collections, other types of sources[51] are now gaining relevance in funding schemes, which might be of great interest for HNR.[52]

While digitization obviously improves access to sources, it does, however, not necessarily help with the discovery of relevant information within them. Transforming digital surrogates into machine readable texts is a process comparable to that of palaeographic transcriptions. But in order to efficiently transcribe a large amount of digitized sources, the process has to be automated.[53] For printed texts, this problem has been solved to a great extent with Optical Character Recognition (OCR), albeit the fact that the general principle that the older the text, the lower the quality of text recognition seems to remain valid. Automated recognition of handwritten texts (HTR), however, still is an interdisciplinary research challenge. Recent developments,[54] using, among others, artificial neuronal networks[55] and deep learning, seem promising. Proximity might have come closer to a goal desired (maybe) by many: to feed a computer with the digital image of a historical manuscript and to receive, without human interaction, machine readable data of the text that the manuscript mediates. Consequences for historical research would hardly be less immense: millions of handwritten historical records could be made accessible, not only for reading at one's home desk but far more, for automatically searching within, annotating, linking and analysing them in any thinkable way with the aid of computer-based procedures. Especially for the field of HNR, this would open vast new terrain.

Another aspect of digitization as a societal phenomenon of the present time will have future impact on HNR: "Born digital" data are becoming more and more relevant for historical research. For instance, with the advent of networked computers in the 1960s, electronic communication such as email has been gaining importance (and is now prevailing). It is already – or will soon be – an object of study for historians too. Surprisingly, unwanted characteristics of historical data, such as lost records or questions of authenticity, seem to have survived in the digital realm in a similar manner. Archiving these records as well as other media of the World Wide Web such as websites is still an open issue. Due to its ephemeral nature, much past electronic communication will not be available as historical sources in the future.[56]

With the help of automated text recognition, digitization opens historical research to a new paradigm in which the accurate reading of a historical source is not necessarily central for understanding the past any more. Instead, macroscopic analyses of a vast amount (and maybe also variety) of sources might prevail under this paradigm. In literary studies, which build far more on printed texts hence easily made machine-readable than historical studies, this paradigm has been (controversially) discussed for a couple of years under the headline "distant reading."[57] For many applications of this "distant reading" that are currently discussed, semantic enrichment of data and its in-depth annotation, essential for HNR, is, however, not a key requirement. For HNR, further steps, especially those of annotating information, are necessary.

Automatic annotation

While digitization is improving, the availability of historical sources and automated text recognition is making them processable on a computer. Furthermore, information has to be extracted and formalized systematically from these sources about historical actors, as well as their attributes and relations, in order to prepare for HNR.[58] Depending on the research design, this process can happen on the level of metadata (e.g. in a correspondence network: who wrote whom)[59], on the level of content (e.g. which two persons are mentioned in the same context), or in a combination of both (e.g. who wrote about about whom). Generally, the content-level is more challenging to address than the meata-data-level when it comes to automating this process.

For HNR, procedures to support this process lie within the realm of Historical Information Science. They encompass:

(1) (Named) entity recognition (NER); that is annotating a textual string as a signifier for a specific type of information e.g. a person's name;
(2) Entity resolution; that is disambiguating this string in order to identify a unique member of that type, e.g. a specific person;
(3) Complex relation extraction; that is extracting from the text whether and how two persons (or other types of entities) relate to each other; and
(4) Contextualising data historically in space, time, and semantics.

All these steps underlie general characteristics of historical sources such as variety, ambiguity, vagueness, fuzziness, and scarcity or incompleteness. They have been studied extensively since the 1970s at the latest[60] but have not yet been solved satisfactorily, at least not on a generic level. Progress in data cleaning and annotation of this kind, particularly employing methods and tools from Computational Linguistics, has been made in non-historical applications of network research of Digital Humanities. However, data from other disciplines is often less complex than historical data. In literary studies, for instance, extracting actors and their co-occurrence on stage can be automated more easily as the number of actors is limited, their names are known, stage instructions that define certain relations exist, and data is complete.[61] Again, one might think of a future machine that is fed with all available sources in a machine-readable form together with a research question such as that of an ego-network of historical person X or of a trade network within a particular period. Yet, by annotating data in the way outlined earlier, profound historical understanding already plays a central role. The key to such historical understanding is contextualisation and interpretation based on human experience (Karl Schlögel), which cannot be delegated to algorithms uncritically. Hence, concerns should be expressed about attempts to automatically generate historical networks. To follow the so-called "Big Data" paradigm[62] and its "end-of-theory" branch[63] is topical and tempting but, for the time being at least, rather misleading as long as it is not connected to theory.[64] The strength of automation in HNR lies, so far, mainly in supporting data preparation, and any progress made here will be of benefit for HNR.

Various methods are currently being tested on historical texts, structured historical data, and research literature. These attempts are mostly situated in the realm of Computer Linguistics and Natural Language Processing and include supervised and unsupervised Machine Learning.[65] Approaches close to HNR encompass automated coding of occupational strings,[66] place name disambiguation,[67] entity recognition, and resolution in cultural heritage collections[68] and relationship extraction from biographical information systems.[69] This kind of annotated data carries information about historical actors that may be useful for network research.

Speaking of relationship extraction in particular, however, one must be aware that several types of such relations exist for conceptualizing a network.[70] They range semantically from permanent genealogical to temporary relations such as that of employer-employee or an affair up to short-term interactions or punctual meetings, and they encompass complex and disputable notions as such of friendship as well as simple and less disputable co-occurrences of names within a certain range of text.

Currently, there is no general set of tools or process pipeline available to undertake annotating tasks for HNR. As research questions and source material are too heterogeneous[71] and a broad variety exists concerning how such relations between people are expressed in the source text (if at all made explicit), it is probably not sensible to develop similar tools at all. However, it is foreseeable that computer-supported annotation can be used in HNR if tailored to specific types of sources and questions.[72] This kind of generalization on a meso level, that transcends single problems but does not promise "all-inclusive" solutions, can again be understood as a future research task for Historical Information Science and will significantly push forward HNR if transferred to a broader scholarly community.

Data linkage

Another benefit that annotated data offer is their capability of instantly creating links to other data sources. With such a network of data, the scope of an HNR research question can potentially be extended. Furthermore, efforts to collect and annotate data can be shared among many. Hence, HNR can benefit from a general trend to make research and other data openly available in a way that relations between data points can be followed across the internet (Linked Open Data, LOD). At the same time, HNR researchers can also contribute to this idea of LOD by providing their data and annotations to the public.[73]

Some considerations have to be made, though. For LOD to work effectively, data has to be interchangeable, that is, reusable within a different context. This is usually solved technically (and legally). For LOD to work efficiently (and to an extent, automatically), however, data has to be interoperable.[74] For instance, data points that have the same identifier must refer to the same entity (e.g. historical person) across the whole LOD-universe and vice versa; one entity cannot have different identifiers in different data sources.

Questions of interoperability and interchange of data have long been studied in Digital Humanities. Especially the "flagship" data model of DH, that of the Text Encoding Initiative (TEI), a set of recommendations of tag and attribute names and their usages, has addressed them since the mid-1980s.[75] TEI provides mechanisms to annotate (textual) data so that interchange among different data sources is facilitated. Efficient interoperability is still a challenge, and promises seem, so far, to be unfulfilled.[76] Interoperability grounded, as a prerequisite, on a shared understanding and conceptualization of data, information, and knowledge[77] and a consistent formalization among potentially heterogeneous sources is difficult if not impossible to attain.

The same can be said about employing the ideas of LOD for HNR. LOD certainly facilitates data interchange, and researchers can manually or semi-automatically draw data from other sources to include them in their own networks. It is still unclear, however, to what extent one can make use of data interoperability in HNR. As mentioned earlier, a shared conceptualization of data is a prerequisite. Evidently, unique and cross-platform identification of persons is a central part of this shared conceptualization, as one has to be sure that the same person X is meant in different data sources. This problem has found its technical solution in authority files.[78]

An example of how interoperability based on authority files may work beneficially for HNR is given by the Deutsche Biographie; a network of biographical data of about 740,000 persons from 19 distributed data sources through this mechanism. Thinking this concept to its end, historical relations could be detected and used for HNR that would not be otherwise visible. If, for instance, one historical source witnesses a relation between A and B and another source, a relation between B and C, then a relation between A and C (with B as a broker) can be inferred even if neither the first source witnesses anything about C nor the second about A. The critical evaluation of this (logical) inference is still the historian's task, though.

Reaching a shared understanding to uniquely identify persons is still a far easier task than that of reaching a shared understanding of what a relation is and what quality it has. Given the various types of possible relations (see earlier example) and different, often subjective notions of relations such as "friendship" or different levels of relevance of a particular relation within different research questions, a general agreement will hardly ever be reached.

It is an open issue for the HIS-branch of Historical Network Research to deal with these problems. Questions need to be addressed such as how to cope with different kinds of types for the same data that would lead to inconsistencies within the network; how to deal with a categorisation used in an external data source that does not fit one's own research design; how to assess the quality of external data sources and what mechanisms exist in order to improve them or tailor them to specific needs; among different (heterogeneous) data sources, what is the greatest common divisor of a shared conceptualization? Some of these questions are also discussed within the TEI,[79] and a joint discussion would be of mutual benefit, in particular as TEI-encoded data might serve as data sources for HNR.

Visualization and publication

Digitization also brings new forms for communicating scholarly findings. They range from traditional publication formats within new business models (journal articles in Open Access) to rather new formats such as blogging and micro blogging. Scholarly work is beginning to be understood as a process that addresses the public not only at its very end. Findings are also being published in preliminary stages, and experimental formats such as collaborative writing[80] or social editing[81] are being developed. The discussion about the future of scholarly publication and communication, another current topic of DH, also encompasses research data for later re-use as a scholarly output.

Scholars of HNR are also confronted with the possibilities (by some seen as opportunities, by others as threats) that new publication paradigms provide. The HNR-platform[82] is an example of a usage for communication and community building. For publishing historical findings, however, it can be observed that the traditional formats of scholarly articles and monographs including static graphics[83] are mostly chosen. As an emerging research paradigm, HNR itself is confronting a traditional scholarly community, that of historians, with a new and disputable methodology. It might be easier to gain acceptance from this community if new methodology is not combined with fresh forms of publication.

As an analogy to HNR, digital scholarly editing, one of the core areas of Digital Humanities (and Digital History), is about to complete its transformation into the Humanities' methodological canon. As a consequence, digital editions might lose their attribute "digital" in the near future as it has become accepted and natural to create and publish scholarly editions digitally. This wide acceptance, however, has come at the cost of compromise. Most digital editions mimic their printed predecessors more than they could (and should). The medium of publication is different (internet instead of a printed book), but the form remains more or less the traditional one: the one that the long-standing community has been used to using, hence the one that demands less effort for overcoming resistance.

This compromise has excluded vast potential that computational methods offer for publication, e.g. interactive, dynamic visualizations. Peter Shillingsburg's statement from 2006 still seems to be valid: "we have not fully understood or exploited the capabilities of electronic texts."[84] This statement can be transferred to HNR. Especially in the area of Visual Analytics, HNR has the potential to make a large contribution to digital scholarship as well as to historical research when (thoughtfully) exploiting these capabilities, not only as a research tool but also as a way to communicate with the scientific community and lay people. What is required, however, is a progressive conversation about forms of scholarly communication and publication that go beyond academic papers and monographs.

Visualization and Visual Analytics are current trends in digital scholarship; they are not new, but computational methods offer a fresh quality to them. One must be aware, however, that in HNR, what is visualised is never the historical reality but rather a hypothetical model of it.[85] As this model is interpretative and subjective, an endless number of potential other models of that reality exists. For instance, a historian can be uncertain as to whether a relation between actors A and B existed. Traditionally, it is expected from the researcher that a decision is made, based on accepted historical theory and methodology or that the answer is left open to the question and that this undecidability is documented. In HNR, however, a formalized statement is required in any case. Within the data, this could be reached by either including the relationship between A and B, by omitting it, or by attributing the relationship in question with a "certainty weight" of, say, 0.5, meaning that the likelihood that this relationship existed is estimated by the researcher with 50%.[86] Data formalized in this way is processable by a computer and can also be used for visualization (and inference).

This example of making explicit statements about the data is comparable to that of constituting a particular reading of a text in a scholarly edition. But what happens if the reader of an edition or in this case, the viewer of the visualization, does not agree with the editor's or researcher's decision? Interpreting the relationship between A and B differently might lead to a different network and consequently to a different model and hypothesis of the past. But which methods exist to test the researcher's hypothesis or make up one's own?

In digital scholarly editing, attempts have been made to use dynamic forms of textual presentations in which the reader is empowered to follow his or her own reading paths, highlight particular aspects of the textual data (at the cost of others) and so on, hence having an interactive tool at hand to verify the editor's decisions as well as to make one's own hypotheses.[87] Such forms have not yet reached a critical mass and broad acceptance as a consequence of the aforementioned compromise. Naturally, they are disputable, especially because the traditional role model between editor and reader, between researcher and recipient, seem to vanish, and responsibilities for research results are disavowed. At the time of this writing, I would argue that this discussion as to whether such new forms of scholarly communication brings scholarship forward or are rather steps backwards is undecided. But as a natural part of research and progress, these questions need to be actively discussed, and HNR can contribute greatly to this debate.

Conclusion

In this chapter, an attempt has been made to characterise Historical Network Research as a topical research paradigm that is located in the scholarly landscape as an integral part of the realms of Digital Humanities, Digital History, and Historical Information Science. Such an interdisciplinary intersection can be employed for mutual benefit. On the one hand, HNR may profit from computer-based practices, procedures, and tools that are being developed and made ready for general use within the area of Historical Information Sciences. On the other hand, HNR itself offers significant potential to improve digital and otherwise formalized methods that can then be used in other scholarly disciplines under the umbrella of the Digital Humanities. The methodological issues that HNR opens up and attempts to answer may contribute on a general level to the theoretical foundation of computer-based approaches in the Humanities.

HNR as a paradigm also contributes to the methodological canon of scholarship in History. By providing complementary approaches to historical questions, it helps to expand the boundaries of research, offers new perspectives to old questions, and potentially opens up new terrain. An anchoring of HNR within the theories of History, as well as those of computer-based Humanities, has yet to be consolidated.

Notes

1 Commonly, the year 1947, when Roberto Busa started his seminal work on a word concordance of the works of Tomas Aquinas aided by IBM calculators and punch cards, is regarded as a starting point for what is now known as Digital Humanities (Susan Hockey, "The History of Humanities Computing," in *A Companion to Digital Humanities*, ed. Susan Schreibman, Ray Siemens and John Unsworth, Blackwell Companions to Literature and Culture 26, pp. 3–19 (Oxford: Blackwell, 2004)). Computational approaches in the Humanities, however, can be traced back further, at least to the 19th century when statistical stylometric calculations where undertaken manually (Manfred Thaller, "Geschichte der Digital Humanities," in Jannidis; Kohle; Rehbein, *Digital Humanities*, pp. 3–12).

2 See for instance Matthew K. Gold and Lauren F. Klein, eds., *Debates in the Digital Humanities* (Minneapolis: University of Minnesota Press, 2012) and Matthew K. Gold and Lauren F. Klein, eds., *Debates in the Digital Humanities 2016* (Minneapolis:

University of Minnesota Press, 2016). See also Helle Porsdam, "Digital Humanities: On Finding the Proper Balance between Qualitative and Quantitative Ways of Doing Research in the Humanities," *Digital Humanities Quarterly* 7, no. 3 (2013).

3 Surprisingly, however, the communities of Digital Humanities and of Digital History do not mix very well. The Call for Papers for the DH conference of the Nordic Countries, published in August 2017, describes and addresses this phenomenon, which can be observed on an international as well on a national level (e.g. Germany) as follows: "While the number of researchers describing themselves as digital historians is increasing, computational approaches to history have rarely captured the attention of those without innate interest in digital humanities. To address this, we particularly invite presentations of historical research whose use of digital methods advances the overall methodological basis of the field" http://eadh.org/news/2017/09/01/cfp-digital-humanities-nordic-countries-0 (accessed 17 December 2018).

4 A lot of older as well as recent publications try to outline and illustrate the scope of Digital Humanities, such as Susan Schreibman, Ray Siemens and John Unsworth, eds., *A Companion to Digital Humanities*, Blackwell Companions to Literature and Culture 26 (Oxford: Blackwell, 2004), Willard McCarty, *Humanities Computing* (Basingstoke, England, New York: Palgrave Macmillan, 2005), and John Unsworth, Raymond G. Siemens and Susan Schreibman, eds., *A New Companion to Digital Humanities*, Blackwell Companions to Literature and Culture 93 (Chichester, West Sussex, UK, Malden, MA, USA: John Wiley & Sons Ltd., 2016). For a brief overview see also Malte Rehbein, "Was sind Digital Humanities?," *Akademie Aktuell* 56, no. 1 (2016). As Digital Humanities has begun to appear on the universities' curricula especially in the German speaking countries, a textbook introduces basic DH-methods: Fotis Jannidis, Hubertus Kohle and Malte Rehbein, eds., *Digital Humanities: Eine Einführung* (Stuttgart: J.B. Metzler Verlag, 2017).

5 Manfred Thaller, "Grenzen und Gemeinsamkeiten: Die Beziehung zwischen der Computerlinguistik und den Digital Humanities" (DHd 2014, Passau, 27 March 2014).

6 Fundamental for discussing the connotation of related terms: Onno Boonstra, Leen Breure and Peter Doorn, "Past, Present and Future of Historical Information Science," *Historical Social Research/Historische Sozialforschung* 29, no. 2 (2004). Particularities of historical research within the realm of Digital Humanities are discussed in Stephen Robertson, "The Differences between Digital Humanities and Digital History," in *Debates in the Digital Humanities 2016*, ed. Matthew K. Gold and Lauren F. Klein (Minneapolis: University of Minnesota Press, 2016). See also Peter Haber, *Digital Past: Geschichtswissenschaft im digitalen Zeitalter* (München: Oldenbourg, 2011) for a wider contextualization of Digital History within the landscape of historical scholarship.

7 Lorna Hughes, Panos Constantopoulos and Costis Dallas, "Digital Methods in the Humanities: Understanding and Describing Their Use across the Disciplines," in Unsworth; Siemens; Schreibman, *A New Companion to Digital Humanities*, pp. 150–70. The authors categories network analysis as one of seven "research activities" of the type "analysis" (ibid., p. 157).

8 I argue elsewhere for the notion of DH as a "transferring science" (Transferwissenschaft), which "mission" is to evaluate, adopt, develop, and generalize procedures and tools until they are ready-to-use within the Humanities and can be transferred into their methodological canon. This includes also setting up DH curricula and training programmes Malte Rehbein, "Geschichtsforschung im digitalen Raum. Über die Notwendigkeit der Digital Humanities als historische Grund- und Transferwissenschaft," in *Papstgeschichte im digitalen Zeitalter: Neue Zugangsweisen zu einer Kulturgeschichte Europas*, ed. Klaus Herbers and Viktoria Trenkle, Beihefte zum Archiv für Kulturgeschichte Heft 85, pp. 19–44 (Köln, Weimar, Wien: Böhlau Verlag, 2017). This includes also setting up DH curricula and training programmes.

9 Especially in German, Russian, Dutch as Historische Fachinformatik, istoricheskaya informatika, historische informatiekunde.

10 Boonstra, Breure and Doorn, "Past, Present and Future of Historical Information Science," p. 20.

11 Ibid., p. 10.

12 This model is inspired by Patrick Sahle's "spheres" (Patrick Sahle, "DH studieren! Auf dem Weg zu einem Kern- und Referenzcurriculum der Digital Humanities," *DARIAH-DE Working Papers 1* (2013), http://webdoc.sub.gwdg.de/pub/mon/dariah-de/dwp-2013-1.pdf). It is not yet complete.

13 Boonstra, Breure and Doorn, "Past, Present and Future of Historical Information Science," p. 20.

14 Cf. especially Marten Düring, Ulrich Eumann, Martin Stark and Linda von Keyserlingk, eds., *Handbuch Historische Netzwerkforschung: Grundlagen und Anwendungen*, Schriften des Kulturwissenschaftlichen Instituts Essen (KWI) zur Methodenforschung 1 (Berlin: LIT-Verlag, 2016).

15 Manuel Castells, *The Rise of the Network Society*, The Information Age: Economy, Society and Culture 1 (Cambridge, MA, Oxford: Wiley-Blackwell, 1996); Jan van Dijk, *The Network Society: Social Aspects of New Media* (London, Thousand Oaks, CA: Sage Publications, 1999).

16 Ian Goldin, *Divided Nations: Why Global Governance Is Failing, and What We Can Do about It* (Oxford: Oxford University Press, 2013), p. 5.

17 Currently, the term "network" seems overloaded and used in an inflationary manner, also in history (Marten Düring et al., "Einleitung," in Düring; Eumann; Stark; Keyserlingk, *Handbuch Historische Netzwerkforschung*, pp. 5, 5–10). There are cases, however, in which the metaphoric is turned around: the North-American network of people and facilities to help escaped slaves to freedom in Canada since the late 18th century was called "the underground railroad," an expression coined in the 1830s. Here, "railroad" is used as a metaphoric expression to illustrate the activities of the network: "Various routes were lines, stopping places were called stations, those who aided along the way were conductors and their charges were known as packages or freight" (History.com staff, "Underground Railroad" (2009), www.history.com/topics/black-history/underground-railroad (accessed 17 December 2018)).

18 There seems to be indeed few restrictions in which to apply this research paradigm, see e.g. Milan Janosov, "Network Science Predicts Who Dies Next in Game of Thrones," https://cns.ceu.edu/article/2017-07-08/network-science-predicts-who-dies-next-game-thrones (accessed 17 December 2018).

19 In a common definition by James Clyde Mitchell, founding member of the International Network for Social Network Analysis, a social network is a "specific set of linkages among a defined set of persons, with the additional property that the characteristics of these linkages as a whole may be used to interpret the social behaviour of the persons involved" (James C. Mitchell, "The Concept and Use of Social Networks," in *Social Networks Urban Situations: Analyses of Personal Relationships in Central African Towns*, ed. James C. Mitchell, p. 2 (Manchester: Manchester University Press, 1969)).

20 I am referring to general definitions in Claire Lemercier, "Formale Methoden der Netzwerkanalyse in den Geschichtswissenschaften: Warum und Wie?," *Österreichische Zeitschrift für Geschichtswissenschaft* 23, no. 1 (2012): pp. 16–41 and Marten Düring and Linda von Keyserlingk, "Netzwerkanalyse in den Geschichtswissenschaften. Historische Netzwerkanalyse als Methode für die Erforschung von historischen Prozessen," in *Prozesse: Formen, Dynamiken, Erklärungen*, ed. Rainer Schützeichel and Stefan Jordan, pp. 337–50 (Wiesbaden: Springer VS, 2015).

21 This is supported by Thomas S. Kuhn's understanding of a research paradigm: a consensus-based set of procedures that defines what objects to be studied, the kindof questions and hypotheses that are to be asked or tested, how these questions are to be asked, and how findings shall be interpreted. In context of HNR, cf. Wilhelm Heinz Schröder, who has characterised Historical Social Research

as a transdisciplinary research paradigm (Wilhelm H. Schröder, "Die Buchreihe 'Historisch-sozialwissenschaftliche Forschungen' als publizistisches 'Flaggschiff' der quantitativen Historischen Sozialforschung in der 'Pionierzeit'," *Historical Social Research/Historische Sozialforschung*, Supplement No. 18 (2006): pp. 14, 6–23).

22 This is generally true for the wider field of historical computing (Boonstra, Breure and Doorn, "Past, Present and Future of Historical Information Science," p. 18).

23 The programme of the international "Digital Humanities" conference in Montréal 2017 lists 15 presentations and posters that discuss phenomena of literary, social, or historical networks. The keyword "network" is listed 47 times, albeit including technical terms such as neuronal networks or network infrastructure (McGill University & Université de Montréal, *Digital Humanities 2017: Conference Abstracts* (2017)). The expressions "network research" or "network study" do, however, not appear. Networks were not addressed in the fundamental DH-Companions (Schreibman, Siemens and Unsworth, eds., *A Companion to Digital Humanities*) and (Unsworth, Siemens and Schreibman, eds., *A New Companion to Digital Humanities*).

24 http://republicofletters.stanford.edu/.

25 Charlotte Schubert, "Die Visualisierung von Quellennetzwerken am Beispiel Plutarchs," *Digital Classics Online* 2, no. 1 (2016).

26 Malte Rehbein, "Reconstructing the Textual Evolution of a Medieval Manuscript," *Literary & Linguistic Computing* 24 (2009): pp. 319–27.

27 Further examples include: Bente Opheim, "Political Networks and Factions: Online Prosopography of Medieval Scandinavian Sagas," *History and Computing* 12, no. 1 (2000): pp. 43–57, David M. Brown, Adriana Soto-Corominas and Juan L. Suárez, "The Preliminaries Project: Geography, Networks, and Publication in the Spanish Golden Age," *Digital Scholarship in the Humanities* 32, no. 4 (December 2017): pp. 709–32, Cornell Jackson, "Using Social Network Analysis to Reveal Unseen Relationships in Medieval Scotland," *Digital Scholarship in the Humanities* 32, no. 2 (June 2017): pp. 336–43, F. Kimura et al., "Visualization of Relationships among Historical Persons from Japanese Historical Documents," *Literary and Linguistic Computing* 28, no. 2 (2013): pp. 271–8, and Marcus Bingenheimer, Jen-Jou Hung and Simon Wiles, "Social Network Visualization from TEI Data," *Literary and Linguistic Computing* 26, no. 3 (2011).

28 See e.g. Peer Trilcke, "Social Network Analysis (SNA) als Methode einer textempirischen Literaturwissenschaft," in *Empirie in der Literaturwissenschaft*, ed. Philip Ajouri, Katja Mellmann and Christoph Rauen, pp. 201–47 (Münster: mentis Verlag, 2013), Laura Mandell, "How to Read a Literary Visualisation: Network Effects in the Lake School of Romantic Poetry," *Digital Studies / Le champ numérique* 3, no. 2 (2012), and Chloe Edmondson, "An Enlightenment Utopia: The Network of Sociability in Corinne," *Digital Humanities Quarterly* 11, no. 2 (2017).

29 Fotis Jannidis, "Netzwerke," in Jannidis; Kohle; Rehbein, *Digital Humanities*, pp. 147–61.

30 Other examples in which graph-based networks with a different semantic annotation are used for Humanities research include and illustrate the variety of this usage: Tom J. Lynch, "Social Networks and Archival Context Project: A Case Study of Emerging Cyberinfrastructure," *Digital Humanities Quarterly* 8, no. 3 (2014), Jamshid Tehrani, Quan Nguyen and Teemu Roos, "Oral Fairy Tale or Literary Fake? Investigating the Origins of Little Red Riding Hood using Phylogenetic Network Analysis," *Digital Scholarship in the Humanities* 31, no. 3 (2016): pp. 611–36; Daniel Gamermann et al., "The Small-World of 'Le Petit Prince': Revisiting the Word Frequency Distribution," *Digital Scholarship in the Humanities* (32, no. 2 (June 2017): pp. 301–311; Maciej Eder, "Visualization in Stylometry: Cluster Analysis Using Networks," *Digital Scholarship in the Humanities* 32, no. 1 (2017): pp. 50–64; Valentina Bartalesi et al., "Towards a Semantic Network of Dante's Works and Their Contextual Knowledge," *Digital Scholarship in the Humanities* 30, Supplement 1 (December 2015):

pp. i28–i35; J. Hu and N. Wang, "Complex Network Perspective on Graphic Form System of Hanzi," *Literary and Linguistic Computing* 28, no. 4 (2013): pp. 660–7; Florian Kräutli and Matteo Valleriani, "CorpusTracer: A CIDOC Database for Tracing Knowledge Networks," *Digital Scholarship in the Humanities* 33, no. 2 (June 2018): pp. 336–46.

31 For a methodological overview, cf. Markus Gmür, "Co-Citation Analysis and the Search for Invisible Colleges: A Methodological Evaluation," *Scientometrics* 57, no. 1 (2003): pp. 27–57. As an example for its implementation cf. Markus Eckl, "Von Forschungsteams zur Wissenschaftscommunity: Eine soziale Netzwerkanalyse der wissenschaftlichen Co-Autorenschaften in der Disziplin der Sozialen Arbeit zwischen 1980 und 2014," *Soziale Passagen* 8 (2016).

32 Such as the website "Historical Network Research" maintained by members of the community: "a platform for scholars to present their work, enable collaboration and provide those new to network analysis with some helpful first information" (http://historicalnetworkresearch.org/).

33 In the much larger area of Digital Humanities, it is still discussed whether DH shall form a scholarly discipline on its own or remain rather a "community of practice" with a low degree of institutionalization (for instance understood so by the sub community "Digital Medievalist", http://digitalmedievalist.org (accessed 17 December 2018)). "The digital humanities is what digital humanists do" Robertson, "Robertson 2016" might be a sufficient definition of such a community.

34 Something that seems to have been quite successful in the area of digital editing and more so (but still on its way) in the computational method of authorship attribution (Patrick Juola, "Authorship Attribution," *Foundations and Trends in Information Retrieval* 1, no. 3 (2007): pp. 233–334). See also the discussion in the final section of this chapter.

35 Cf. Boonstra, Breure and Doorn, "Past, Present and Future of Historical Information Science," p. 19.

36 Barry Wellman, "Doing It Ourselves: The SPSS Manual as Sociology's Most Influential Recent Book," in *Required Reading: Sociology's Most Influential Books*, ed. Dan Clawson, pp. 71–8, 73 (Amherst: University of Massachusetts Press, 1998).

37 Ibid., p. 74. In his famous quote: "L'historien de demain sera programmeur ou il ne sera plus" (Emmanuel Le Roy Ladurie, *Le territoire de l'historien*, Bibliothèque des histoires (Paris: Gallimard, 1973), p. 14), in which he referred to quantitative history only, Emmanuel Le Roy Ladurie did not foresee this development. For him, it was clear that only a deep understanding of the mechanisms behind software will empower the historian to fully exploit its potential and to avoid failures (cf. Manfred Thaller, "Geschichte der Digital Humanities," in Jannidis; Kohle; Rehbein, *Digital Humanities*, pp. 3–12).

38 I argue in more detail elsewhere: Rehbein, "Geschichtsforschung im digitalen Raum".

39 Matthias Bixler and Daniel Reupke, "Von Quellen zu Netzwerken," in Düring; Eumann; Stark; Keyserlingk, *Handbuch Historische Netzwerkforschung*, pp. 116–17, 101–22.

40 Cf. Martin Stark, "Netzwerkberechnungen. Anmerkungen zur Verwendung formaler Methoden," in Düring; Eumann; Stark; Keyserlingk, *Handbuch Historische Netzwerkforschung*, pp. 156, 155–71, Malte Rehbein, "Digitalisierung braucht Historiker/innen, die sie beherrschen, nicht beherrscht," *H-Soz-Kult*, 27 November 2015. In general terms, as digital methods not only are those of HNR arising and getting more widely used, skills required to apply (and to assess and evaluate!) them have to be included in curricula of history, cf. e.g. Rehbein, "Geschichtsforschung im digitalen Raum," Malte Rehbein and Patrick Sahle, "Digital Humanities lehren und lernen: Modelle, Strategien, Erwartungen," in *Evolution der*

Informationsinfrastruktur: Kooperation zwischen Bibliothek und Wissenschaft, ed. Heike Neuroth, Norbert Lossau and Andrea Rapp, pp. 209–28 (Glückstadt: vwh Hülsbusch, 2013).

41 This aspect cannot be overestimated. Anyone using graphics to communicate research results, a core element of HNR, must be aware of the effect on the recipient of symbols and their arrangement. Cf. Katja Mayer, "Netzwerkvisualisierungen. Anmerkungen zur visuellen Kultur der Historischen Netzwerkforschung," in Düring; Eumann; Stark; Keyserlingk, *Handbuch Historische Netzwerkforschung*, pp. 147, 139–53. For a general introduction see Edward R. Tufte, *The Visual Display of Quantitative Information* (Cheshire, CT (Box 430, Cheshire 06410): Graphics Press, 1983), for information visualization within DH: Malte Rehbein, "Informationsvisualisierung," in Jannidis; Kohle; Rehbein, *Digital Humanities*, pp. 328–42.

42 Katja Mayer, "Netzwerkvisualisierungen. Anmerkungen zur visuellen Kultur der Historischen Netzwerkforschung," in Düring; Eumann; Stark; Keyserlingk, *Handbuch Historische Netzwerkforschung*, pp. 139, 139–53.

43 Cf. Marten Düring and Florian Kerschbaumer, "Quantifizierung und Visualisierung. Anknüpfungspunkte in den Geschichtswissenschaften," in Düring; Eumann; Stark; Keyserlingk, *Handbuch Historische Netzwerkforschung*, pp. 38, 31–43 with references to Robert W. Fogel, Konrad Jarausch, and William O. Aydelotte. Arguing from the readers' perspective: N. Thomson, "How to Read Articles which Depend on Statistics," *Literary and Linguistic Computing* 4, no. 1 (1989): pp. 6–11.

44 Following Thomas S. Kuhn's notion of science (Thomas S. Kuhn, *The Structure of Scientific Revolutions* (Chicago, IL: University of Chicago Press, 1962)), as well as Friedrich Nietzsche's understanding of truth (cf. Wiebrecht Ries, *Nietzsche zur Einführung*, 7th ed. (Hamburg: Junius Verlag, 2004), pp. 1020–4) and Lothar Kolmer, *Geschichtstheorien* (Stuttgart: UTB, 2008), p. 68).

45 For the latter cf. Düring, Eumann, Stark and Keyserlingk, eds., *Handbuch Historische Netzwerkforschung*, p. 7.

46 Matthias Bixler, "Die Wurzeln der Historischen Netzwerkforschung," in Düring; Eumann; Stark; Keyserlingk, *Handbuch Historische Netzwerkforschung*, pp. 59, 45–61.

47 Information problems of historical sources, of relationships between sources, of historical analysis, and of the presentation of sources or analysis Boonstra, Breure and Doorn, "Past, Present and Future of Historical Information Science," pp. 20–1.

48 Cf. Christian Marx, "Forschungsüberblick zur Historischen Netzwerkforschung. Zwischen Analysekategorie und Metapher," in Düring; Eumann; Stark; Keyserlingk, *Handbuch Historische Netzwerkforschung*, pp. 65, 63–84, (Robert Gramsch, "Zerstörte oder verblasste Muster? Anwendungsfelder mediävistischer Netzwerkforschung und das Quellenproblem," in Düring; Eumann; Stark; Keyserlingk, *Handbuch Historische Netzwerkforschung*, pp. 69, 85–99), Matthias Bixler and Daniel Reupke, "Von Quellen zu Netzwerken," in Düring; Eumann; Stark; Keyserlingk, *Handbuch Historische Netzwerkforschung*, pp. 107, 101–22.

49 As well as it has been with the realms of library and information science and archival studies.

50 So-called retro-digitization of printed edition as well as of secondary texts have to be mentioned here, too, as both are of importance for HNR.

51 To be mentioned, among others, are historical newspapers and serial sources such as account books and matriculation registers. As an example: the diocese of Passau, Germany, is digitizing its complete series of parochial records from the 16th to the 20th century and hence opening them up for genealogical research, social history, and, potentially, HNR (cf. Robert Gramsch, "Zerstörte oder verblasste Muster? Anwendungsfelder mediävistischer Netzwerkforschung und das Quellenproblem," in Düring; Eumann; Stark; Keyserlingk, *Handbuch Historische Netzwerkforschung*,

pp. 85–99). As pioneering in terms of systematization, completeness and transparency to be mentioned: Monasterium.net (http://monasterium.net/, accessed 17 December 2018) and Matricula (http://data.matricula-online.eu/de/, accessed 17 December 2018).

52 Within a pilot "Archivalische Quellen" (archival sources), funded by the German Research Council DFG, efficient procedures have been developed, tested, and should be ready for broad application.

53 Apart from automation, crowdsourcing has also been tried to transcribe (and annotate) historical records (Melissa Terras, "Crowdsourcing in the Digital Humanities," in Unsworth; Siemens; Schreibman, *A New Companion to Digital Humanities*, pp. 420–38). There are manifold examples, including the "Transcribe Bentham" project (blogs.ucl.ac.uk/transcribe-bentham/) or the data entry project of the Danish censuses 1787–1880 (Nanna F. Clausen, "The Danish Demographic Database: Principles and Methods for Cleaning and Standardisation of Data," in *Population Reconstruction*, ed. Gerrit Bloothooft et al., 1st ed. 2015, pp. 3–22 (Cham, s.l.: Springer International Publishing, 2015)).

54 The forthcoming results from the EU-funded project READ should be observed (https://read.transkribus.eu/, accessed 17 December 2018).

55 A method of machine learning, motivated from biology as well as another example of using the word "network."

56 Cf. Terry Kuny, "A Digital Dark Ages? Challenges in the Preservation of Electronic Information," *63rd IFLA Council and General Conference*, 1997.

57 Franco Moretti, *Graphs, Maps, Trees: Abstract Models for a Literary History*, 1. publ (London: Verso, 2005).

58 Cf. Matthias Bixler and Daniel Reupke, "Von Quellen zu Netzwerken," in Düring; Eumann; Stark; Keyserlingk, *Handbuch Historische Netzwerkforschung*, pp. 104, 101–22.

59 The Kalliope database of meta data on personal papers and manuscripts is an example in which such metadata can be used for rudimentary network analysis (http://kalliope.staatsbi bliothek-berlin.de/, accessed 17 December 2018).

60 Boonstra, Breure and Doorn, "Past, Present and Future of Historical Information Science," p. 35.

61 This way, in an exemplary case study, 465 German dramas between 1731 and 1929 have been automatically analysed in order to create and visually compare networks of characters over time (Frank Fischer et al., "Distant Reading Showcase: 200 Jahre deutsche Dramengeschichte auf einen Blick" (DHd 2016, Leipzig, 9 March 2016), http://dhd2016.de/sites/default/files/dhd2016/files/Distant_Reading_-Showcase__465_German_Dramas__DHd2016_Poster.pdf (accessed 17 December 2018)).

62 For a discussion of the use of the term "Big Data" in the Humanities cf. Christof Schöch, "Big? Smart? Clean? Messy? Data in the Humanities," *Journal of Digital Humanities* 2, no. 3 (2013).

63 „Out with every theory of human behavior. [… The traditional] approach to science – hypothesize, model, test – is becoming obsolete" (Chris Anderson, "The End of Theory: The Data Deluge Makes the Scientific Method Obsolete," *Wired Magazine*, 16 July 2008).

64 One might, for instance, simply ask why different data sets yield different results for the same research questions. I am discussing this in greater detail on biographical data elsewhere (Rehbein, "Geschichtsforschung im digitalen Raum").

65 An overview of state-of-the-art methods in automated processing of structured historical data on the example of genealogy provides Gerrit Bloothooft, Peter Christen, Kees Mandemakers and Marijn Schraagen, eds., *Population Reconstruction*, 1st ed. 2015 (Cham, s.l.: Springer International Publishing, 2015).

66 Ibid., Kevin Schürer, Tatiana Penkova and Yanshan Shi, "Standardising and Coding Birthplace Strings and Occupational Titles in the British Censuses of 1851 to 1911," *Historical Methods* 48, no. 4 (2015): pp. 195–213.
67 Ian Gregory et al., "Geoparsing, GIS, and Textual Analysis: Current Developments in Spatial Humanities Research," *International Journal of Humanities and Arts Computing* 9, no. 1 (2015): pp. 1–14.
68 Seth van Hooland, Max de Wilde, Ruben Verborgh, Thomas Steiner and Rik van de Walle, "Exploring Entity Recognition and Disambiguation for Cultural Heritage Collections," *Literary and Linguistic Computing* 30, no. 2: pp. 262–79.
69 Matthias Reinert et al., "From Biographies to Data Curation: The Making of www.deutsche-biographie.de," *Proceedings of the First Conference on Biographical Data in a Digital World*, 2015: pp. 13–19. The project currently focusses on genealogical relations and those of the type teacher-student.
70 For a thorough and theoretical discussion from the viewpoint of sociology, cf. Mustafa Emirbayer, "Manifesto for a Relational Sociology," *The American Journal of Sociology* 103, no. 2 (1997): pp. 281–317.
71 Christian Marx, "Forschungsüberblick zur Historischen Netzwerkforschung. Zwischen Analysekategorie und Metapher," in Düring; Eumann; Stark; Keyserlingk, *Handbuch Historische Netzwerkforschung*, pp. 66, 63–84.
72 Cf. Robert Gramsch, "Zerstörte oder verblasste Muster? Anwendungsfelder mediävistischer Netzwerkforschung und das Quellenproblem," in Düring; Eumann; Stark; Keyserlingk, *Handbuch Historische Netzwerkforschung*, pp. 85–99 for a proposal on how to categorise HNR research questions in mediaeval studies.
73 This touches also the question of managing research data, which is a general topic in DH and information science. Solutions from there can be reused also by HNR.
74 For a definition (and distinction) between interchangeability and interoperability, cf. Syd Bauman, "Interchange vs. Interoperability," *Proceedings of Balisage: The Markup Conference 2011* (2011), www.balisage.net/Proceedings/vol7/print/Bauman01/BalisageVol7-Bauman01.html (accessed 17 December 2018).
75 Nancy Ide et al., "Proposal for Funding for an Initiative to Formulate Guidelines for the Encoding and Interchange of Machine-Readable Texts" (1988), www.tei-c.org/Vault/SC/scg02.html (accessed 17 December 2018).
76 Desmond Schmidt, "Towards an Interoperable Digital Scholarly Edition," *Journal of the Text Encoding Initiative* 7 (2014).
77 Shared conceptualization is the idea behind "ontologies," proposed by Thomas Gruber (Thomas Gruber, "A Translation Approach to Portable Ontology Specifications," *Knowledge Acquisition* 5, no. 2 (1993): pp. 199–220). Paired with semantic web technologies, ontologies are one of the current research areas in DH (Malte Rehbein, "Ontologien," in Jannidis; Kohle; Rehbein, *Digital Humanities*, pp. 162–76).
78 Mostly commonly the Virtual International Authority File (VIAF), initiated by the Library of Congress and the German National Library and the Gemeinsame Normdatei (GND), maintained by the German National Library.
79 John Unsworth, "Computational Work with Very Large Text Collections: Interoperability, Sustainability, and the," *Journal of the Text Encoding Initiative* 1, no. 1 (2011), http://jtei.revues.org/215.
80 E.g. Terry Eagleton, *Marxism and Literary Criticism*, 2nd ed. (New York: Routledge, 2002), p. 63.
81 Kenneth M. Price, "Social Scholarly Editing," in Unsworth; Siemens; Schreibman, *A New Companion to Digital Humanities*, pp. 137–49.
82 http://historicalnetworkresearch.org/ (accessed 17 December 2018).
83 Cf. Katja Mayer, "Netzwerkvisualisierungen. Anmerkungen zur visuellen Kultur der Historischen Netzwerkforschung," in Düring; Eumann; Stark; Keyserlingk, *Handbuch Historische Netzwerkforschung*, pp. 149, 139–53.

84 Peter L. Shillingsburg, *From Gutenberg to Google: Electronic Representations of Literary Texts* (Cambridge: Cambridge University Press, 2006), p. 88.
85 Johanna Drucker, "Humanities Approaches to Graphical Display," *Digital Humanities Quarterly* 5, no. 1 (2011).
86 A similar mechanism is implemented in the TEI with the @cert-attribute.
87 As an example see Rehbein, "Reconstructing the Textual Evolution," pp. 319–27 and the earlier discussion.

Bibliography

Ajouri, Philip, Katja Mellmann, and Christoph Rauen, eds. *Empirie in der Literaturwissenschaft*. Münster: mentis Verlag, 2013.

Anderson, Chris. "The End of Theory: The Data Deluge Makes the Scientific Method Obsolete." *Wired Magazine* 16, no. 7 (2008).

Bartalesi Valentina, Carlo Meghini, Paola Andriani, and Mirko Tavoni. "Towards a Semantic Network of Dante's Works and Their Contextual Knowledge." *Digital Scholarship in the Humanities* 30, Supplement 1 (2015): pp. i28–i35.

Bauman, Syd. "Interchange vs. Interoperability." *Proceedings of Balisage: The Markup Conference 2011*, 2011. www.balisage.net/Proceedings/vol7/print/Bauman01/BalisageVol7-Bauman01.html (accessed 17 December 2018).

Bingenheimer, Marcus, Jen-Jou Hung, and Simon Wiles. "Social Network Visualization from TEI Data." *Literary and Linguistic Computing* 26, no. 3 (2011): pp. 271–8.

Bixler, Matthias. "Die Wurzeln der Historischen Netzwerkforschung." In Düring; Eumann; Stark; Keyserlingk, *Handbuch Historische Netzwerkforschung*, pp. 45–61.

Bixler, Matthias, and Daniel Reupke. "Von Quellen zu Netzwerken." In Düring; Eumann; Stark; Keyserlingk, *Handbuch Historische Netzwerkforschung*, pp. 101–22.

Bloothooft, Gerrit, Peter Christen, Kees Mandemakers, and Marijn Schraagen, eds. *Population Reconstruction*. 1st ed. Cham, s.l.: Springer International Publishing, 2015.

Boonstra, Onno, Leen Breure, and Peter Doorn. "Past, Present and Future of Historical Information Science." *Historical Social Research/Historische Sozialforschung* 29, no. 2 (2004).

Brown, David M., Adriana Soto-Corominas, and Juan Luis Suárez. "The Preliminaries Project: Geography, Networks, and Publication in the Spanish Golden Age." *Digital Scholarship in the Humanities* 32, no. 4 (2017): pp. 709–32.

Castells, Manuel. *The Rise of the Network Society*. The Information Age: Economy, Society and Culture 1. Cambridge, MA: Oxford, 1996.

Clausen, Nanna Floor. "The Danish Demographic Database: Principles and Methods for Cleaning and Standardisation of Data." In *Population Reconstruction*. Edited by Gerrit Bloothooft et al. 1st ed., pp. 3–22. Cham, s.l.: Springer International Publishing, 2015.

Clawson, Dan, ed. *Required Reading: Sociology's Most Influential Books*. Amherst: University of Massachusetts Press, 1998.

Drucker, Johanna. "Humanities Approaches to Graphical Display." *Digital Humanities Quarterly* 5, no. 1 (2011).

Düring, Marten, and Florian Kerschbaumer. "Quantifizierung und Visualisierung. Anknüpfungspunkte in den Geschichtswissenschaften." In Düring; Eumann; Stark; Keyserlingk, *Handbuch Historische Netzwerkforschung*, pp. 31–43.

Düring, Marten, and Linda von Keyserlingk. "Netzwerkanalyse in den Geschichtswissenschaften. Historische Netzwerkanalyse als Methode für die Erforschung von historischen Prozessen." In *Prozesse: Formen, Dynamiken, Erklärungen*. Edited by Rainer Schützeichel and Stefan Jordan, pp. 337–50. Wiesbaden: Springer VS, 2015.

Düring, Marten, Ulrich Eumann, Martin Stark, and Linda von Keyserlingk. "Einleitung." In Düring; Eumann; Stark; Keyserlingk, *Handbuch Historische Netzwerkforschung*, pp. 5–10.

———, eds. *Handbuch Historische Netzwerkforschung: Grundlagen und Anwendungen*. Schriften des Kulturwissenschaftlichen Instituts Essen (KWI) zur Methodenforschung 1. Berlin: LIT-Verlag, 2016.

Eagleton, Terry. *Marxism and Literary Criticism*. 2nd ed. New York: Routledge, 2002.

Eckl, Markus. "Von Forschungsteams zur Wissenschaftscommunity: Eine soziale Netzwerkanalyse der wissenschaftlichen Co-Autorenschaften in der Disziplin der Sozialen Arbeit zwischen 1980 und 2014." *Soziale Passagen* 8 (2016).

Eder, Maciej. "Visualization in Stylometry: Cluster Analysis Using Networks." *Digital Scholarship in the Humanities* 32, no. 1 (2017): pp. 50–64.

Edmondson, Chloe. "An Enlightenment Utopia: The Network of Sociability in Corinne." *Digital Humanities Quarterly* 11, no. 2 (2017).

Emirbayer, Mustafa. "Manifesto for a Relational Sociology." *The American Journal of Sociology* 103, no. 2 (1997): pp. 281–317.

Fischer, Frank, Andreas Vogel, Mathias Göbel, Peer Trilcke, Dario Kampkaspar, and Christopher Kittel. "Distant Reading Showcase: 200 Jahre deutsche Dramengeschichte auf einem Blick." *DHd 2016* (9 March 2016), Leipzig. http://dhd2016.de/sites/default/files/dhd2016/files/Distant_Reading_Showcase__465_German_Dramas__DHd2016_Poster.pdf (accessed 17 December 2018).

Gamermann, Daniel, Carmen Moret-Tatay, Esperanza Navarro-Pardo, and Pedro Fernandez de Córdoba Castellá. "The Small-World of 'Le Petit Prince': Revisiting the Word Frequency Distribution." *Digital Scholarship in the Humanities* (2016).

Gmür, Markus. "Co-Citation Analysis and the Search for Invisible Colleges: A Methodological Evaluation." *Scientometrics* 57, no. 1 (2003): pp. 27–57.

Gold, Matthew K., and Lauren F. Klein, eds. *Debates in the Digital Humanities*. Minneapolis: University of Minnesota Press, 2012.

———, eds. *Debates in the Digital Humanities 2016*. Minneapolis: University of Minnesota Press, 2016. http://dhdebates.gc.cuny.edu/debates/2 (accessed 17 December 2018).

Goldin, Ian. *Divided Nations: Why Global Governance Is Failing, and What We Can Do about It*. Oxford: Oxford University Press, 2013.

Gramsch, Robert. "Zerstörte oder verblasste Muster? Anwendungsfelder mediävistischer Netzwerkforschung und das Quellenproblem." In Düring; Eumann; Stark; Keyserlingk, *Handbuch Historische Netzwerkforschung*, pp. 85–99.

Gregory, Ian, Christopher Donaldson, Patricia Murrieta-Flores, and Paul Rayson. "Geoparsing, GIS, and Textual Analysis: Current Developments in Spatial Humanities Research." *International Journal of Humanities and Arts Computing* 9, no. 1 (2015): pp. 1–14.

Gruber, Thomas. "A Translation Approach to Portable Ontology Specifications." *Knowledge Acquisition* 5, no. 2 (1993): pp. 199–220.

Haber, Peter. *Digital Past: Geschichtswissenschaft im digitalen Zeitalter*. München: Oldenbourg, 2011.

Herbers, Klaus, and Viktoria Trenkle, eds. *Papstgeschichte im digitalen Zeitalter: Neue Zugangsweisen zu einer Kulturgeschichte Europas*. Beihefte zum Archiv für Kulturgeschichte Heft 85. Köln, Weimar, Wien: Böhlau Verlag, 2017.

History.com staff. "Underground Railroad." (2009). www.history.com/topics/black-history/underground-railroad (accessed 17 December 2018).

Hockey, Susan. "The History of Humanities Computing." In *A Companion to Digital Humanities*. Edited by Susan Schreibman, Ray Siemens and John Unsworth, pp. 3–19. Blackwell Companions to Literature and Culture 26. Oxford: Blackwell, 2004.

Hu, J., and N. Wang. "Complex Network Perspective on Graphic Form System of Hanzi." *Literary and Linguistic Computing* 28, no. 4 (2013): pp. 660–7.

Hughes, Lorna, Panos Constantopoulos, and Costis Dallas. "Digital Methods in the Humanities: Understanding and Describing Their Use across the Disciplines." In Unsworth; Siemens; Schreibman, *A New Companion to Digital Humanities*, pp. 150–70.

Ide, Nancy, C.M. Sperberg-McQueen, Robert Amsler, Donald Walker, Susan Hockey, and Antonio Zampolli. "Proposal for Funding for an Initiative to Formulate Guidelines for the Encoding and Interchange of Machine-Readable Texts." (1988). www.tei-c.org/Vault/SC/scg02.html (accessed 17 December 2018).

Jackson, Cornell. "Using Social Network Analysis to Reveal Unseen Relationships in Medieval Scotland." *Digital Scholarship in the Humanities* (2016).

Jannidis, Fotis. "Netzwerke." In Jannidis; Kohle; Rehbein, *Digital Humanities*, pp. 147–61.

Jannidis, Fotis, Hubertus Kohle, and Malte Rehbein, eds. *Digital Humanities: Eine Einführung*. Stuttgart: J.B. Metzler Verlag, 2017.

Janosov, Milan. "Network Science Predicts Who Dies Next in Game of Thrones." (8 July 2017). https://cns.ceu.edu/article/2017-07-08/network-science-predicts-who-dies-next-game-thrones (accessed 17 December 2018).

Juola, Patrick. "Authorship Attribution." *Foundations and Trends in Information Retrieval* 1, no. 3 (2007): pp. 233–334.

Kimura, F., T. Osaki, T. Tezuka, and A. Maeda. "Visualization of Relationships among Historical Persons from Japanese Historical Documents." *Literary and Linguistic Computing* 28, no. 2 (2013): pp. 271–8.

Kolmer, Lothar. *Geschichtstheorien*. Stuttgart: UTB, 2008.

Kräutli, Florian, and Matteo Valleriani. "CorpusTracer: A CIDOC Database for Tracing Knowledge Networks." *Digital Scholarship in the Humanities* (2017).

Kuhn, Thomas S. *The Structure of Scientific Revolutions*. Chicago, IL: University of Chicago Press, The University of Chicago Press, 1962.

Kuny, Terry. "A Digital Dark Ages? Challenges in the Preservation of Electronic Information." *63rd IFLA Council and General Conference*, Copenhagen, Denmark, August 31 - September 5, 1997.

Lemercier, Claire. "Formale Methoden der Netzwerkanalyse in den Geschichtswissenschaften: Warum und Wie?" *Österreichische Zeitschrift für Geschichtswissenschaft* 23, no. 1 (2012): pp. 16–41.

Le Roy Ladurie, Emmanuel. *Le territoire de l'historien*. Bibliothèque des histoires. Paris: Gallimard, 1973.

Lynch, Tom J. "Social Networks and Archival Context Project: A Case Study of Emerging Cyberinfrastructure." *Digital Humanities Quarterly* 8, no. 3 (2014).

Mandell, Laura. "How to Read a Literary Visualisation: Network Effects in the Lake School of Romantic Poetry." *Digital Studies/Le champ numérique* 3, no. 2 (2012).

Marx, Christian. "Forschungsüberblick zur Historischen Netzwerkforschung. Zwischen Analysekategorie und Metapher." In Düring; Eumann; Stark; Keyserlingk, *Handbuch Historische Netzwerkforschung*, pp. 63–84.

Mayer, Katja. "Netzwerkvisualisierungen. Anmerkungen zur visuellen Kultur der Historischen Netzwerkforschung." In Düring; Eumann; Stark; Keyserlingk, *Handbuch Historische Netzwerkforschung*, pp. 139–53.

McCarty, Willard. *Humanities Computing*. Basingstoke, England, New York: Palgrave Macmillan, 2005. http://site.ebrary.com/lib/alltitles/docDetail.action?docID=10254194 (accessed 17 December 2018).

Mitchell, James Clyde. "The Concept and Use of Social Networks." In *Social Networks Urban Situations: Analyses of Personal Relationships in Central African Towns*. Edited by James C. Mitchell. Manchester: Manchester University Press, 1969.

———, ed. *Social Networks Urban Situations: Analyses of Personal Relationships in Central African Towns*. Manchester: Manchester University Press, 1969.

Moretti, Franco. *Graphs, Maps, Trees: Abstract Models for a Literary History*. 1st publ. London: Verso, 2005.

Neuroth, Heike, Norbert Lossau, and Andrea Rapp, eds. *Evolution der Informationsinfrastruktur: Kooperation zwischen Bibliothek und Wissenschaft*. Glückstadt: vwh Hülsbusch, 2013.

Opheim, Bente. "Political Networks and Factions: Online Prosopography of Medieval Scandinavian Sagas." *History and Computing* 12, no. 1 (2000): pp. 43–57.

Porsdam, Helle. "Digital Humanities: On Finding the Proper Balance between Qualitative and Quantitative Ways of Doing Research in the Humanities." *Digital Humanities Quarterly* 7, no. 3 (2013). www.digitalhumanities.org/dhq/vol/7/3/000167/000167.html (accessed 17 December 2018).

Price, Kenneth M. "Social Scholarly Editing." In Unsworth; Siemens; Schreibman, *A New Companion to Digital Humanities*, pp. 137–49.

Rehbein, Malte. "Reconstructing the Textual Evolution of a Medieval Manuscript." *Literary & Linguistic Computing* 24 (2009): pp. 319–27.

———. "Digitalisierung braucht Historiker/innen, die sie beherrschen, nicht beherrscht." *H-Soz-Kult* 27, no. 11 (2015).

———. "Was Sind Digital Humanities?" *Akademie Aktuell* 56, no. 1 (2016): pp. 13–17.

———. "Geschichtsforschung im digitalen Raum. Über die Notwendigkeit der Digital Humanities als historische Grund- und Transferwissenschaft." In *Papstgeschichte im digitalen Zeitalter: Neue Zugangsweisen zu einer Kulturgeschichte Europas*. Edited by Klaus Herbers and Viktoria Trenkle, pp. 19–44. Beihefte zum Archiv für Kulturgeschichte Heft 85. Köln, Weimar, Wien: Böhlau Verlag, 2017.

———. "Informationsvisualisierung." In Jannidis; Kohle; Rehbein, *Digital Humanities*, pp. 328–42.

———. "Ontologien." In Jannidis; Kohle; Rehbein, *Digital Humanities*, pp. 162–76.

Rehbein, Malte, and Patrick Sahle. "Digital Humanities lehren und lernen: Modelle, Strategien, Erwartungen." In *Evolution der Informationsinfrastruktur: Kooperation zwischen Bibliothek und Wissenschaft*. Edited by Heike Neuroth, Norbert Lossau and Andrea Rapp, pp. 209–28. Glückstadt: vwh Hülsbusch, 2013.

Reinert, Matthias, Maximilian Schrott, Bernhard Ebneth, and Malte Rehbein. "From Biographies to Data Curation: The Making of www.deutsche-biographie.de." *Proceedings of the First Conference on Biographical Data in a Digital World*, Amsterdam, The Netherlands, April 9, 2015, pp. 13–19.

Rhian Lewis, Cecily Raynor, Dominic Forest, Michael Sinatra, Stéfan Sinclair: *Digital Humanities 2017, DH 2017, Conference Abstracts*, McGill University & Université de Montréal, Montréal, Canada, August 8-11, 2017

Ries, Wiebrecht. *Nietzsche zur Einführung*. 7th ed. Hamburg: Junius Verlag, 2004.

Robertson, Stephen. "The Differences between Digital Humanities and Digital History." In *Debates in the Digital Humanities 2016*. Edited by Matthew K. Gold and Lauren F. Klein. Minneapolis: University of Minnesota Press, 2016.

Sahle, Patrick. "DH studieren! Auf dem Weg zu einem Kern- und Referenzcurriculum der Digital Humanities." *DARIAH-DE Working Papers* 1 (2013). http://webdoc.sub.gwdg.de/pub/mon/dariah-de/dwp-2013-1.pdf (accessed 17 December 2018).

Schmidt, Desmond. "Towards an Interoperable Digital Scholarly Edition." *Journal of the Text Encoding Initiative* 7 (2014).

Schöch, Christof. "Big? Smart? Clean? Messy? Data in the Humanities." *Journal of Digital Humanities* 2, no. 3 (2013). http://journalofdigitalhumanities.org/2-3/big-smart-clean-messy-data-in-the-humanities/ (accessed 17 December 2018).

Schreibman, Susan, Ray Siemens, and John Unsworth, eds. *A Companion to Digital Humanities*. Blackwell Companions to Literature and Culture 26. Oxford: Blackwell, 2004.

Schröder, Wilhelm Heinz. "Die Buchreihe 'Historisch-sozialwissenschaftliche Forschungen' als publizistisches 'Flaggschiff' der quantitativen Historischen Sozialforschung in der 'Pionierzeit'." *Historical Social Research/Historische Sozialforschung* 18 (2006): pp. 6–23.

Schubert, Charlotte. "Die Visualisierung von Quellennetzwerken am Beispiel Plutarchs." *Digital Classics Online* 2, no. 1 (2016).

Schürer, Kevin, Tatiana Penkova, and Yanshan Shi. "Standardising and Coding Birthplace Strings and Occupational Titles in the British Censuses of 1851 to 1911." *Historical Methods* 48, no. 4 (2015): pp. 195–213.

Schützeichel, Rainer, and Stefan Jordan, eds. *Prozesse: Formen, Dynamiken, Erklärungen*. Wiesbaden: Springer VS, 2015.

Shillingsburg, Peter L. *From Gutenberg to Google: Electronic Representations of Literary Texts*. Cambridge: Cambridge University Press, 2006. www.loc.gov/catdir/enhancements/fy0733/2006497553-b.html (accessed 17 December 2018).

Stark, Martin. "Netzwerkberechnungen. Anmerkungen zur Verwendung formaler Methoden." In Düring; Eumann; Stark; Keyserlingk, *Handbuch Historische Netzwerkforschung*, pp. 155–71.

Tehrani, Jamshid, Quan Nguyen, and Teemu Roos. "Oral Fairy Tale or Literary Fake? Investigating the Origins of Little Red Riding Hood Using Phylogenetic Network Analysis." *Digital Scholarship in the Humanities* 31, no. 3 (2016): pp. 611–36.

Terras, Melissa. "Crowdsourcing in the Digital Humanities." In Unsworth; Siemens; Schreibman, *A New Companion to Digital Humanities*, pp. 420–38.

Thaller, Manfred. "Grenzen und Gemeinsamkeiten: Die Beziehung zwischen der Computerlinguistik und den Digital Humanities." *DHd 2014* (27 March 2014), Passau.

———. "Geschichte der Digital Humanities." In Jannidis; Kohle; Rehbein, *Digital Humanities*, pp. 3–12.

Thomson, N. "How to Read Articles Which Depend on Statistics." *Literary and Linguistic Computing* 4, no. 1 (1989): pp. 6–11.

Trilcke, Peer. "Social Network Analysis (SNA) als Methode einer textempirischen Literaturwissenschaft." In *Empirie in der Literaturwissenschaft*. Edited by Philip Ajouri, Katja Mellmann and Christoph Rauen, pp. 201–47. Münster: mentis Verlag, 2013.

Tufte, Edward R. *The Visual Display of Quantitative Information*. Cheshire, CT (Box 430, Cheshire 06410): Graphics Press, 1983.

Unsworth, John. "Computational Work with Very Large Text Collections: Interoperability, Sustainability, and the." *Journal of the Text Encoding Initiative* 1, no. 1 (2011).

Unsworth, John, Raymond George Siemens, and Susan Schreibman, eds. *A New Companion to Digital Humanities*. Blackwell Companions to Literature and Culture 93. Chichester, West Sussex, UK, Malden, MA, USA: John Wiley & Sons Ltd, 2016. http://onlinelibrary.wiley.com/book/10.1002/9781118680605 (accessed 17 December 2018).

van Dijk, Jan. *The Network Society: Social Aspects of New Media*. London, Thousand Oaks, CA: Sage Publications, 1999. http://search.ebscohost.com/login.aspx?direct=true&scope=site&db=nlebk&db=nlabk&AN=309855 (accessed 17 December 2018).

van Hooland, Seth, Max de Wilde, Ruben Verborgh, Thomas Steiner, and Rik van de Walle. "Exploring Entity Recognition and Disambiguation for Cultural Heritage Collections." *Literary and Linguistic Computing* (2013). http://llc.oxfordjournals.org/content/early/2013/11/29/llc.fqt067.full.pdf (accessed 17 December 2018).

Wellman, Barry. "Doing It Ourselves: The SPSS Manual as Sociology's Most Influential Recent Book." In *Required Reading: Sociology's Most Influential Books*. Edited by Dan Clawson, pp. 71–8. Amherst: University of Massachusetts Press, 1998.

6 Glossary

Actor, node A (socially) acting entity. A distinction is made between individual actors (people) and collective actors (e.g. companies or organisations). In graph theory or in a visualisation based on graph algorithms an actor is typically represented by a node.

Attributes, edge attributes, node attributes An attribute is additional information about actors (nodes) or their relationships (edges). For the interpretation of a given network structure it might be important to collect attribute data. Edge attributes describe a certain aspect of an edge, for example how often two actors speak to each other or how much money was given from a creditor to a debtor. Node attributes describe individual aspects of an actor, e.g. age, gender, profession or religion.

Bimodal network (also bipartite network, two-mode network) A social network in which actors are only linked via a second type of actor. For example, two people may be linked by an organisation they work for or two authors may use the same term. Unlike unimodal networks, the relationships are always between different types of actors and not between actors of the same type.

Brokerage, broker, bridge, structural hole Brokerage is about connecting two separate parts of a network. The broker is an actor who indirectly relates otherwise unconnected actors and is therefore in a structurally favourable position to potentially control the flow of information inside the network structure. A bridge is the relationship between two actors that actually connects different and otherwise separate parts of a network. Without the bridge there might be a structural hole in the network, leaving actors unconnected and not able to interact with one another.

Centrality measure Centrality measures are used to calculate the potential influence of actors in a network. There are various types of centrality measure:

Degree centrality (in-/out-degree centrality in the case of directed networks) is the number of direct relationships an actor has. A hub is an actor with a far higher degree centrality than the average in a given network.

Betweenness is a measure of centrality; it gives the highest score to the node that appears most often on the shortest path between all other nodes.

Closeness centrality is measured with the average path distance of an actor to all other actors in a network, including indirect relationships.

Component, cluster, clique, *k*-core, community Components or clusters are parts of a network. In a strict sense, cliques are maximal connected parts of a network where all actors are directly related to another. In a more loose sense, cliques are also often regarded as parts of a network in which the actors have a greater average connection to one another compared with other parts of the network. *K*-cores are maximal connected parts of the network where each actor has at least *k* relations inside the core, e.g. 2-cores or bi-components are subnetworks in which actors inside the component are connected at least twice. Communities are parts of a network in which the actors are densely connected with each other but have few relationships with actors outside the community. This definition of a community is very close to the loose definition of a clique.

Complete network (also whole network) A social network in which one or more defined actor types are connected to each other via one (uniplex) or more (multiplex) defined relationship types, e.g. friendship relationships between pupils within a class or trade agreements and peace treaties between nations. Complete networks can be unimodal, bimodal or multimodal. Ego networks are the opposite of complete networks.

Data Data are a set of individual pieces of information that are typically machine-readable.

Density Density is a measure that describes the structural properties of a network. It is calculated by comparing the proportion of existing relationships in a network with all the potential relationships if all the actors were connected to each other.

Ego network, egocentric network, personal network A social network in which the direct social relationships between one actor (ego) and other actors (alteri) are represented. Often the relationships among the alteri are also included in the representation. Ego networks are the opposite of complete networks. Personal networks are sometimes just comprised of alteri-alteri connections.

Graph algorithm, layout algorithm Graph algorithms arrange and optimise the visual representation of a network graph according to an underlying mathematical principle. Many different algorithms exist and can be fine-tuned with additional parameters. Layout algorithms help reveal patterns in network data. Force-directed algorithms are often used: they consider edges to be "springs" that attract well-connected nodes and seek to avoid crossings between ties. The same data can look very different with different layout algorithms, and even the same algorithm may produce a different visualisation each time it is run; there is not one "right" way to visualise a network.

Graph theory Graph theory is a branch of mathematics. Graphs are mathematical objects consisting of nodes and edges. Social network analysis is applied graph theory in which actors and their relationships are operationalised as nodes and edges.

Historical Network Research (HNR), historical social network analysis, historical network analysis Historical network research is a transdisciplinary research paradigm that attempts to transfer social network analysis theories and methods to historical questions and data. Because of the subject matter, HNR is a mixed-method approach, combining quantitative methods with traditional, more qualitative humanities methods.

Isolate, dyad, triad An isolate is an actor that has no relationships with other actors. A dyad is a group of two related actors. A triad is a group of three interconnected actors and also the smallest network possible.

Multi-level network, multi-layer network A multi-level network is a social network in which several actor classes (a bimodal/multimodal network), assigned to different aggregation levels, are connected to each other via different relationship types (multiplex), both within a single level and between levels.

Multimodal network A social network in which more than two different classes of actor are connected via relationships, e.g. authors, terms or publications. The relationships are always between different categories of actors and not between actors in the same category. Multimodal networks are complete networks.

Multiplex network A social network is considered to be multiplex if several types of relationship between actors are represented in it (e.g. friendship, kinship, neighbourhood, or cooperation).

Network A (social) network consists of a precisely defined (externally delimited) set of at least three actors (nodes) and their relationships (edges) with each other. There may be different classes of actors in a network. The relationships of these actors may be uniplex or multiplex. Different network types include ego networks, complete networks and unimodal, bimodal or multimodal networks.

Network data Network data is typically machine-readable data that contains information on actors and their relationships, edge and node attributes, and the network itself.

Network metaphor The most common way to refer to networks in the humanities. A network metaphor typically describes the observation that social relationships have an effect on something or somebody without specifying it further.

Network visualisation Visualisation is the optical representation of networks. Although in principle it is possible to draw networks by hand or to represent them in the form of a matrix, this quickly becomes impracticable for larger and more complex networks. For this reason, special software incorporating graph and layout algorithms is generally used for the visual representation of networks. The most commonly used visualisation algorithm is the spring embedder algorithm. Network visualisation based on algorithms is considered as formal social network analysis; the basic assumptions of the algorithms used should always be taken into account when interpreting the visualised networks.

Path distance, diameter, average path distance The geodetic distance or path distance is the shortest path between two actors in a network, e.g. actors with a direct relationship have a path distance of one. The diameter of a network is the longest possible path distance between two actors in a network; it's a structural measure to describe the network as a whole. The average path distance is an indicator of the general reachability of actors in a network.

Relationship, edge, directed, undirected A (social) relationship is a connection between actors. This relationship may be directed and originate only from one actor (e.g. lending), or it may be undirected, i.e., reciprocal (e.g. friendship) or weighted via edge attributes. In graph theory or in a visualisation based on graph algorithms a relationship is represented by an edge.

Social network analysis Social network analysis is the formalised study of interactions and relationships between actors and the resulting network structure based on graph theory.

Software for social network analysis Helps to create, compute, visualise and modify network data. Many different tools have been developed for different purposes. Gephi, Visone, UCINET and Pajek are among the most well-known standard network analysis tools. Humanists with no training in formal methods might use tools like Palladio, Vistorian or Nodegoat as a starting point for their studies.

Unimodal network, one-mode network, unipartite network A social network in which one type of actor is directly connected to another via one (uniplex) or more (multiplex) relationships. Unimodal networks are complete networks.

Uniplex network A social network is uniplex if it represents only one type of relationship between the actors involved.

Index

Note: Page locators in *italic* represent a figure. Page locators in **bold** represent a table

For Product Safety Concerns and Information please contact our EU representative GPSR@taylorandfrancis.com
Taylor & Francis Verlag GmbH, Kaufingerstraße 24, 80331 München, Germany

www.ingramcontent.com/pod-product-compliance
Ingram Content Group UK Ltd.
Pitfield, Milton Keynes, MK11 3LW, UK
UKHW020016200926
475702UK00014B/1912
* 9 7 8 1 0 3 2 2 3 6 6 7 4 *